Chaos and Structures in Nonlinear Plasmas

Chaos and Structures in Nonlinear Plasmas

W. Horton
Univ. of Texas at Austin
USA

Y-H Ichikawa
Chubu Univ.
Japan

World Scientific
Singapore • New Jersey • London • Hong Kong

Published by

World Scientific Publishing Co Pte Ltd
P O Box 128, Farrer Road, Singapore 912805
USA office: Suite 1B, 1060 Main Street, River Edge, NJ 07661
UK office: 57 Shelton Street, Covent Garden, London WC2H 9HE

Library of Congress Cataloging-in-Publication Data
Chaos and structures in nonlinear plasmas / editors, W. Horton, Y. H. Ichikawa.
    p.   cm.
Includes bibliographical references.
ISBN 9810226365
    1. Plasma dynamics. 2. Chaotic behavior in systems. 3. Nonlinear
theories.   I. Horton, C. W. (Claude Wendell), 1942-   .
II. Ichikawa, Yoshi H.
QC718.5.D9C43   1996
530.4'4--dc20   96-33752
                                                                       CIP

British Library Cataloguing-in-Publication Data
A catalogue record for this book is available from the British Library.

Copyright © 1996 by World Scientific Publishing Co. Pte. Ltd.

All rights reserved. This book, or parts thereof, may not be reproduced in any form or by any means, electronic or mechanical, including photocopying, recording or any information storage and retrieval system now known or to be invented, without written permission from the Publisher.

For photocopying of material in this volume, please pay a copying fee through the Copyright Clearance Center, Inc., 222 Rosewood Drive, Danvers, MA 01923, USA.

Printed in Singapore.

Preface

The purpose of this book is to help scientists and engineers explore the nonlinear dynamics of plasmas while developing their proficiency with the tools of nonlinear science. Nonlinear science owes much of its origin to the early studies of plasma physicists. For example, two key plasma physics problems from the nineteen sixties led first to what is now the standard estimate for the criterion for the onset of stochasticity in a Hamiltonian system and secondly to the inverse scattering method of solving the initial value problem for integrable nonlinear wave equations. The method for gauging the onset of a significant volume of stochasticity in the phase space arose from the study of the structures of the magnetic fields in toroidal plasma confinement vessels. The problem of determining the confinement time for charged particles in the magnetic mirror trap led physicists to the standard map. The standard map, as its name implies, is a universal description of Hamiltonian chaos showing the beautifully intricate mixtures of integrable and chaotic particle trajectories. For the closed two degree-of-freedom system the surface of section technique is used to develop an equivalent two-dimensional area preserving map from which the integrable structures and chaotic sea in the phase space can be found. These methods and numerous examples are developed in Chapters 1 to 4.

The concept of robust, localized intrinsically nonlinear solutions of a class of nonlinear wave equations which appear in many fields of science and engineering was developed by the celebrated team of Martin Kruskal and Norman Zabusky in the late nineteen sixties. From Zabusky's precise computer simulations of the elastic interactions of the soliton solutions of the Korteweg-de Vries (KdV) equation and Kruskal's mathematical description of the dynamics observed in the computer simulations emerged the paradigm for future studies of integrable nonlinear wave equations. These studies led to the invention of the inverse scattering transform (IST) method for solving the initial value problem giving precisely the number of solitons and the wave field components that arise from a given initial disturbance. From this seminal method for integrating the KdV equation other nonlinear wave equations, such as the cubic Schrödinger equation and the Kadomtsev-Petviashvili equation, were found to be integrable by extending the IST method developed for the KdV equation. These methods are developed in Chapter 5. For a complete, modern development of the analysis of integrable nonlinear wave equations the reader may wish to consult *Solitons, Nonlinear Evolution and Inverse Scattering* by Ablowitz and Clarkson (1991). Numerous other books devoted to soliton theory are cited in Chapter 5.

The plasma physics contributions, coupled with the developments by mathematicians and other scientists and engineers, have led to the development of a new field of research simply called Nonlinear Science. There are already numerous excellent books in the area of nonlinear science. We felt the need for a new work for plasma science students and researchers focused on the physical problems in plasma science. Thus Nonlinear Science encompasses many areas beyond the scope of the present text. Here we are content to present the most central aspects of nonlinear plasma physics found to occur in laboratory and space plasmas.

The book assumes the reader has a thorough knowledge of classical mechanics and electromagnetic wave theory. Some knowledge of plasma physics will be required to appreciate the origin and importance of the problems addressed in the book. The authors believe that the work is sufficiently self-contained that a graduate student with no previous studies of plasma physics will be able to follow the nonlinear science aspects of the subject without recourse to a plasma physics text. For those wishing to supplement their reading with a plasma physics text we suggest the text *Physics of High Temperature Plasmas* by G. Schmidt and *Plasma Physics* by Ichimaru as useful companions. For a confinement and tokamak physics companion the reader may consult *Plasma Confinement* by Hazeltine and Meiss (1992) or *Theory of Tokamak Plasmas* by White (1989).

In Chapters 1 and 2 we develop the theory of nonlinear oscillations, parametric instabilities, and stochastic trajectories from overlapping resonances. In Chapter 3 we develop the understanding of nonlinear motions for the central examples of nonlinear dynamics and the diffusion approximation for the chaotic trajectories by working through the Hamiltonian motions of charged particles in the classic electromagnetic field configurations for space and laboratory plasmas.

Chapter 4 develops the theory of the standard map with its single parameter K controlling the nonlinearity. The method of understanding the structure of phase space by studying the stability of fixed points with bifurcations analysis is developed. The stochastic orbits are studied and described by the diffusion coefficient $D(K)$, and the accelerator orbits are found and characterized. The generalization of the theory to two parameter maps (K, β) with β controlling the nonlinear phase advance is developed by introducing the relativistic standard map.

In Chapter 5 we derive the classic nonlinear wave equations known as the KdV, modified KdV, the cubic nonlinear Schrödinger equation and others. The theory of the modulational instability is developed. Its final state leads to localized nonlinear solutions. In the second part of Chapter 5 the theory of the integrability of the KdV equation is developed with the inverse scattering transform method to integrate the initial value problem. The integrability of the nonlinear cubic Schrödinger equation and other nonlinear wave equations is discussed.

In Chapter 6 the theory of the two-dimensional nonlinear drift wave vortices is developed, and the fascinating duality of the partly wave-like and partly soliton-like properties of the solutions is shown in some detail. There is a direct analog between the plasma drift wave vortices and the geophysical neutral fluid vortex structures in the atmosphere and oceans. The mixed wave and vortex solutions of these two-dimensional nonlinear wave equations correspond to the large scale ($\gtrsim 1000\,\text{km}$) cyclonic (low pressure) and anti-cyclonic (high pressure) weather patterns seen daily in the weather reports. In the geophysical weather context the nonlinear equation is called the Charney (1948) equation. In the magnetized plasma context the equation is called the Hasegawa-Mima (1977) equation. This mixture of waves and vortices is called drift turbulence in the confined, magnetized plasma. Drift wave turbulence appears to be the universal, dominant mechanism for transporting particles and energy-momentum across the magnetic field. Because of the central importance of the drift waves for plasmas and the analog Rossby waves for the dynamics of the atmospheres and oceans, many computer simulations and rotating water tank experiments have been carried out to determine their properties. Only a small fraction of this work is reported here. The 1994, Vol. 4 issue of the journal *CHAOS* is devoted entirely to the study of this type of planetary and plasma physics work on wave-vortex dynamics.

In Chapter 7 the statistical theory of plasma turbulence is developed and applied to the problem of drift wave turbulence. Here the statistical assumptions of turbulence theory are formulated, and the method of closure in the chain of multiple correlations is presented. The method of renormalization equations for the correlation function and the response function or propagator is presented. The properties of the reduced equations are developed for their wavenumber spectrum and transport properties. The closed statistical theory is shown to agree with computer simulations in parameter regimes where the coherent vortex structures of Chapter 6 are a subdominant component of the turbulence.

The authors owe many thanks to their colleagues, especially their co-authors and students, who have helped us achieve some degree of understanding of nonlinear plasma dynamics. The authors especially wish to thank Suzy Mitchell for skillful and tireless word processing. They also thank the many students, including J. Hernandez, X.N. Su, D. Boozer, J.-L. Thiffeault, Y. Nomura, and J.C. Bowman for helping with the final stages of preparing the document.

The work was completed during the sabbatical of the first author at the Institute for Theoretical Physics at the University of California at Santa Barbara. We thank Profs. Ravi Sudan and Steve Cowley for a stimulating and productive plasma physics program. We wish to thank Prof. Jim Langer, Director, Prof. Dan Hones, and the ITP staff for providing a hospitable setting for finishing this book.

Contents

Preface v

1 Nonlinear Oscillations 1
   1.1 Harmonic Generation in Small Amplitude Oscillations 2
   1.2 Amplitude Dispersion from Secular Terms 6
   1.3 Parametric Instabilities 7
      1.3.1 Parametric instability threshold and the growth rate 9
      1.3.2 Mathieu equation and Floquet theory 10
   1.4 *Example 1.1 Parametric Excitation of the Pendulum* 13
   1.5 *Example 1.2 Stability of the Ponderomotive Potential* 15
   1.6 *Example 1.3 Oscillations of a Charged Particle in Two Longitudinal Waves* 18

2 Hamiltonian Dynamics 23
   2.1 Measure-Preserving Flows and the Hamiltonian 23
   2.2 Poincaré Surface of Section 25
   2.3 Fixed Points and Invariant Curves 26
   2.4 KAM Theory 27
   2.5 KAM Torii, the Golden Mean and Leaky Barriers 28
   2.6 KAM Theory in Laboratory Plasmas 31
   2.7 Visualization of the Magnetic Flux Surfaces by Electron Beam Mapping 31
   2.8 Magnetic Islands and the Unstable-Chaotic Trajectories of Field Lines in Tokamaks 36
   2.9 Measure of the Resonant Rational Surfaces 44
      2.9.1 The number of irreducible fractions 44
      2.9.2 Measure of the resonant surfaces 45
   2.10 *Example 2.1 How a Swing Behaves* 47
   2.11 Invariant Tori and the KAM Theorem 48

3 Stochasticity Theory and Applications in Plasmas 51
   3.1 Trajectories in Straight, Nonuniform Magnetic Fields 53

		3.1.1	Integrable orbits and invariants of charged particle motion 54

- 3.1.2 Nonlinear oscillator for weakly inhomogeneous magnetic field 59
- 3.1.3 Nonlinear orbits in the field reversed configuration or sheet pinch 60
- 3.2 Adiabatic Invariants 62
  - 3.2.1 Small gyroradius limit $(2m\,E_\perp \ll p_y^2)$ 63
  - 3.2.2 Small gyroradius and a weak electrostatic potential 64
  - 3.2.3 Current sheet invariant 65
- 3.3 Onset of Chaos from a Transverse Electric Field 69
  - 3.3.1 $1\tfrac{1}{2}$-D Hamiltonian 70
  - 3.3.2 Melnikov-Arnold integral 71
  - 3.3.3 The Whisker map and the standard map 72
  - 3.3.4 The standard map 73
- 3.4 Onset of Chaos from a Normal Magnetic Field Component 74
  - 3.4.1 2D nonlinear coupled oscillations 75
  - 3.4.2 Two degrees-of-freedom and degeneracy 77
  - 3.4.3 Resonance conditions and the surface of section 78
- 3.5 $\mathbf{E} \times \mathbf{B}$ Motion in Two Low-Frequency Waves and the Diffusion Approximation 84
  - 3.5.1 Test particle motion in drift waves 85
  - 3.5.2 Action-angle variables for $\mathbf{E} \times \mathbf{B}$ motion 87
  - 3.5.3 The onset of stochasticity in two drift waves 91
  - 3.5.4 *Example 3.1:* $\mathbf{E} \times \mathbf{B}$ *drift islands in the cylindrical plasma with sheared flow* 96
  - 3.5.5 The diffusion approximation 99
  - 3.5.6 *Example 3.2: Onset of* $\mathbf{E} \times \mathbf{B}$ *diffusion from two equal amplitude waves* 100
- 3.6 Renormalized $\mathbf{E} \times \mathbf{B}$ Diffusion Coefficients 102
- 3.7 $\mathbf{E} \times \mathbf{B}$ Motion in a Sheared Magnetic Shear 107
- 3.8 Hamilton's Equations of Motion in Non-Canonical Coordinates 109

4 Phase Space Structures in Hamiltonian Systems 117
- 4.1 The Standard Map 117
- 4.2 Maps for the Motion of a Charged Particle in an Infinite Spectrum of Longitudinal Waves 120
- 4.3 Fixed Points, Accelerator Orbits and the Tangent Map 122
- 4.4 Stochastic Motion and the Diffusion Coefficient 125
  - 4.4.1 Probability distribution functions for the orbits 126
  - 4.4.2 Evolution of the momentum distribution function 131

	4.4.3 Standard map at large values of A 131
	4.4.4 Standard map at small values of A 134
	4.4.5 Discussion of physics for the diffusion in the standard map . . . 134
4.5	Regular Motion in the Relativistic Standard Map 136
4.6	Symmetries of the Relativistic Standard Map 140
4.7	Stability of the Periodic Orbits 145
4.8	Poincaré-Birkhoff Multifurcation for the Period-4 Orbit 155
4.9	Concluding Remarks on the Standard and Relativistic Maps 160
4.10	*Example 4.1 Fermi Acceleration* 161

5 Solitons in Plasmas   165

5.1	Nonlinear Coherent Modes in Vlasov Plasmas 166
5.2	Amplitude Modulation of Nearly Monochromatic Waves and the Modulational Instability 169
5.3	Modulational Instability and Nonlinear Landau Damping 178
	5.3.1 NLS solitons with nonlinear Landau damping 180
5.4	Reductive Perturbation Analysis of Nonlinear Wave Propagation . . . 183
	5.4.1 Nonlinear ion acoustic wave 183
	5.4.2 Parallel propagating right and left polarized Alfvén waves . . . 187
	5.4.3 General nonlinear equation for oblique waves 188
5.5	Birth of the Soliton 190
5.6	The Inverse Scattering Transformation Method 191
5.7	Gelfand-Levitan Equation for the Scattering Potential 194
5.8	Inverse Scattering Transform (IST) for the NLS Equation 196
5.9	Generalization of the Integrability Conditions 198
5.10	Alfvén Soliton 201
5.11	One-Dimensional Soliton Gas Models 206
5.12	*Example 5.1 — Elastic Collisions of Two Solitons* 206
	5.12.1 Petviashvili drift wave solitons and the soliton gas 209
5.13	*Example 5.2 — 3 × 3 Matrix Formalism of the Inverse Scattering Transformation* 219

6 Vortex Structures in Hydrodynamic and Vlasov Systems   223

6.1	The Drift Wave–Rossby Wave Analogy 223
6.2	The Drift Wave Mechanism and Vortex 228
	6.2.1 Nonlinear drift wave equation and passive scalar equation . . . 230
	6.2.2 Conservation laws 231
	6.2.3 Solitary monopolar structures and the trapping condition . . . 234
6.3	Solitary Dipolar Vortex Solutions 235
6.4	Drift Wave-Ion Acoustic Wave Equations 236
	6.4.1 Conditions for the conservation of energy and enstrophy . . . 239

	6.4.2	Ertel's conservation theorem 240
	6.4.3	Solitary drift wave vortices: shearless case 240
	6.4.4	Larichev-Reznik dipole vortices 242
	6.4.5	Maximum amplitude and size relations 243
	6.4.6	Energy and enstrophy of the dipole vortex 244
6.5	Experimental Features and Computer Simulations of the Dipole Vortices 246	
	6.5.1	Plasma drift wave experiments 246
	6.5.2	Drift wave vortex collisions and coalescence 248
	6.5.3	Anomalous transport during inelastic vortex collisions 257
6.6	Intermittent Transport from Vortex Collisions 260	
6.7	Interaction Energies and Kurtosis in Distributions of Vortices and Waves 263	
6.8	Fluctuation Spectrum from a Gas of Dipole Vortices 266	
6.9	Monopolar Vortices Produced by Sheared Flows 268	
	6.9.1	Monopole solutions due to the combined action of the Poisson bracket and KdV nonlinearities 268
	6.9.2	Dipole vortex splitting into monopoles 272
6.10	Discussion and Conclusions 275	

7 Statistical Properties and Correlation Functions for Drift Waves   277

7.1	Introductory Remarks 277	
7.2	Nonlinear Drift Wave Equation 280	
	7.2.1	Nonlinear drift wave equation in configuration space 281
	7.2.2	Nonlinear drift wave equation in $\mathbf{k}$ space 284
	7.2.3	Volume contraction in the phase space 285
	7.2.4	Statistical turbulence theory 288
	7.2.5	Two-time correlation functions 293
	7.2.6	Properties of correlation functions for $\mathbf{k}_1 + \mathbf{k}_2 = 0$ 294
	7.2.7	Space correlation function and the wavenumber spectrum . . . 295
	7.2.8	Effect of dissipation on space-time evolution 296
7.3	Renormalized, Markovianized Spectral Equations 297	
	7.3.1	Incoherent fluctuation source 299
	7.3.2	Dynamics of the correlation function in renormalized turbulence theory 299
	7.3.3	Properties of $T_\mathbf{k}(\{E\})$ — the nonlinear transfer operator . . . 300
7.4	Local, Isotropic Approximation 301	
	7.4.1	High-wavenumber spectral balance 303
	7.4.2	Quasilinear transport from the renormalized turbulence theory spectrum 305
	7.4.3	Relation to the Hasegawa-Wakatani drift-wave equation 306

7.5	Low-Order Wave Coupling		308
	7.5.1	Three-wave coupling and strange attractors	308
	7.5.2	Integrable limit of the three-wave equations	311
	7.5.3	Dissipative dynamics and the strange attractor	312
	7.5.4	Random phase approximation	314

Bibliography   **321**

Index   **335**

Chapter 1

Nonlinear Oscillations

To appreciate the advances made in nonlinear plasma physics during the 1980s it is important to know the classical aspects of the theory of nonlinear oscillations. In addition the principal theoretical methods and the physical phenomena found in the study of the one-dimensional nonlinear oscillator are important aspects of many nonlinear problems in physics and engineering. A list of phenomena produced in nonlinear oscillations is given in Table 1.1. The work *Nonlinear Oscillations* by Minorsky (1962) presents a systematic development of the techniques developed for nonlinear ordinary differential equations by mathematicians, engineers, and physicists up to 1960. A more accessible introduction to the properties of nonlinear oscillators can be found in Chapter 3 of the textbook *Newtonian Dynamics* by Baierlein (1983). At a more advanced level the Landau and Liftschitz (1958) *Classical Mechanics* gives a concise treatment of nonlinear oscillators, while the standard reference for understanding the behavior of coupled nonlinear oscillators is the review article *A Universal Instability of Many-Dimensional Oscillator Systems* by Chirikov (1979).

In this Chapter we develop the small amplitude expansion for the one degree-of-freedom nonlinear oscillator deriving the formulas for the harmonic generation and the amplitude dependence of the oscillator's frequency. While the derivation is simple, the technique is important since the same problem of secular terms from higher order phase-coherent harmonics arises in the much more complicated Vlasov-Poisson system treated in Chapter 5, and in a more general sense in the drift wave-Rossby wave turbulence treated in Chapters 6 and 7. The second major topic in Chapter 1 is the condition for the parametric instability of a linear oscillator in which a parameter varies periodically. In terms of nonlinear physics this problem arises when one degree-of-freedom in an $N \geq 2$ degree-of-freedom system has sufficiently high energy (the pump) to appear as a fixed periodic driving force in one, or several, of the other degrees-of-freedom. The stable-unstable domains of such a system are fairly intricate and are known in terms of the properties of the two-parameter Mathieu equation.

For nonlinear oscillators coupled to an external force (or degree-of-freedom) through the time periodic variation of a parameter of the system, important resonances do occur. Within certain resonance domains the linear parametrically driven oscillator is unstable to exponential growth. Outside these domains the system has a periodic nonsinusoidal oscillation. In Sec. 1.3 we give the calculation of the parametric instability and the stability boundaries. In Secs. 1.4, 1.5, and 1.6 we work through three examples that display the nonlinear effects presented in Chapter 1.

1.1 Harmonic Generation in Small Amplitude Oscillations

We begin by introducing one of the most universal solution procedures, namely the small amplitude expansion. The small amplitude expansion is a cornerstone in the methods for many complicated problems which have no exact solution, and thus it is useful to begin here in our description of nonlinear phenomena. We start with a simple nonlinear oscillator which may be solved exactly by finding the energy integral and working out the resulting quadrature in terms of elliptic functions.

Table 1.1. Nonlinear Phenomena

Oscillation amplitude $= a$
Oscillation angular frequency $= \omega_0$
Harmonic generation $\omega_0 \to 2\omega_0, 3\omega_0, \ldots, n\omega_0$
Period doubling $\omega_0 \to \omega_0/2, \omega_0/4, \ldots$
Shift of equilibrium $x_0 = 0 \to x_0 = \alpha a^2 + \cdots$
Frequency a function of amplitude $\omega_0 \to \omega(a^2)$
Amplitude dispersion for waves $\omega_0 \to \omega(k, a^2)$
Hysteresis in dissipative system $\to$ state depends on path taken

Let a be the characteristic amplitude of the oscillation $x(t)$. We seek solutions in the form of an expansion

$$x(t) = x^{(0)} + x^{(1)} + x^{(2)} + \cdots + x^{(n)} \tag{1.1}$$

where each successive term is one order higher in the small amplitude

$$\frac{x^{(\ell+1)}}{x^{(\ell)}} \sim a \ll 1. \tag{1.2}$$

1.1 Harmonic Generation in Small Amplitude Oscillations

Use of expansion (1.1) in a nonlinear equation produces an equation with a large number of terms. If the highest nonlinearity is of degree x^M, then there are terms from zeroth order $x^{(0)}$ to degree $\left(x^{(n)}\right)^M$ with magnitude a^{nM}. When a is small the only way to balance the equation is to have all terms of the small order a^ℓ sum to zero. Thus, in this method we convert a single nonlinear equation of degree M into a system of nM linear equations.

Often it is possible to classify the terms at order ℓ so that the dominant terms, at large t for example, are a simple subset of all the order ℓ terms. In this case one looks for ways to sum the subset of dominant terms to all orders. This selective summation of dominant terms to all orders is a second standard procedure that is used in plasma physics to generate so-called renormalized series for propagators and renormalized response functions. The same type of selective summation of dominant terms is used with fields to develop renormalized turbulence theory in Sec. 7.2.

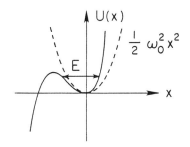

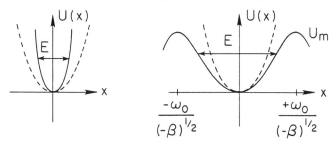

Fig. 1.1. The potential energy $U(x)$ for three types of nonlinear forces. (a) the quadratic nonlinearity αx^2 shifting the equilibrium and giving a first harmonic; (b) the hardening cubic nonlinearity $\beta > 0$ increasing the oscillation frequency; and (c) the softening cubic nonlinearity $\beta < 0$.

Now let us consider how the small amplitude expansion works for the nonlinear (often called anharmonic) oscillator given by

$$\frac{d^2x}{dt^2} + \omega_0^2 x = -\alpha x^2 - \beta x^3. \tag{1.3}$$

For $\beta > 0$ there is a cubic nonlinear restoring force which increases the frequency of the oscillator at finite amplitudes, and the α effect shifts the mean value of oscillator coordinate to the side $x\alpha < 0$ due to the de-symmetrization of the restoring force. These two effects are shown in Fig. 1.1 by constructing the potential energy $U(x)$ for Eq. (1.3). For $\beta < 0$ the energy of the nonlinearity weakens the linear restoring force, and the oscillator energy must be less than $\omega_0^4/4|\beta|$ for stable motion. These features are readily seen from the energy integral and potential $U(x)$ shown in Fig. 1.1. In Eq. (1.3) the dynamical variable $x(t)$ is the shift from the equilibrium position given by $x^{(0)} = 0$ in Eq. (1.1) so that the amplitude expansion begins at first order with the solution

$$x^{(1)}(t) = a\cos(\omega_0 t + \zeta) \to a\cos(\omega_0 t)$$

by a suitable choice of the reference time of $t = 0$.

Substituting expansion (1.1) into Eq. (1.3) and demanding that terms of the same degree a^m of each order in a balance identically, we obtain the following system of n linear equations:

$$\ddot{x}^{(1)} + \omega_0^2 x^{(1)} = 0$$

$$\ddot{x}^{(2)} + \omega_0^2 x^{(2)} = -\alpha(x^{(1)})^2$$

$$\ddot{x}^{(3)} + \omega_0^2 x^{(3)} = -2\alpha x^{(1)} x^{(2)} - \beta(x^{(1)})^3 \tag{1.4}$$

$$\vdots$$

$$\ddot{x}^{(n)} + \omega_0^2 x^{(n)} = -2\alpha x^{(1)} x^{(n-1)} \cdots - 3\beta(x^{(1)})^2 (x^{(n-2)})$$

$$= f_n\left(x^{(1)}, x^{(2)} \cdots x^{(n-1)}\right).$$

Now, in principle it is straightforward to solve the linear system (1.4) to construct the series $\Sigma_n x^{(n)}(t)$.

The second order solution $x^{(2)}(t)$ is obtained by reducing $\alpha(x^{(1)})^2 = \alpha a^2 \cos^2(\omega_0 t)$ to its average and its $2\omega_0$ harmonic component to obtain

$$\ddot{x}^{(2)} + \omega_0^2 x^{(2)} = -\frac{\alpha a^2}{2}(1 + \cos 2\omega_0 t) \tag{1.5}$$

1.1 Harmonic Generation in Small Amplitude Oscillations

with the particular solution

$$x^{(2)}(t) = x_0^{(2)} + x_{2\omega_0}^{(2)} \cos(2\omega_0 t) \tag{1.6}$$

where $x_0^{(2)}$ is the shift of the equilibrium position of the oscillator

$$x_0^{(2)} = -\frac{\alpha\, a^2}{2\omega_0^2}$$

and $x_{2\omega_0}^{(2)}$ is the amplitude of the second harmonic components

$$x_{2\omega_0}^{(2)}(t) = +\frac{\alpha\, a^2}{6\omega_0^2} \cos(2\omega_0 t).$$

In third order the driving term on the right-hand side is again decomposed into its Fourier harmonics using $\cos\omega_0 t \cos 2\omega_0 t = (\cos 3\omega_0 t + \cos\omega_0 t)/2$ to obtain

$$\ddot{x}_0^{(3)} + \omega_0^2\, x^{(3)} = \left(\tfrac{5}{6}\frac{\alpha^2 a^3}{\omega_0^2} - \tfrac{3}{4}\beta a^3\right)\cos(\omega_0 t) + \left(-\frac{\alpha^2 a^3}{6\omega_0^2} - \tfrac{1}{4}\beta a^3\right)\cos(3\omega_0 t). \tag{1.7}$$

The important qualitatively new feature that arises here is that the nonlinearity has taken the beating of the lower order oscillations, in this case $2\omega_0 \pm \omega_0 \to 3\omega_0$ and ω_0, to produce a resonant driving force giving rise to the secular growth of the solution

$$x_{\omega_0}^{(3)}(t) = \frac{f_3 t}{2\omega_0}\sin(\omega_0 t). \tag{1.8}$$

where

$$f_3 = \tfrac{5}{6}\frac{\alpha\, a^3}{\omega_0^2} - \tfrac{3}{4}\beta\, a^3. \tag{1.9}$$

The higher harmonic component $x_{3\omega_0}^{(3)}(t) \propto a^3 \cos(3\omega_0 t)$ has a relatively small amplitude because the frequency $3\omega_0$ is far off resonance. The third order nonresonant contribution from Eq. (1.7) is

$$x_{3\omega_0}^{(3)}(t) = \frac{a^3}{8\omega_0^2}\left(\frac{\alpha^2}{6\omega_0^2} + \tfrac{1}{4}\beta\right)\cos(3\omega_0 t).$$

The solution in Eq. (1.8) shows the important phenomena of secular growth from resonant nonlinear interactions. The solution is correct for finite t and sufficiently small amplitudes such that $|x^{(3)}/x^{(1)}| < 1$. The method for removing such a short time limit restriction is to allow for nonlinear shifts in the oscillation frequency $\omega \to \omega_0(a^2)$. In this case $x^{(1)} + x_{\omega_0}^{(3)}$ becomes

$$x(t) = a\cos\left[\omega_0(1 + \kappa a^2)t\right]$$

with
$$\kappa = \tfrac{3}{8}\frac{\beta}{\omega_0^2} - \tfrac{5}{12}\frac{\alpha^2}{\omega_0^4}.$$

In the context of coherent waves in Chapter 5 and turbulence in Chapter 7 the allowance for this type of amplitude dependent frequency shift in the wave-particle resonance is required to remove secular terms in the small amplitude expansion of the particle distribution function $f(x,v,t)$. In the plasma beat wave accelerator and other nonlinear resonant wave coupling problems, the secular growth of the coupling product $\omega_1 + \omega_2 = \omega_3$, is used to drive up the amplitude of the product wave as occurs in Eq. (1.8). In all these cases, where the total energy of the system is limited, there is a limiting time (the pump depletion time) beyond which the secularly growing solution is not valid. Study of this limit for various practical problems is an important area of nonlinear plasma physics. A mechanical analog is given in Example 1.1 in Sec. 1.4 at the end of this chapter.

1.2 Amplitude Dispersion from Secular Terms

In the present example the constancy of the energy integral of the nonlinear oscillator,
$$E = \tfrac{1}{2}\dot{x}^2 + \tfrac{1}{2}\omega_0^2 x^2 + \tfrac{1}{3}\alpha\, x^3 + \tfrac{1}{4}\beta\, x^4, \tag{1.10}$$
clearly stops the secular growth. The method of removing the secular growth is to recognize that at each order of the small amplitude expansion these dominant terms continue to arise. The sum of the set of secular terms at order $a^{2n+1}t^n$ form a summable series that is contained in the following nonlinear frequency shift:

$$a\cos\left[\omega_0(1+\kappa a^2)t\right] = a\cos(\omega_0 t)\left(1 - \tfrac{1}{2}\omega_0^2\,\kappa^2 a^4\,t^2 + \cdots\right)$$

$$- a\sin(\omega_0 t)\left(\omega_0\kappa\, a^2 t - \tfrac{1}{6}\omega_0^3\,\kappa^3\,a^6\,t^3\cdots\right). \tag{1.11}$$

Thus, the important lesson is that at each order in perturbation theory, the secular terms must be recognized and allowed for in the perturbation expansion. In the present case by allowing for a frequency shift $\omega_0 \to \omega_0(1+\kappa\,a^2)$ the set of secular terms of order $a^{2n+1}\,\omega_0^n\,t^n$ can be summed to all orders. The amplitude dependence of the nonlinear frequency $\omega_0 \to \omega(a^2)$ is called amplitude dispersion and is a key feature of nonlinear oscillations and nonlinear waves. This amplitude dependence for nonlinear frequency is intimately connected with their ability to self-trap into localized, solitary wave structures as we will show in Chapters 5 and 6.

To systematically introduce the amplitude dispersion in the small amplitude perturbation expansion one proceeds as follows. We recognize that there is a well defined, but unknown, period $2\pi/\Omega$ of the nonlinear oscillator due to the energy integral in Eq. (1.10). Thus, we introduce the phase variable

$$\theta = \Omega\, t \tag{1.12}$$

and transform the oscillator equation to

$$\Omega^2 \frac{d^2x}{d\theta^2} + \omega_0^2 x(\theta) = f(x). \quad (1.13)$$

Now the small amplitude expansion is extended to include both $x(t)$ and $\Omega(a)$

$$\begin{aligned}x(\theta) &= x^{(1)} + x^{(2)} + x^{(3)} + \cdots \\ \Omega &= \omega^{(0)} + \omega^{(1)} + \omega^{(2)} + \cdots,\end{aligned} \quad (1.14)$$

and the new freedom provided by the parameters $\{\omega^{(n)}\}$ is used at each order in the expansion to eliminate secularities. The new expansion in Eq. (1.14) is called a secular perturbation theory since it eliminates the secularly growing terms that occur at each order, for example in Eq. (1.8) at third order.

With the expansions in Eq. (1.14) substituted into Eq. (1.13) the theory proceeds as before with the first and second order giving $\omega^{(0)} = \omega_0$ and $\omega^{(1)} = 0$. At third order the equation becomes

$$2\omega_0 \omega^{(2)} a \cos\theta + \omega_0^2 \ddot{x}^{(3)} + \omega_0^2 x^{(3)} = f^{(3)}_{\omega_0} \cos\theta + f^{(3)}_{3\omega_0} \cos 3\theta \quad (1.15)$$

balancing the $\cos\theta$-term gives the frequency shift

$$\omega^{(2)} = \frac{f^{(3)}_{\omega_0}}{2\omega_{0a}} = \frac{\omega_0}{2}\left(-\frac{5}{6}\frac{\alpha^2}{\omega_0^4} + \frac{3}{4}\frac{\beta}{\omega_0^2}\right)a^2 \quad (1.16)$$

and the $\cos 3\theta$ term gives

$$x^{(3)} = a\left(\frac{\alpha^2 a^2}{48\omega_0^4} + \frac{\beta a^2}{32\omega_0^2}\right)\cos(3\theta). \quad (1.17)$$

Now the approximate solution $\sum_1^3 x^{(n)}(\theta = \Omega t)$ is uniformly convergent in t with error of order a^4 for all θ.

Thus, the important lesson is to recognize that nonlinearity introduces the qualitatively new behavior of amplitude dependent frequency shifts that must either be incorporated into the expansions at the outset, or recognized and summed to all orders as the dominant secular terms in the perturbation series.

1.3 Parametric Instabilities

An important class of nonlinear excitations that lead to exponential growth or decay of the oscillations, either in waves or particle orbits depending on the application, is the parametric instability. The name "parametric" describes the fact that in its final form the problem reduces to an oscillator equation in which a parameter has an explicit periodic oscillation, or modulation, in space or time. Thus, the reduced system is a linear non-autonomous (open) system.

In the study of nonlinear waves the situation arises in many nonlinear wave coupling problems (Kruer, 1988). A strong, high frequency wave $E_0 \cos(\omega_0 t)$ in a plasma typically modulates a parameter such as the plasma density $\delta n/n$ which then leads to a resonant decay of the primary oscillation into lower frequency oscillations at ω_p when $\omega_0 = n\omega_p$ with $n = 2, 3, 4 \ldots$. In wave theory the interpretation is given that a higher frequency quantum of excitation energy ω_0 decays into two lower frequency quanta when $\omega_0 = \omega_p + \omega_p$.

For plasma waves with $\omega_p = (4\pi n e^2/m_e)^{1/2}$ the parametric instability problem reduces to an equation of the form

$$\partial_t^2 \xi + \omega_p^2 \left(1 + \frac{\delta n}{n} \cos(\omega_0 t)\right) \xi = 0 \tag{1.18}$$

easily derived using $\omega^2 = \omega_p^2(1 + \delta n/n)$ for modulated plasma waves and using $\delta n \to \delta n \cos(\omega_0 t)$ for the driver. For single particle motion, equations of the form of Eq. (1.18) occur due to the modulation of the system parameters such as the magnetic field curvature in the alternating gradient focusing in a synchrotron. Generically, in these types of problems the orbit around an elliptic fixed point is coupled to another degree-of-freedom.

In terms of the anharmonic oscillator governed by Eq. (1.3) in the previous section, the parametric decay instability can arise when a driving force $f = f_\Omega \cos(\Omega t)$ with $\Omega \simeq 2\omega_0$ is applied to the oscillator. The direct linear response is nonresonant and given by

$$x_\Omega(t) = \frac{f_\Omega \cos(\Omega t)}{-\Omega^2 + \omega_0^2} \simeq -\frac{f_\Omega}{3\omega_0^2} \cos(\Omega t). \tag{1.19}$$

Now the driven oscillation induces a modulation through the quadratic nonlinearity $\alpha x^2 = \alpha(x_\Omega + x(t))^2 \cong \alpha x_\Omega^2 + 2\alpha\, x_\Omega(t) x(t)$. Keeping only the first order $x(t)$ motion gives the modulational oscillator equation

$$\ddot{x} + \omega_0^2 \left(1 - \frac{2\alpha f_\Omega}{3\omega_0^4} \cos(\Omega t)\right) x = 0. \tag{1.20}$$

The strength of the modulation, in this example, is given by the combination of parameters $2\alpha f_\Omega/3\omega_0^4$ which we may define as h. Landau and Lifschitz (1958) and Minorsky (1962) give more examples.

Equations of the form (1.18) and (1.20) are called Mathieu equations or, for more general periodic modulations, Hill equations. Here we first calculate the parametric instability threshold and the growth rate by the small amplitude expansion and then give the general properties of the Mathieu equation.

1.3.1 Parametric instability threshold and the growth rate

Let us calculate the characteristic exponents of the near resonance modulations at $\Omega \cong 2\omega_0$ from the beating of the oscillations $x_\omega(t)$ and $x_\Omega(t)$ to produce the difference frequency $\Omega - \omega_0 \cong \omega_0$ to form a resonant oscillation. The equation

$$\ddot{x} + \omega_0^2(1 + h \cos \Omega t) x(t) = 0 \tag{1.21}$$

for small h is solved by expanding the solution as

$$x(t) = a(t) \cos\left(\omega_0 + \frac{\delta\omega}{2}\right) t + b(t) \sin\left(\omega_0 + \frac{\delta\omega}{2}\right) t \tag{1.22}$$

assuming that $da/dt \ll \omega_0 a$ and $db/dt \ll \omega_0 b$. Here we let $\Omega = 2\omega_0 + \delta\omega$ defining $\delta\omega$ as the mismatch between Ω and $2\omega_0$. The beat frequencies in (1.21) and (1.22) arise from the products of $\cos(2\omega_0 + \delta\omega)t \cos(\omega_0 + \delta\omega_0/2)t = 1/2 \cos(3\omega_0 + 3/2\, \delta\omega)t + 1/2 \cos(\omega_0 + \delta\omega/2)t$, since $(\omega_0 + \delta\omega/2) - (2\omega_0 + \delta\omega) = -(\omega_0 + \delta\omega/2)$ where $2\omega_0 + \delta\omega$ is the driver frequency Ω. Substitute Eq. (1.22) into (1.21) and using $\ddot{a} \ll 2\omega_0 \dot{b}$, $\ddot{b} \ll 2\omega_0 \dot{a}$, the system reduces to

$$\begin{pmatrix} \dot{a} \\ \dot{b} \end{pmatrix} = \begin{pmatrix} 0 & -\frac{\delta\omega}{2} - \frac{\omega_0 h}{4} \\ \frac{\delta\omega}{2} - \frac{\omega_0 h}{4} & 0 \end{pmatrix} \begin{pmatrix} a \\ b \end{pmatrix}. \tag{1.23}$$

It is easy to generalize the system to include weak growth or damping terms γ_1, γ_2 in the diagonal position of the matrix in Eq. (1.23). The solution of Eq. (1.23) is

$$\begin{pmatrix} a(t) \\ b(t) \end{pmatrix} = \begin{pmatrix} a_0 \\ b_0 \end{pmatrix} e^{st} \tag{1.24}$$

with

$$s^2 = \frac{\omega_0^2}{4}\left(\frac{h^2}{4} - \frac{\delta\omega^2}{\omega_0^2}\right). \tag{1.25}$$

Thus, the fixed point at $x = \dot{x} = 0$ is destabilized when the modulation amplitude h exceeds the frequency mismatch by

$$h > h_{\text{crit}} = 2\left|\frac{\delta\omega}{\omega_0}\right|. \tag{1.26}$$

At the critical value $h = h_{\text{crit}}$ (where $s^2 = 0$) there are two linearly independent periodic solutions (ce_1 and se_1) of the equation. These correspond to the eigenvalue problem given by Eq. (1.21) with the boundary condition of a periodic solution $x(\theta + \pi) = x(\theta)$ with $\theta = \Omega t/2$. The fundamental periodic solutions are called $ce_1(\theta, q)$ and $se_1(\theta, q)$ with $q = -ha/2$ and $a = 4\omega_0^2/\Omega^2$ in Abramowitz and Stegun (1970) (20.2.30) and are summarized in Table 1.2. For $q > 1$ higher order resonances denoted by the integer r are important with the eigenvalues $a_r(q)$ and $b_r(q)$ given in Table 1.2. For $h < h_{\text{crit}}$ the solutions are stable quasiperiodic oscillations given by Eq. (1.22) with $a(t), b(t)$ given by the real parts of the system (1.24)–(1.25). In this case the oscillation has a low frequency

$$\nu = is = \tfrac{1}{2}\sqrt{\delta\omega^2 - h^2\,\omega_0^2/4} \tag{1.27}$$

modulation of its amplitude given by $\cos(\nu t)\cos(\omega_0 + \delta\omega/2)t$.

Table 1.2. Mathieu Function Properties

Small q
$a_1(q) = 1 + q - q^2/8$
$b_1(q) = 1 - q - q^2/8$
$a = a_1(q)\quad x(t) = ce_1(t,q) \cong \cos t - \tfrac{q}{8}\cos 3t$
$a = b_1(q)\quad x(t) = se_1(t,q) \cong \sin t - \tfrac{q}{8}\sin 3t$
Large q
$a_r(q) \sim b_{r+1}(q) \simeq -2q + \sqrt{q}\,(2r+1)$

These motions are described as stable or unstable about the fixed point of Eq. (1.21) by introducing the phase plane with $(x, y) \equiv (x, \dot{x})$ and noting that $(x, y) = (0, 0)$ is a fixed point or exact equilibrium solution. For finite amplitude solutions near $(0, 0)$ the motion diverges exponentially along the unstable manifold when condition (1.26) is satisfied and rotates around $(0, 0)$ with the angular frequency ν when $h < h_{\text{crit}}$.

The condition that $x(\theta)$ be periodic gives the eigenvalue condition that $s\pi = -i\nu\pi = -in\pi$ with $n = 0, \pm 1, \pm 2, \ldots$. The eigenvalue condition determines a banded structure of the frequency shift $\delta\omega = \Omega - 2\omega_0$ as a function of h (or $q = -ha/2$).

For the work of calculating the behavior of nonlinear dynamical systems the use of a computer is indispensable. Fortunately, relatively inexpensive personal computers along with reference books such as the *Handbook of Mathematical Functions* (Abramowitz and Stegun, 1970) allows the researcher to make advanced calculations. The second valuable tool for nonlinear research is the use of software packages such as *Mathematica* (Wolfram, 1991). We will show the details of some of these more typical calculations using these tools in a few examples.

1.3.2 Mathieu equation and Floquet theory

It is useful to know a little about the higher order modulational decay solutions both for the case of stronger drivers and for the problem of the spectrum

1.3 Parametric Instabilities

of eigenmodes in toroidal confinement systems with periodic boundary conditions. A standard dimensionless form (Abramowitz and Stegun, 1970) for the parametric equation (1.21) is the Mathieu equation

$$\ddot{y} + (a - 2q\cos 2t)y = 0. \tag{1.28}$$

In the case of Eq. (1.21) $a = (2\omega_0/\Omega)^2$ and $2q = -ha$. The periodic solutions, which correspond to the critical condition $h = h_{\text{crit}}$ above, are found along the eigenvalues $a = a_r(q)$ and $a = b_r(q)$ where $r = 0, 1, 2, 3\ldots$ with cosine-like eigenfunctions for $a_r(q)$ and sine-like eigenfunctions for the $b_r(q)$ eigenvalues. The stable (S) and unstable (US) domains in (a, q) parameter space are shown in Fig. 1.2a, and the principal first order resonance is given in detail in Fig. 1.2b.

For small and large q the eigenvalues for periodic solutions are given by

$$a_r \simeq b_{r+1} \simeq -2q + \sqrt{q}(2r+1) \qquad q > 1$$

and for $r > 1$ (1.29)

$$a_r \simeq b_r = r + q^2/2(r^2 - 1) \qquad q < 1$$

The diagram for the stability in the a-q-plane is shown in Fig. 1.2 following Abramowitz and Stegun (1970) (p. 724). The periodic solutions along the marginal stability curves are defined as $y(t) = ce_r(t,q) \sim \cos(rt)$ and $y(t) = se_r(t,q) \sim \sin(rt)$ where the $q \ll 1$ limits are indicated. Inside the resonant regions the solutions are exponentially growing (marked US for unstable), and outside the resonant domains (marked S for stable) they are amplitude modulated oscillatory solutions.

The generalization of these results to an equation of the form

$$\ddot{x} + \omega^2(t)x = 0 \tag{1.30}$$

with

$$\omega(t + \tau) = \omega(t) \tag{1.31}$$

is called Floquet Theory and is an important area of study in the stability theory of periodic cycles. Briefly, the properties of the solutions of Eq. (1.30) are as follows:

If $x(t)$ is a solution of Eq. (1.29) then so is $x(t + \tau)$, so that if $x_1(t)$ and $x_2(t)$ are two linearly independent solutions of the equation, then there is a matrix A_{ij} such that

$$\begin{pmatrix} x_1(t+\tau) \\ x_2(t+\tau) \end{pmatrix} = \begin{pmatrix} A_{11} & A_{12} \\ A_{21} & A_{22} \end{pmatrix} \begin{pmatrix} x_1(t) \\ x_2(t) \end{pmatrix}. \tag{1.32}$$

The eigenvalues of A are the characteristic exponents of the system as given for the special case of the Mathieu equation by Eqs. (1.24)–(1.25). In terms of the eigenvalues λ_1 and λ_2 and eigenvectors we have

$$x_1(t+\tau) = \lambda_1 x_1(t)$$
$$x_2(t+\tau) = \lambda_2 x_2(t), \tag{1.33}$$

and it can be shown that $\lambda_1 \lambda_2 = 1$ by using the constancy of the Jacobian or Wronskian $x_1 \dot{x}_2 - x_2 \dot{x}_1$.

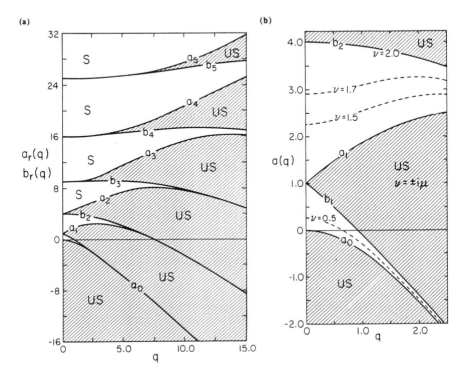

Fig. 1.2. The stability domains and the eigenvalues for periodic solutions of the Mathieu equation. The stable domains are marked S, and the unstable domains that expand out of the resonances at $a_n = n^2$ for $q \to 0$ are marked US. Along the neutral curves $a_n(q)$ and $b_n(q)$ the even and odd periodic solutions occur. (a) is the large scale stability diagram showing $n = 1, 2, 3, 4$ and 5 resonances, and (b) is the enlargement of the primary, first order $n = 1$ resonance region, giving the parametric decay $\Omega = \omega_0 + \omega_0 = 2\omega_0$ in Sec. 1.3.1. Figures 1.2 are adopted from Fig. 20.8 of M. Abramowitz and Stegun (1970).

If λ_1 is real, then $\lambda_1 = e^{\mu\tau}$ and $\lambda_2 = e^{-\mu\tau}$ and the solutions are unstable. If λ_1 is complex, then $\lambda_1 = e^{i\nu\tau}$ and $\lambda_2 = e^{-i\nu\tau}$ and the solutions are amplitude modulated with the time variation given by $e^{\pm i\nu t} P(t)$ where $P(t + \tau) = P(t)$. The critical conditions for periodic solutions are that $\lambda_1 = \pm 1$ or that $\nu\tau/\pi = 0, 1, 2 \ldots$. For the Mathieu equation (1.27) the period is $\tau = \pi$ and the values of ν are indicated in Fig. 1.2(b). The unstable eigenvalues $\lambda = e^{\mu\pi}$ are given in Fig. 1.3 for the first order resonance region.

1.4 Example 1.1 Parametric Excitation of ...

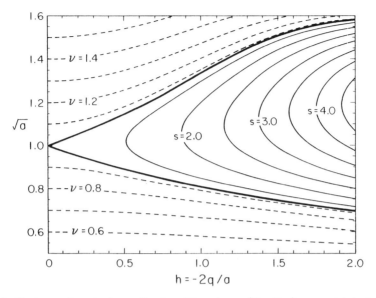

Fig. 1.3. Contours of constant growth rates of the primary ($\Omega = 2\omega_0$) parametric instability in the $h = -2q/a$ versus $\sqrt{a}$ plane. The solid lines inside the unstable domain give the amplification factor $s = e^{i\nu\pi} = e^{\gamma\pi}$ per period, and the dashed lines in the stable domain give the frequency shift ν.

1.4 Example 1.1 Parametric Excitation of the Pendulum

For a pendulum of length ℓ with a pivot point that oscillates vertically as $z(t) = \Delta z \cos(\Omega t)$ find the critical amplitude for the destabilization of the stable fixed point corresponding to small amplitude oscillations at $\omega_0 = \sqrt{(g/\ell)}$. Estimate the maximum of the finite amplitude oscillations after the period doubling bifurcation.

Based on physical reasoning one may recognize that the effective gravity for the blob of mass m is $g_\text{eff} = g + \ddot{z}(t)$ due to the acceleration of the pivot and write down the equation of the nonlinear, modulated pendulum of length ℓ

$$\ddot{\theta} + \frac{1}{\ell}(g + \ddot{z}(t))\sin\theta = 0.$$

Alternatively, this equation may be derived by using the Lagrangian $L = T - U$ with $T = (m/2)\left[(\ell\dot{\theta}\cos\theta)^2 + (\ell\dot{\theta}\sin\theta + \dot{z})^2\right]$ and $U = -mg\ell\cos\theta + z(t)$. For $z = z_0 + \Delta z \cos(\Omega t)$ the effective frequency is $\omega_0^2(1 - h\cos(\Omega t))$ with $h = \Omega^2 \Delta z/g$ and $\omega_0^2 = g/\ell$. The first threshold for destabilization of the stable fixed point is at the resonance $\Omega \cong 2\omega_0 = 2(g/\ell)^{1/2}$ where for $h > h_{\rm crit}$ given in Eq. (1.9) the oscillations are unstable exponentially spiralling out of the $(\theta, \dot{\theta}) = (0,0)$ fixed point in the phase plane. The width of the unstable frequency mismatch domain given by Eq. (1.26) is

$$|\Omega - 2\omega_{\rm osc}| < \tfrac{1}{2}\omega_0 h = \frac{\omega_0}{2}\left(\frac{\Omega^2 \Delta z}{g}\right) \qquad (1.34)$$

corresponding to the shaded region defined by $b_1 < a(q) < a_1$ in Fig. 1.2b.

For finite $\theta(t) = \theta_m \cos(\omega_0 t)$ the oscillator frequency according to Eq. (1.16) becomes

$$\omega = \omega_0(1 - \theta_m^2/16)$$

since $\alpha = 0$ and $\beta = -\omega_0^2/6$ for the small θ expansion,

$$\ddot{\theta} + \omega_0^2(\theta - \theta^3/6) = 0,$$

of the pendulum oscillator ($\ddot{\theta} + \omega_0^2 \sin\theta = 0$). We estimate that the steady state is reached when the frequency shift results in the critical condition in Eq. (1.26) being satisfied. This estimate gives the amplitude of the solution that bifurcates from the parametric instability as

$$|\Omega - 2\omega_{\rm osc}| \simeq 2\delta\omega^{n\ell} = 2\omega_0 \,\theta_m^2/16 = h\omega_0/2.$$

Using $h = \Omega^2 \Delta z/g$, the maximum oscillation amplitude θ_m is given approximately by

$$\theta_m^2 = 4(\Delta z\, \Omega^2/g).$$

To demonstrate the use of the formulas developed here for the parametric instability and the nonlinear oscillator frequency shift, we give the numerical solution for this example. For $\Omega/\omega_0 = 2$ and $h = \Omega^2 \Delta z/g = 1/4$ the small amplitude solution, which is in the unstable parameter domain, grows as shown in Fig. 1.4 with $\theta(t) \cong \theta_0 \exp(\omega_0 t/16)\cos(\omega_0 t)$ as given by Eqs. (1.22)–(1.25). The exponential growth continues until the amplitude is such that nonlinear frequency shift reaches the value $2\delta\omega^{n\ell} = \theta_m^2/8 = h/2 = 1/8$, which is the critical value for $s^2 = 0$ in Eq. (1.25). The numerical solution peaks at $\theta_m \cong 1.3$ and then returns to a low amplitude. This type of nonlinear amplitude limit of the parametric instability is a common phenomena in plasma dynamics. An imortant example occurs in the plasma beat wave accelerator of Tajima and Dawson (1979).

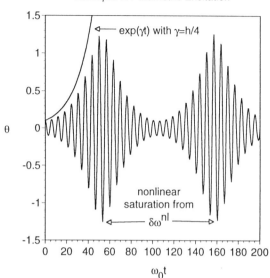

Fig. 1.4. Time dependence of the angle $y_1 = \theta(t)$ of the parametrically driven pendulum in Example 1.1. The function shows the growth due to the parametric resonance at small $|\theta(t)|$ and the saturation of the growth due to nonlinear frequency shift away from resonance at $|\theta| \lesssim \pi/2$.

1.5 Example 1.2 Stability of the Ponderomotive Potential

The motion $x(t)$ of a charged particle in a high frequency ω electromagnetic field can often be described by the motion of its average $x(t) = \langle x \rangle \to X(t)$ position in an effective potential $U(X)$ called the ponderomotive potential. We show a simple example where the ponderomotive potential description fails when the amplitude of the potential exceeds a limit determined from the Mathieu equation.

Consider the Hamiltonian for a particle in the presence of equal amplitude left and right running waves as given by

$$H = \frac{p^2}{2m} - \frac{e\varphi_0}{2}\cos(k(x+v_{ph}t)) - \frac{e\varphi_0}{2}\cos(k(x-v_{ph}t))$$

$$= \frac{p^2}{2m} - e\varphi_0 \cos(kx)\cos(\omega t) \qquad (1.35)$$

where $\omega = kv_{ph}$.

The Hamiltonian equations of motion $\dot{x} = \partial H/\partial p = p/m$ and $\dot{p} = -\partial H/\partial x$ reduce to
$$\ddot{x} = -\frac{ek\varphi_0}{m}\sin(kx)\cos(kv_{ph}t) = f(x)\cos(\omega t). \tag{1.36}$$
For sufficiently large ω we may expand the force as
$$f(x = X + \delta x) = f(X) + \delta x(\partial f/\partial X), \tag{1.37}$$
and for $\delta\ddot{x} \cong f(X)\cos(\omega t)$ which integrates to give the "quiver" motion
$$\delta x(t) = -\frac{f(X)}{\omega^2}\cos(\omega t). \tag{1.38}$$
Retaining the small term $\delta x\,\partial f/\partial X$ in Eq. (1.37) and using Eq. (1.38) we obtain for the average motion
$$\ddot{X} = \langle\ddot{x}\rangle = \left\langle \delta x\frac{\partial f}{\partial X}(X)\cos(\omega t)\right\rangle = -f(X)\frac{\partial f}{\partial X}\frac{\langle\cos^2(\omega t)\rangle}{\omega^2}$$
$$= -\frac{\partial}{\partial X}\left(\frac{f^2(X)}{4\omega^2}\right) = -\frac{\partial U}{\partial X} \tag{1.39}$$
where the ponderomotive potential $U(X) = f^2(X)/4\omega^2$, and in the case of Eq. (1.36), has stable minima at the positions $kX_n = n\pi$. Thus the ponderomotive force description leads to stable motion at $X_n = n\pi/k$.

Now we consider the full equation of motion (1.36) noting that $\dot{x} = 0$, $x_n = n\pi/k$ are fixed points. Consider the motion $\xi_n(t) = x(t) - n\pi/k$ about these fixed points in the linear approximation. We obtain from the linearization of Eq. (1.36) with $\omega t = 2\theta$ the Mathieu equation
$$\left[\frac{\omega^2}{4}\frac{d^2}{d\theta^2} - \omega_b^2(-1)^n\cos(2\theta)\right]\xi_n(\theta) = 0$$
where $\omega_b \equiv k(e\varphi_0/m)^{1/2}$ is the bounce frequency for motion in the bottom of the sinusoidal potential. The (a,q) parameters of the Mathieu equation are
$$a = 0 \quad\text{and}\quad q = (-1)^n\frac{2\omega_b^2}{\omega^2}. \tag{1.40}$$
From Fig. 1.2 we see that the solution $\xi_n(\theta)$ is exponentially growing for $q = q_c \geq 0.91$. The nonlinear analysis (Schmidt and Bialek, 1982) shows that at this point where
$$k\left(\frac{e\varphi_0}{m}\right)^{1/2}_c = \omega_b = \left(\frac{q_c}{2}\right)^{1/2}\omega \tag{1.41}$$
the stable elliptic points at $(X_n, \dot{X}) = (n\pi/k, 0)$ bifurcate into two neighboring elliptic points surrounded by a chaotic figure-eight boundary as shown in Chapter 3.

Thus the ponderomotive potential is not qualitatively correct for $\omega_b/\omega \gtrsim 0.67$. Of course, the original idea is based on ω being sufficiently large so this result quantifies the meaning of the high frequency limit on ω.

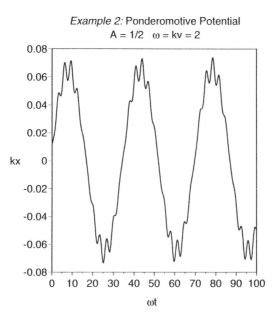

Fig. 1.5. Coordinate of the particle in the standing wave in Example 1.2 showing the small, high frequency oscillations in $\delta x(t)$ about the low frequency motion $X(t)$ of the oscillation center governed by the ponderomotive potential $U(X)$ in Eq. (1.39).

In Fig. 1.5 we show the numerical solution of Eq. (1.36) in the dimensionless form $\dot{x} = p; \dot{p} = -A \sin x \cos vt$. The long-time scale oscillations for $q = 2A/v^2 = 2(0.5)/(2)^2 = 1/4$ are due to the trapping of the particle in the ponderomotive potential $U(X) = (A^2/2v^2) \sin^2 X$ (Eq. (1.39)) and have the period $T = 2\pi v/A \cong 8\pi$ for small (X). The small ripples on the trajectory in Fig. 1.5 are due to the standing waves with $\omega = kv_{ph} = 2$ in Eq. (1.36). By decreasing the driving wave frequency the unstable motion of the particle is found.

1.6 Example 1.3 Oscillations of a Charged Particle in Two Longitudinal Waves

Now we generalize Eq. (1.35) to the case of two waves with different amplitudes (φ_1, φ_2), wavenumbers (k_1, k_2) and frequencies (ω_1, ω_2). The two-wave Hamiltonian is

$$H = \frac{1}{2}v^2 - \frac{e\varphi_1}{m}\cos(k_1 x - \omega_1 t) - \frac{e\varphi_2}{m}\cos(k_2 x - \omega_2 t + \alpha). \quad (1.42)$$

Clearly, the important physical consideration is the difference in the phase velocities of the wave $\Delta v_{1,2} = \omega_1/k_1 - \omega_2/k_2$ compared with the trapping velocity $v_{tr} \equiv (e\varphi/m)^{1/2}$ of the larger of the two waves. When $\Delta v_{1,2} \gg \max(v_{tr})$ the situation is similar to that in Example 2 with the particle experiencing a high frequency $\omega = k\Delta v_{1,2}$ oscillation unless it is near the trapping region of one of the two waves.

For the analysis of the motion one transforms to a reference frame in which the larger of the two waves, say $\varphi_1 > \varphi_2$, is at rest by introducing $x' = x - (\omega_1/k_1)t$. In this frame the phase velocity of wave φ_2 is $\Delta v_{2,1} = \omega_2/k_2 - \omega_1/k_1$. Now using the wavelength of the larger wave to measure distance $k_1 x' \to x''$ and scaling time $k_1(\omega_2/k_2 - \omega_1/k_1)t = k_1 \Delta v_{2,1} t \to t'$, the velocities are measured in $dv/dt \to \Delta v_{2,1}(dx'/dt')$. In these dimensionless variables the canonical form (Escande and Doveil, 1981) of the two wave problem is given by the Hamiltonian

$$H = \frac{1}{2}v^2 - M\cos x - P\cos[k(x-t)] \quad (1.43)$$

where $M = e\varphi_1/m(\Delta v_{1,2})^2$, $P = e\varphi_2/m(\Delta v_{1,2})^2$ and $k = k_2/k_1$. The transformation to these dimensionless variables, where the primes on x, t have been dropped, identifies the three-dimensionless parameters (M, P, k) of the two-wave problem.

A typical phase space portrait of the system is shown in Fig. 1.6. For $P = 0$ the M-wave separatrix between passing and trapped motion occurs for $H = M$ which from Eq. (1.43) gives $(\dot{x})_{sx} = \pm 2 M^{1/2} \cos(x/2)$. For $M = 0$ the P-wave separatrix is $(\dot{x})_{sx} = 1 \pm 2P^{1/2} \cos(kx/2)$. For finite M and P the motion becomes stochastic when the combined widths of these trapping resonance regions given by

resonance overlap parameter s

$$s = 2M^{1/2} + 2P^{1/2} \quad (1.44)$$

approaches the width of the region between the phase velocities $\Delta v_{1,2}$ of the waves, which is unity by the choice of normalized (x,t) variables in Eq. (1.43). The resonance criterion $s \cong 1$ for stochasticity only applies when the amplitudes M and P are of comparable magnitude. The curve (1.44) roughly separates the region for stochastic behavior of the particle and is shown in Fig. 1.7 by the curve $2M^{1/2} + 2P^{1/2} = 1$ labelled S. Escande and Doveil (1981) and Escande (1983, 1985) give a renormalization map procedure for computing the *critical curve* that determines the boundary in M, P, k space for the onset of stochasticity. A few elements of their lengthy, but informative analysis are given here.

1.6 Example 1.3 Oscillations of a Charged Particle in ...

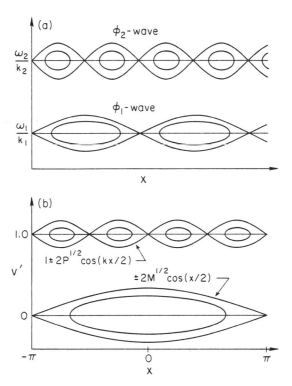

Fig. 1.6. The single wave trapping domains drawn on the particle phase space ($x, v = p/m$) for the two longitudinal waves in Example 1.3. The reduction to the dimensionless variables (M, P, k) in the rest frame of the bigger wave is indicated in going from frame (a) to frame (b).

In the limit $M \gg P$ it is useful to evaluate the effect of the P-wave along orbits of the integrable M-wave motion. For this purpose one introduces the elliptic integrals given in Table 1.3 by letting $m = \sin(x_0/2) = (E+M)^{1/2}/(2M)^{1/2}$ so that $\dot{x}^2 = 4M(\sin^2(x_0/2) - \sin^2(x/2))$. Then transforming to

$$\sin\left(\frac{x(t)}{2}\right) = \sin\left(\frac{x_0}{2}\right) \sin\varphi(t) \tag{1.45}$$

yields

$$u = M^{1/2} t = \int_0^\varphi \frac{d\varphi'}{\sqrt{1 - m \sin^2 \varphi'}} = \operatorname{sn}^{-1}(\varphi, m) \tag{1.46}$$

or $\operatorname{sn}(u, m) = \sin\varphi$. The right-hand side of Eq. (1.46) is the definition of the first of the tripolar Jacobi functions sn, cn and dn.

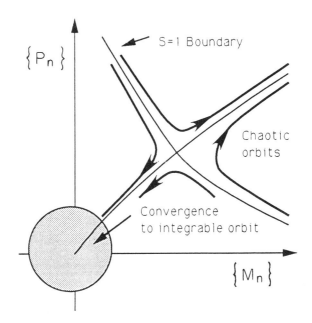

Fig. 1.7. The two-parameter plane (M, P) of the two-wave problem in Example 1.3 showing how the sequence of renormalization transformations maps the sequence of $\{M_n, P_n\}$ parameters either to the convergence region at the origin or to a divergence region depending on the position relative to the resonance overlap parameter condition $s(M, P) = 1$ defined in Eq. (1.44).

Now φ is a periodic function of u with one quarter of the period being $\int_0^{\pi/2} (1 - m \sin^2 \varphi)^{-1/2} d\varphi \equiv K(m)$. Thus we may seek the Fourier series expansion of φ and any function $F(\varphi)$ in terms of $\cos(\ell\, 2\pi u/4K(m))$ with $\ell = 0, 2, 4 \ldots 2n$ for even functions.

In particular the x-dependence of the P-wave along the orbit from the M-wave can be expanded in the Fourier series. Obtaining the coefficient $V_n(m)$ of the Fourier series is a difficult problem in analysis. The result given by Escande and Doviel (1981) is

$$P \cos[k(x - t)] = P \sum_{n=-\infty}^{+\infty} V_n(m) \cos[(k + n)\theta - kt] \qquad (1.47)$$

from which we see that the P-wave acts as a sum of waves at the velocities $d\theta/dt = k/(k + n)$ with strength $V_n(m)$. Calculation of $V_n(m)$ is difficult and we refer the reader to Escande and Doviel (1981) and Hatori and Irie (1987).

The secondary waves, (1.47), as seen in the motion of the particle along the primary wave, now act to produce a chain of islands. The renormalization map procedure is to recognize that once the most important of the two new waves from the spectrum in Eq. (1.47) are identified, a second set of transformations from (M, P, k) to (M', P', k') can be carried out to form a new two-wave problem. The new two-wave problem is then subjected to the same analysis, and the subsequent sequence of transformations to each new two-wave problem defines the renormalization map $R(M, P, k)$. When the sequence of transformations converges to the fixed point at the origin of the M-P parameter space, that is the sequence of transformed $M, P \to 0$ as indicated in Fig. 1.7, the system is integrable. When the iterations of the renormalization map $R(M, P, k)$ fail to converge to a fixed point at the origin, the system has global stochasticity. We return to the study of the stochasticity in this generic type of $1\frac{1}{2}$-dimensional Hamiltonian system in Chapters 2 and 3.

Table 1.3. Nonlinear Oscillations of the Pendulum
$H(p, \theta) = \frac{1}{2} p^2 - U \cos \theta$ with $H(p, \theta) = E = \text{const}$
Jacobi Elliptic Function Ref. Abramowitz-Stegun

Rotation $E > U$	Oscillations $E < U$
Amplitude $= \phi(t)$	Amplitude $= \psi(t)$
$\phi(t) = \frac{1}{2} \theta(t) = \text{am}\left[\left(\dfrac{E+U}{2}\right)^{1/2} t, k\right]$	$\sin\left(\dfrac{\theta}{2}\right) = \sin\left(\dfrac{\theta_0}{2}\right) \sin \psi(t)$
$k = m^{1/2} = \left(\dfrac{2U}{E+U}\right)^{1/2} < 1$	$k = m^{1/2} = \left(\dfrac{E+U}{2U}\right)^{1/2}$
$u = \text{phase} = \left(\dfrac{E+U}{2}\right)^{1/2} t$	$u = \text{phase} = U^{1/2} t$
$\dfrac{d\phi}{du} = \text{dn}(u, k)$	$\sin(\theta/2) = m^{1/2} \sin \psi = m^{1/2} \text{sn}(u, m)$
$\sin \phi = \text{sn}(u, k)$	$\cos \psi = \text{cn}(u, m)$
$\cos \phi = \text{cn}(u, k)$	$\dfrac{d\text{sn}}{du} = \dfrac{m^{1/2} \text{cn}(u, m)}{\text{dn}(u, m)} \dfrac{d\psi}{du}$
Energy conservation	Energy conservation
$\left(\dfrac{d\phi}{du}\right)^2 + k^2 \sin^2 \phi = 1$	$\left(\dfrac{d\psi}{du}\right)^2 + m \sin^2 \psi = 1$

Experimental realization of the onset of stochastic velocity diffusion at the overlap condition $s \approx 1$ for the two-wave problem in Example 3 has been demonstrated by Fasoli et al. (1993). Using laser-induced florescence to optically tag barium ions in a Q-machine, the experimenters are able to directly measure the parallel velocities of the ions. The velocity distribution shows the sharp onset of stochastic diffusion at the point where the overlap criterion $s \approx 1$ is satisfied. The collisional diffusion below the onset of stochasticity is one order of magnitude smaller than the two-wave induced diffusion above the threshold in Eq. (1.44) for the onset of chaotic orbits.

Chapter 2

Hamiltonian Dynamics

An important class of dynamical systems occurs when the flow of the system points conserves a measure such as phase space volume. Such measure preserving flows have distinctly different properties and theorems from the volume contracting flows that occur in dissipative systems. The constraint on the flow $\dot{x}_i = f_i(x)$ of conserving volume in the x_i-vector space is a strong restriction that allows the dynamics to be derived from the gradients of a scalar function called the Hamiltonian $H(p,q,t)$.

The topic of this Chapter, Hamiltonian dynamics, is a broad field of research itself. Fortunately, many excellent texts exist on Hamiltonian dynamics, and the reader undoubtedly has some experience with the subject and his or her own favorite books on the topic. Here we develop only those concepts and results that are now a core part of the tools of plasma physics. The reader will need to consult texts such as Arnold, *Mathematical Methods of Classical Mechanics* (1978), Lichtenberg and Lieberman's *Regular and Chaotic Motions* (1991), or the comprehensive work by the plasma physicists who played a key role in developing the subject, *Nonlinear Physics*, by Sagdeev, Usikov, and Zaslavsky (1988). Other relevant works are cited in the course of the Chapter.

2.1 Measure-Preserving Flows and the Hamiltonian

In simple cases of a Euclidean vector space for x_i the measure preserving condition can be expressed as conservation of phase space volume by the requirement that

$$\sum_{i=1}^{n} \frac{\partial \dot{x}_i(x)}{\partial x_i} = 0. \tag{2.1}$$

Equation (2.1) guarantees the conservation of the volume integrals $\int dx_1\, dx_2, \ldots dx_n$ taken along the flow, called the Poincaré "integral invariants," to distinguish such time-moving invariants from the usual invariants such as energy and momentum in mechanics. General geometrical forms of the integral invariants of Poincaré are found in Sec. 9.44 of Arnold (1978). For $n = 2$ and 3 the constraint (2.1) is the same as the condition of incompressible flow in hydrodynamics. In fact 2D incompressible hydrodynamic flows provide one of the simplest realizations of Hamiltonian dynamics in the laboratory. In this case the constraint in Eq. (2.1) is the necessary and sufficient condition for the existence of the stream function ψ giving $(x = x_1(t), y = x_2(t))$ the flow

$$\dot{x} = -\frac{\partial \psi}{\partial y}(x, y, t)$$
$$\dot{y} = \frac{\partial \psi}{\partial x}(x, y, t) \tag{2.2}$$

which is the $1\frac{1}{2}$-D Hamiltonian system with $x = p, y = q$ and the Hamiltonian $H = \psi$. System (2.2) is a very important special case in both hydrodynamics and plasma physics, and the theory is developed in Secs. 3.5. Complete, systematic mathematical developments of the subject are to be found in a text of Arnold (1978), in the review article by Chirikov (1979), and in a text by Lichtenberg and Lieberman (1991).

A general dynamical system is described in canonical pairs of coordinates called $p_i(t)$ and $q_i(t)$ for each degree-of-freedom of the system. For N degrees-of-freedom the $2N$ vector space of $\mathbf{p}, \mathbf{q}$ is defined as the phase space of the system. The condition of volume preservation for the flow is satisfied by the existence of the scalar Hamiltonian function $H(\mathbf{p}, \mathbf{q}, t)$ as the generator of the equations of motion

$$\frac{dp_i}{dt} = -\frac{\partial H}{\partial q_i}$$
$$\frac{dq_i}{dt} = \frac{\partial H}{\partial p_i}. \tag{2.3}$$

The time derivative of any function $F(p, q, t)$ is then given by the Poisson brackets $\{H, F\}$ since

$$\frac{dF}{dt} = \sum_{i=1}^{N} \frac{\partial F}{\partial q_i}\frac{dq_i}{dt} + \frac{\partial F}{\partial p_i}\frac{dp_i}{dt} + \frac{\partial F}{\partial t}$$

$$= \sum_{i=1}^{N}\left[\frac{\partial F}{\partial q_i}\frac{\partial H}{\partial p_i} - \frac{\partial F}{\partial p_i}\frac{\partial H}{\partial q_i}\right] + \frac{\partial F}{\partial t} \tag{2.4}$$

$$: = \{H, F\} + \frac{\partial F}{\partial t}. \tag{2.5}$$

As a special case we choose $F = H$ to find that

$$\frac{dH}{dt} = \frac{\partial H}{\partial t}(p, q, t), \tag{2.6}$$

and thus for a closed system in which H is only a function of p, q and independent of time the Hamiltonian $H(p,q) = E$ is constant. The value of the constant E is in most cases the physical energy of the system. In the case where $H(p,q) = E =$ constant the dynamics is restricted to a $2N-1$ dimensional surface in the $2N$ dimensional phase space.

A most common and well-studied example occurs for the closed two degree-of-freedom problem. Here $N = 2$ and the energy surface $H = E$ constrains the flow to evolve in a $d = 3$ vector space in the 4-dimensional phase space. The topology of the Hamiltonian dynamics in the 3-dimensional vector space has been studied exhaustively over the past decade. Having a physical problem reduce to this $d = 3$ system, also called the $1\frac{1}{2}$ degree-of-freedom problem, is tantamount to having the problem solved.

All the microscopic equations of classical physics have this Hamiltonian structure or can be cast into the Hamiltonian structure (Kandrup and Morrison (1993)).

Of course, we must keep in mind that Newton's laws of motion are more general than Hamiltonian dynamics. Newton's laws apply to both conservative and dissipative forces such as friction. The dissipative forces produce a volume contracting flow in the phase space which attempts to shrink all the phase space structures to lower dimensional regions called attractors. The complexity of the attractors and strange attractors is shown in the simple example of the driven-damped pendulum. The richness of the chaotic motion of this dissipative $1\frac{1}{2}$ degree-of-freedom problem is so great that the entire monograph of Baker and Gollub (1990) is devoted to describing the dynamics and phase space attractors for the driven-damped pendulum.

The phase space structures for Hamiltonian systems differ qualitatively from those of dissipative systems. The principle difference is in the long-time limit the chaotic solutions in the dissipative systems fill only a subspace of dimension $d_f < d$ while the chaotic solutions of the measure preserving (Hamiltonian) system fill a finite fraction of the full phase space d. The dimension d_f is typically a fractal for the dissipative system.

2.2 Poincaré Surface of Section

The measure preserving flow property (2.1) of the Hamiltonian system provides a natural way of analyzing the dynamics in a system with bounded trajectories. A surface in the phase space is introduced, and intersections of a system's trajectory with the surface are recorded as points. Only intersections or punctures of the surface in one direction (from, for example, the negative to the positive side of the surface) are recorded. When only the information given by the points of intersection of the trajectory with this surface of section is considered, the complex flow from the differential equations is reduced to a measure preserving mapping of points on the surface. This technique of constructing one or more surfaces of section and studying the mapping of the points on the surface may be the single most important tool we have for analyzing the character of the nonlinear dynamics. As we shall see it is used over and over and is even realized experimentally to find the confinement properties in toroidal magnetic plasma confinement devices.

The first systems to show chaotic behavior are those with a $d = 3$ phase space. Since an N-degree-of-freedom problem has a $d = 2N$ phase space these systems are traditionally called a $1\frac{1}{2}$-D degree-of-freedom problems. These $1\frac{1}{2}$-D problems abound in the study of dynamics in engineering and physics. The surface of section is a two-dimensional surface in the three-dimensional phase space.

2.3 Fixed Points and Invariant Curves

The mapping of the points on the surface of section created by the flow from the equations of motion is called T and after n passes through the surface the points on the surface of section, sometimes abbreviated as SoS, are mapped by T^n. A fixed point of T^n is a point which is mapped onto itself after n applications of T, that is, by T^n. Such a point, $x_i = T^n(x_i)$, is a period-n fixed point. Generally, there are both points (fixed points) and curves on the surface of section that map by T onto themselves. A curve C that maps onto itself is called an invariant curve.

In the systems we consider there is always at least one fixed point and often a large number of fixed points. In the vicinity of the fixed points the mapping can be linearized, and the behavior of the neighboring flow is used to characterize the fixed point as stable or unstable. For stable fixed points the eigenvalues of the linearized map lie on the unit circle in the complex plane and occur in complex conjugate pairs.

For unstable fixed points the eigenvalues λ of the linear map move off the unit circle in pairs (λ_1, λ_2) with the area preserving property requiring that $\lambda_1 \lambda_2 = 1$. The change of the stability character of the fixed point with the variation of a system parameter μ is called a bifurcation, and the study of the bifurcations of the system with system parameters is one of the ubiquitous studies in science and engineering.

The trajectories in the neighborhood of an unstable fixed point are hyperbolic curves diverging away from the focus at the fixed point. The trajectories in the neighborhood of a stable fixed point are elliptic curves rotating around the focus at the fixed point.

In Chapter 1 the Mathieu Eq. (1.21) provides the standard example of a linear map of the $(x, \dot{x})$-plane by the flow in the $d = 3$ phase space defined by $(x, \dot{x}, \phi = \Omega t)$. The phase space is toroidal in structure with ϕ the angle around the major axis of the torus and $(x, \dot{x}) \rightarrow (r, \theta)$ giving the motion about the mirror axis of the torus defined by the angles (θ, ϕ). The linear mapping is parameterized by $T(a, q)$, and the fixed point is $(0, 0)$. The stability of the fixed point changes from a stable, elliptic fixed point for a, q values inside the domains marked stable (S) in Fig. 1.2 to unstable, hyperbolic in the regions marked unstable (US) in Fig. 1.2. The bifurcations are given by the critical curves $a(q)$ analyzed in Sec. 1.3.1.

In systems of coupled nonlinear oscillators the flow around the fixed point is analyzed for its stability just as in the example described above with the Mathieu equation. About the fixed point there is an $N \times N$ periodic matrix with oscillation frequencies from the ω_ℓ given by the underlying oscillators. What we have learned from the stability diagram of the Mathieu equation is that even when the asymptotic or WKB theory, which applies when $a \gg q$, suggests the existence of an adiabatic invariant(s) $\oint p_\ell \, dq_\ell$ the system has resonances when

$$\sum_{\ell=1}^{p} n_\ell \omega_\ell = 0 \tag{2.7}$$

for which the fixed points are unstable and the flows diverge in hyperbolic trajectories. Here the integers $\{n_\ell\}$ describe the order of the resonance e.g. in Eq. (1.28) $2\omega_0 \pm n\Omega = 0$. These hyperbolic fixed points are the key to the exponential divergence of neighboring orbits in unstable systems. In Sec. 2.5 the Lyapunov exponents are introduced to measure quantitatively the strengths of the unstable divergence of neighboring orbits.

2.4 KAM Theory

Mathematicians have long studied the properties of nonlinear maps T and the mathematical physicists Kolmogorov, and mathematicians Arnold and Moser recognized the importance of applying the properties of the nonlinear maps to the dynamical systems to determine the nature of the stability in the nonlinear motions.

The two-dimensional map

$$\begin{pmatrix} x' \\ y' \end{pmatrix} = T(x, y) \tag{2.8}$$

produced from the flows of systems such as given by Eqs. (2.2) are common in plasma physics. For the theorems of Kolmogorov, Arnold, and Moser, called KAM theory for convenience, to apply the maps need to possess the "twist condition" that produces a monotonically increasing rotation of the points in the surface of section.

The twist condition is that the mapping rotate or twist a line such that

twist condition

$$\frac{\partial y'}{\partial x} > 0 \tag{2.9}$$

for all x. Nontwist maps are developed in del-Castillo-Negrete et al. (1996).

The twist condition guarantees that around a fixed point there is a monotonic increase of the rotation angle $\alpha(r)$ with distance r from the fixed point

$$\alpha' = \frac{d\alpha}{dr} \neq 0. \tag{2.10}$$

In the local (r, θ) coordinates around the fixed point we see that the invariant curves of the $T(x,y)$-mapping are produced by rotation

$$\theta' = \theta + \alpha(r) \tag{2.11}$$

where the rotation number α must be irrational for the images $T^n(x,y)$ to trace out a curve of points as $n \to \infty$. The area of the annulus $r_1 \leq r \leq r_2$ is preserved under the mapping. In sharp contrast are those special situations for which the rotation number is rational $\alpha = (m/n)2\pi$: the map $T^n(x,y)$ returns a point to itself after n iterations of the map. We see that the value of m determines the number of rotations of the image around the fixed point ($r = 0$) during the n iterations of the map. Moser (1962) for example, recognized that it is at these rational values of the rotation angle that the phase space opens up in the presence of a small, generic perturbation into regions containing stable and unstable orbits. From another point of view we may recognize the importance of the rational rotational transforms $\alpha/2\pi = m/n$ as the resonance condition (2.7) found for the onset of the exponentially diverging solution from the Mathieu equation (1.21).

The importance of the twist mapping and the opening up of the invariant curves at the rational values of the rotation number was recognized early in the studies of toroidal plasma confinement systems.

Grad (1967) credits the Moser (1962) theorems in his argument that a 3D perturbation to the integrable 2D system

$$H_0(x,y) \to H_0(x,y) + H_1(x,y,\phi)$$

gives rise to magnetic field line trajectories that are ergodic in an annulus $r_1 < r < r_2$ and thus good magnetic surfaces do not exist in truly 3D toroidal magnetic confinement systems. In Sec. 2.7 we will see the implication of this theorem for the helical/stellarator confinement systems where small magnetic field errors lead to severely degraded confinement times for the charged particles.

2.5 KAM Torii, the Golden Mean and Leaky Barriers

The resonant surfaces defined by $\alpha/2\pi = m/n$ form a dense set since there are rationals m/n arbitrarily close to any irrational value of α. In practice it is only the low order rationals that are dangerous in terms of opening up large ergodic annuli. We will estimate the width of the ergodic regions for small perturbations using the technique of the Melnikov-Arnold integrals and the Whisker maps in Sec. 3.3.

The proof of the KAM theorem is rather technical. Here we begin by discussing some related theorems that convey some of the ideas behind the KAM theorem. After developing the theory for magnetic fields we state and discuss the KAM theorem in Sec. 2.11.

2.5 KAM Torii, the Golden Mean and Leaky Barriers

Ergodic Theorem:

The ergodic theorem states that the long-time average $\bar{f}$ of a function $f(p,q)$ over the phase space is equal to the ensemble average $\langle f \rangle = \int \Pi dp\, dq\, f(p,q) / \int \Pi dp\, dq$ over the accessible region of the phase space defined by $H(p,q) = E$ and any other dynamical constraints.

In the two-dimensional map around the fixed point $r = 0$ on the surface of section the ergodic theorem reduces to the theorem that for a periodic $f(\theta)$ the time average $\bar{f} = \lim_{n \to \infty} (T^n f)/n$ equals the space average $\oint f(\theta)\, d\theta/2\pi$. This ergodic theorem is valid for the dynamics $\theta' = \theta + \alpha(r)$ provided α is irrational for

$$\bar{f} = \lim_{n \to \infty} \frac{1}{n} \sum_{k=0}^{n-1} f(\theta + k\alpha) = \int_0^{2\pi} f(\theta) \frac{d\theta}{2\pi} \equiv f_0. \quad (2.12)$$

The importance of the rationals $\alpha = m/n$ in the statement of the condition for the KAM theorem becomes clear if one considers the rate of convergence of $\bar{f}$ to f_0 with finite n as a function of α. In other words, what is the difference between f_0 and the finite time average $\frac{1}{n} T^n f$? Taking out the average value f_0 from $f(\theta)$, which gives the contribution $n f_0$, we may compute this difference $\delta f(\theta)$ from the Fourier series expansion

$$\delta f(\theta) = \sum_{k=1}^{\infty} \left[a_k e^{ik\theta} + a_k^* e^{-ik\theta} \right]$$

assuming $|a_k| < a_1/k^2$ to eliminate discontinuous functions (square wave and sawtooth functions). Now the time average can be computed. Summing the geometric series gives that

$$\sum_{k=0}^{n-1} \delta f(\theta + k\alpha) = \sum_{k=0}^{\infty} a_k e^{ik\theta} \left(\frac{1 - e^{ikn\alpha}}{1 - e^{ik\alpha}} \right) + \text{c.c.}. \quad (2.13)$$

Clearly, the sum diverges where $k\alpha = 2\pi m$. The sum will converge to a continuous function $g(\theta)$ provided $\alpha/2\pi$ is bounded away from the rationals

$$\left| \frac{\alpha}{2\pi} - \frac{m}{n} \right| > \frac{\epsilon}{n^3}. \quad (2.14)$$

While the rationals have measure zero, the measure of the region excluded by the convergence condition (2.14) is finite. To see this we note that for given n the range of m is 1 to $n-1$ and thus the length of the excluded region is $\epsilon(n-1)/n^3 \cong \epsilon/n^2$ for n terms. As we go to longer times or large n the sum of resonant intervals becomes $\sum \epsilon/n^2 = \epsilon \pi^2/6$ which is a finite fraction of the interval $0 < \alpha/2\pi < 1$.

In the KAM theory the role of δf is the deviation of the invariant curves $\delta r(\theta)$ from their initial, unperturbed positions. The theorem shows that for regions close to rational rotation $\alpha(r)$ or winding numbers $\nu(I)$ there is no convergence, and in fact, a set of island chains of new alternating elliptic and hyperbolic fixed points opens up. This behavior will be clearly seen in the following studies of the standard map for $r' = r'(r, \theta)$, $\theta' = \theta(r, \theta)$.

From the preceding analysis one may expect that the most robust surfaces, resisting the breakup into new chains of elliptic and hyperbolic points correspond to the values of α most difficult to approximate by a rational m/n. Greene (1979) has shown that the last of the invariant surfaces, (or the last KAM surface), to breakup are those with winding number ν, or the rotational number $\alpha = \nu$ for the example in (2.11), given by the reciprocal of the "golden mean" γ. The "most irrational" of all numbers called the "golden mean" is most easily understood by the continued fraction representation of numbers.

The "golden mean" γ, and its reciprocal, are said to be the most irrational of all numbers,

$$\gamma = \frac{1 + \sqrt{5}}{2} \tag{2.15}$$

due to their difficulty to approximate by rationals. The difficulty is due to the fact that the continued fraction representation of ν_{gm},

$$\nu_{\text{gm}} = \frac{1}{\gamma} = 0 + \cfrac{1}{1 + \cfrac{1}{1 + \cfrac{1}{1 + \cdots}}} = [0, 1, 1, 1, \ldots] \tag{2.16}$$

converges slowly compared to other sequences. Equation (2.16) follows from iterating $\nu = 1/(1 + \nu)$ while the positive root is $(\sqrt{5} - 1)/2$.

The high order rational numbers used to bracket the invariant curves are expressed in terms of the Fibonacci numbers F_n. The winding number at order $\ell + 1$ may be written as

$$\nu_{\ell+1} = \frac{F_\ell}{F_{\ell+1}} \tag{2.17}$$

where F_ℓ are from the Fibonacci series obtained from any two initial numbers F_0 and F_1 by simply taking the sum of the two proceeding numbers

$$F_{\ell+1} = F_\ell + F_{\ell-1}. \tag{2.18}$$

MacKay et al. (1984a) has developed an alternative proof of the KAM theorem based on the convergence conditions for a series of maps rather than on the orbits themselves. When the sequence of maps that magnify or "renormalize" the region around the critical orbits converges, the invariant torus (KAM surface) is present. The critical orbits are orbits with winding numbers given by the Fibonacci series (2.17) and (2.18). The renormalization approach, while more abstract, is a powerful tool for studying conditions for the onset of chaos in nonlinear dynamics.

2.6 KAM Theory in Laboratory Plasmas

The abstract phase space of KAM theory discussed in the previous section can be the same as the physical space in two important examples well known in plasma physics. In the first case the motion of the electrons in a toroidal confinement device closely follows the trajectory of the magnetic field lines due to the smallness (submillimeter) of the gyroradius in the strong magnetic field. Using electrons injected parallel to the magnetic field the KAM surfaces and the ergodic regions of the $1\frac{1}{2}$-D Hamiltonian phase space have been mapped out in helical/stellarator magnetic confinement systems in detail. This electron mapping gives a direct picture of the KAM structures. A recent example of using this technique to map out the flux surfaces in the ATF torsatron helical system is described in Sec. 2.7.

The second case where the physical space is the same as the abstract phase space is the $1\frac{1}{2}$-D incompressible drift motion produced by the $\mathbf{E} \times \mathbf{B}$ plasma flows. In certain problems the parallel motion of the particles is unimportant either due to the constancy of the fluctuations along the field line ($k_\parallel \cong 0$) or due to the averaging of the fields by the rapid bounce motion of the electrons along the magnetic field lines. In this $\mathbf{E} \times \mathbf{B}$ drift regime the dynamical system is directly Hamiltonian in the form that the x, y coordinates measuring the distances across the magnetic field $B_0 \hat{z}$ are the canonical p, q coordinates.

2.7 Visualization of the Magnetic Flux Surfaces by Electron Beam Mapping

In toroidal magnetic traps the equilibrium magnetic field is designed to form closed, nested toroidal flux surfaces in $\mathbb{R}^3$ (real $d = 3$ vector space). These real space toroidal surfaces are the KAM surfaces in the $d = 3$ phase space of the Hamiltonian for the magnetic field line trajectories. The winding number in the phase space becomes directly the rotational transform $t(r)$ of the helical confinement system or the so-called "safety factor"[1] $q(r) \equiv 1/t(r)$ of tokamak theory. KAM theory states that it is the magnetic surfaces with rational values of t or q that break up from generic perturbations leading to unstable or chaotic trajectories due to the chains of stable-unstable fixed points introduced along the rational surfaces $q(r) = m/n$ by resonant perturbations.

At $q(r) = m/n$ the field line (or phase space point) returns on itself after m turns around the principal axis of the torus (or m-periods of the time torus) and n-turns around the minor axis. The angle around the major axis is called ϕ and around the minor axis θ.

---

[1] The function $q(r)$ is called the safety factor due to the powerful kink instability that occurs for $q(r) < 1$.

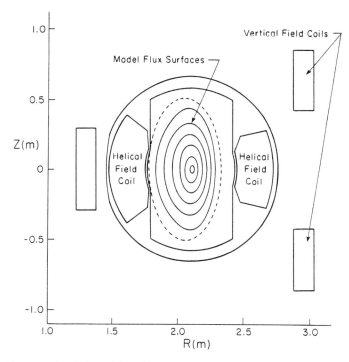

Fig. 2.1. Cross-sectional view of ATF (Advanced Toroidal Facility) at the reference toroidal (or longitudinal angle) $\phi = 0$ showing the external current-carrying coil sets and the calculated magnetic flux surfaces.

The rational magnetic surfaces and the unstable trajectories are directly measured in helical systems using electron beams and a semi-transparent fluorescent screen. In the helical system the equilibrium is intrinsically non-symmetric due to the combination of external helical magnetic fields, varying as $\cos(m\theta - \ell\phi)$ for example, and the toroidal magnetic field $B_\phi = B_0/(1 + r/R \cos\theta)$ where θ and ϕ are the poloidal and toroidal angles, r and R are the minor and major radii of the toroidal device. In the helical toroidal system the plasma currents are small so that the magnetic field is controlled by the external windings, however, the leads feeding current to the helical coils carry large currents and produce significant perturbations to the designed field. In the tokamak the large plasma current producing the poloidal field leads to large perturbed currents from resistive-MHD motions of the plasma. Thus, in both the helical and tokamak confinement systems significant magnetic field perturbations produce layers of magnetic islands surrounded by chaotic magnetic field lines.

2.7 Visualization of the Magnetic Flux Surfaces by ...

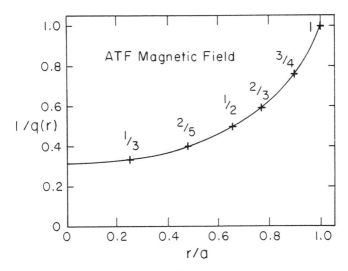

Fig. 2.2. The rotational transform profile for ATF, with the principal resonances indicated. The profile can be varied using the independently programmable vertical field (VF) coils shown in Fig. 2.1.

The importance of magnetic islands is dramatically shown by the large islands produced from (error) magnetic perturbations in the Advanced Toroidal Facility better known as ATF.

In the recent ATF helical system shown in Fig. 2.1 from Harris et al. (1990) relatively large $m = 2$ magnetic islands with widths $W_I \sim 5$ cm were measured and calculated as shown in Fig. 2.2. To construct Fig. 2.2(a) Harris et al. (1990) used a low energy electron beam (energy 80 V and current 5 mA) injected at a single point (r_0, θ_0) in the cross-section at a given ϕ angle. The rotation rate $1/q(r)$ of the magnetic field lines calculated for the model magnetic surfaces in Fig. 2.1 is shown in Fig. 2.2 where the rational values are located. A map is created by placing a highly transparent fluorescent screen across another toroidal location ϕ_s. A small fraction of the beam collides with the screen on each passage leading to a surface of section map with a moderately large number ($N \gtrsim 100$) of crossings. By moving the source point (r_0, θ_0) to different positions the complete surface of section structure of the magnetic fields was mapped out. The measurements immediately revealed a prominent $m = 2$ island of width 5 to 6 cm centered around the $q = 1/t = 2$ surface and smaller islands at the $q = 3, 3/2, 4/3$ and other low-order rational surfaces. The positions of these rational surfaces is shown in Fig. 2.2 giving the rotational transform $t(r)$. Other experiments with plasma in the toroidal vessel showed that the $m = 2$ island directly increased the radial transport losses. Using numerical integration of the field line equations, similar to the calculations described below,

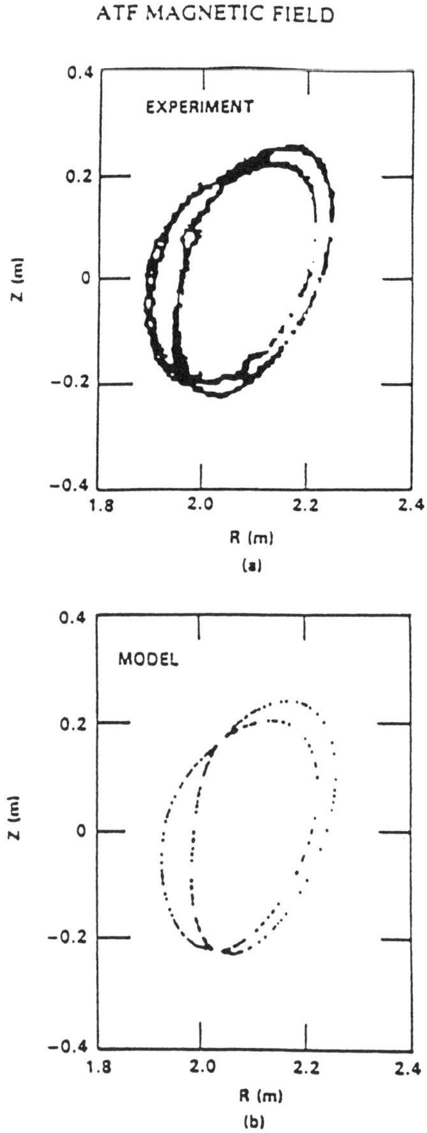

ATF MAGNETIC FIELD

Fig. 2.3. Magnetic islands on the $\iota = \frac{1}{2}$ surface in ATF: (a) experimentally measured with an electron beam and (b) computed using the actual current feed line geometry.

the current feeds and buswork for the helical magnetic field coils were redesigned (with additional external dipole field loops inserted to compensate for the intrinsic feed current loops). With the new design the $m = 2$ island width was reduced to $\lesssim 1\,\text{cm}$ and the $q = 3$ island in the central region ($r \lesssim 5\,\text{cm}$) remained large but can be avoided by operating so that the on-axis rotational transform is maintained above $1/3$.

2.8 Magnetic Islands and the Unstable-Chaotic Trajectories of Field Lines in Tokamaks

In the tokamak system the most dangerous and universal mechanism for creating magnetic islands and the stochastic magnetic field line trajectories arises from the generation of helical plasma currents from unstable resistive MHD motions within the plasma. The external field errors can be kept small due to the symmetry of the system and the intrinsic toroidal ripple $\Delta B/B$ is small due to the discrete toroidal field coils and is a non-resonant ($m = 0$) perturbation of the form $\Delta B \cos(L\phi)$ where L is the number of toroidal field coils ($\sim 18 - 20$).

Electrons with small pitch angles follow the magnetic field lines $\mathbf{B}(\mathbf{x})$ to within a few times the electron gyroradius ($\rho_e = v_e/\omega_{ce}$) in a toroidal confinement device. Thus, they can rapidly map out the magnetic field lines and lead to direct measures of the confinement properties of the magnetic structures. In contrast to the toroidal helical system or stellarator system described in 2.7, the tokamak has no vacuum confinement of single electrons. The confining field $B_\theta(r)$ is produced by the toroidal plasma current $j_\phi(r)$ so the system is an example of a self-consistent, or self-organized, confined plasma state. Nevertheless, high energy, small pitch angle electrons trace out the magnetic field lines in the plasma.

To first order in ρ_e/R, where ρ_e is the electron gyroradius and R the major radius, the electron guiding center follows the trajectory $\mathbf{r}(t)$

$$\frac{d\mathbf{r}}{dt} = \frac{v_\| \mathbf{B}(\mathbf{x})}{|\mathbf{B}(\mathbf{x})|}\bigg|_{\mathbf{x}=\mathbf{r}(t)}. \tag{2.19}$$

To the extent that the motion of the particle has nearly constant $v_\|$, which is valid for small pitch angles $\alpha = \tan^{-1}(v_\perp/v_\|) \ll 1$ where the mirror force $F_\| = -\mu \nabla_\| B$ is weak, the electron particle simply traces out the magnetic field lines given by

$$\frac{d\mathbf{x}}{ds} = \frac{\mathbf{B}(\mathbf{x})}{B(\mathbf{x})} \equiv \mathbf{b}(\mathbf{x}) \tag{2.20}$$

where $ds = v_\| \, dt$ is the element of distance along the magnetic field line. In toroidal magnetic traps there is a fixed strong toroidal magnetic field $\mathbf{B} \cong B\,\mathbf{e}_\phi$ which allows the motion described by Eq. (2.20) to be reduced to a 2-dimensional motion (x, y)

across the magnetic field or in the projection onto the (r, θ) coordinates in the poloidal cross-sectional plane defined by $\phi = $ const.

The magnetic field component perpendicular to the toroidal field $B_\phi \hat{\mathbf{e}}_\phi$ is called the poloidal magnetic field $\mathbf{B}_p$ and is described by the toroidal vector potential A_ϕ or the poloidal magnetic flux function $\Psi = -RA_\phi$ through $\mathbf{B}_p = \nabla \times \mathbf{A}_\phi = \nabla \times (\nabla \phi \, RA_\phi) = \nabla \phi \times \nabla \Psi$. Using these representations and the approximation that $B \simeq B_\phi \gg B_p$ in Eq. (2.20) the field line trajectory reduces to the $1\frac{1}{2}$-D Hamiltonian system

$$\frac{d\theta}{d\phi} = \frac{1}{rB_\phi} \frac{\partial \Psi}{\partial r}(r, \theta, \phi) \tag{2.21}$$

$$\frac{dr}{d\phi} = \frac{-1}{rB_\phi} \frac{\partial \Psi}{\partial \theta}(r, \theta, \phi) \tag{2.22}$$

in which the Hamiltonian is Ψ and the effective time variable is ϕ and with $(p, q) \to (B_\phi r^2/2, \theta)$ the canonically conjugate variables. Thus the $d = 3$ phase space becomes the real $\mathbb{R}^3$ space in the laboratory. The development of the Hamiltonian form of the magnetic field with $H \to \Psi(I, \theta, \phi)$ where the action I is the toroidal flux $I = \int_0^r B_\phi(r') r' dr' \cong B_\phi r^2/2$ and θ are the action-angle variables of Hamiltonian theory given by Boozer (1983).

For $\partial \Psi / \partial \phi = 0$ the trajectories are integrable with $r = r(\theta)$ given by $\Psi(r, \theta) = $ const. This follows by taking the ratio of Eq. (2.22) to Eq. (2.21): we immediately recognize the integral of motion from $dr \frac{\partial \Psi}{\partial r} + d\theta \frac{\partial \Psi}{\partial \theta} = 0$. We also see that $d\theta/d\phi = dH/dI$ in these magnetic action-angle coordinates.

The dynamical equations (2.21)–(2.22) may be written more generally using the contravariant components (B^r, B^θ, B^ϕ) for the magnetic field whereupon (2.21)–(2.22) become $d\theta/d\phi = B^\theta/B^\phi$ and $dr/d\phi = B^r/B^\phi$. Note that the contravariant components B^α do not have the dimensions of $\mathbf{B}$ as usual for contravariant and covariant (tensorial) components of a physical vector. The analysis of the divergence for 3D vector field $\mathbf{B}(x)$ and its associated "natural" coordinate system is a subject in itself. The first Chapter of the "Theory of Tokamak Plasmas," White (1989) is devoted to a general covariant formulation of magnetic coordinates. The recent monograph by D'haeseler et al. (1991) develops this subject in detail.

The unperturbed motion generated from Eqs. (2.21)–(2.22) with equilibrium flux function $\Psi(r)$ gives the rotational transform $t(r)$ or the associated period function $q(r)$ from Eq. (2.21)

$$\frac{d\theta}{d\phi} = \frac{1}{rB_\phi} \frac{d\Psi}{dr} = \frac{RB_\theta}{rB_\phi} = t(r) = 1/q(r) \tag{2.23}$$

and $r = $ const from Eq. (2.22). At each r evolution by $\phi \to \phi + 2\pi$ produces a twist map as described in Eq. (2.23) provided $q(r)$ is monotonic in r. In the disturbed

2.8 Magnetic Islands and the Unstable-Chaotic Trajectories of ...

system the full dynamics of the magnetic field lines is generated by flux function

$$\Psi = \Psi_{0,0}(r) + \sum_{m,n} \psi_{m,n}(r)\cos(m\theta - n\phi) \tag{2.24}$$

where the functions $\psi_{m,n}(r)$ arise from the plasma motions and external field coils. Carreras et al. (1981) take as a simple model for the $\psi_{m,n}(r)$ functions

$$\psi_{m,n}(r) = \widehat{\psi}_{m,n}\left(\frac{r}{r_s}\right)^m \left(\frac{1-r}{1-r_s}\right) \frac{2}{1+\exp(10(r-r_s)/r_s)} \tag{2.25}$$

which are approximations for the resistive MHD eigenfunctions with boundary conditions that $\psi_{m,n}(r \to 0) \to r^m$ and $\psi_{m,n}(r = 1) = 0$. Here we measure r in units of the radius $r = a$ of the conducting wall at which we assume $\delta B_r(r/a = 1) = 0$ and flux Ψ in units of $a^2 B_\phi$. In the resonant approximation used below only the amplitude $\widehat{\psi}_{m,n} = \psi_{m,n}(r = r_s)$ at the rational or singular surface $r = r_s$ is used in the analytic theory. The perturbations $\psi_{m,n}(r)$ are resonant with the equilibrium **B** at the rational surface r_s defined by

$$q(r_s) = \frac{m}{n}. \tag{2.26}$$

For a generic perturbation it is the surfaces with rational winding numbers that break up according to KAM theory for the reason described in Sec. 2.5.

Magnetic perturbations of the form $\psi_{m,n}(r)$ arise from resistive MHD instabilities driven by the radial gradients of the toroidal plasma current $j_\phi(r)$ and the plasma pressure $p(r)$. These magnetic disturbances must be localized to regions of small $k_\parallel^2 v_A^2 = (m - nq(r))^2 (v_A^2/q^2 R^2)$ to minimize the strongly stabilizing shear-Alfvén wave oscillations. The functions given in Eq. (2.25) are models for these types of resistive MHD instabilities (White et al., 1977, 1984) adequate for the purpose of calculating the 3D magnetic structures.

For sufficiently small amplitudes $\widehat{\psi}_{m,n}$ the motion given by Eqs. (2.21)–(2.25) are predominantly nested chains of magnetic islands separated by thin stochastic layers. The island chains are described by an effective pendulum equation. In the vicinity of the resonance in Eq. (2.26) the motion is given by $r = r_s + \Delta r(\phi)$ with the width $\Delta r(\phi)$ controlled by the magnetic shear parameter

$$s = \frac{r_s q'(r_s)}{q(r_s)} \tag{2.27}$$

and the amplitude $\widehat{\psi}_{m,n}$ of the perturbation. The shear or twist rate in Eq. (2.27) acts as the effective mass of the pendulum, and the amplitude $m\psi_{m,n}(r_s)\sin(m\theta - n\phi)$ is the effective restoring force.

A canonical transformation may be used to introduce the local helical phase variable

$$\zeta = \theta - \frac{n}{m}\phi \tag{2.28}$$

and conjugate displacement variable

$$\Delta r = x = r(\phi) - r_s. \tag{2.29}$$

Taking only the *resonant term* the motion near the resonant surface is given by

$$\frac{d\zeta}{d\phi} = t(r) - \frac{n}{m} = \frac{1}{q(r)} - \frac{n}{m} = -\left(\frac{s}{q_s}\right)\frac{x}{r_s} \tag{2.30}$$

$$\frac{dx}{d\phi} = \frac{m}{r_s}\widehat{\psi}_s \sin m\zeta. \tag{2.31}$$

In calculating $d\theta/d\phi$ we use Eq. (2.23) and drop the small oscillatory terms from $\psi'_{m,n}(r)\cos(m\theta - n\phi)$ whereas in calculating $dr/d\phi$ the only contribution comes from the small perturbations in Eq. (2.24).

The general solutions of Eqs. (2.30) and (2.31) are expressed in terms of the Jacobi elliptic functions following the analysis in Sec. 1.6. The Hamiltonian for Eqs. (2.30)–(2.31) is $H = \frac{1}{2}(s/q_s r_s)x^2 - (\widehat{\psi}_s/r_s)\cos(m\zeta)$. The equation for the separatrix is given by

$$\tfrac{1}{2}s(\Delta r)^2 = q_s\,\widehat{\psi}_s\,(1 + \cos(m\zeta)) \tag{2.32}$$

$$= 2q_s\,\widehat{\psi}_s\,\cos^2\left(\frac{m\zeta}{2}\right). \tag{2.33}$$

The separatrix solution $\Delta r(\zeta)$ describes the chain of m stable (s) elliptic fixed points ($\widehat{\psi}_s > 0$) at r_s and $m\zeta_{\text{us}} = 2\pi\ell$ with $\ell = 0, 1, 2\ldots, m-1$. The chain of m unstable (s) hyperbolic fixed points occurs at r_s and $\zeta_{\text{us}} = 2\pi(\ell + 1/2)/m$ with $\ell = 0, 1, 2, \ldots, m-1$. Examples of the two $m = 2$ islands and three $m = 3$ islands are shown in Fig. 2.4. The magnetic island width W_I is defined as the maximum radial excursion of the last orbit trapped within the domain of the associated fixed point which, from Eq. (2.32), is given by

$$W_I = 2\max(\Delta r(\phi)) = 4\left(\frac{q_s}{s}\right)^{1/2}\widehat{\psi}_{m,n}^{1/2}(r_s)$$

$$= r_s\left(\frac{1}{s}\right)^{1/2}\left(\frac{\delta B_r}{m B_\theta}\right)^{1/2}. \tag{2.34}$$

Following Carreras *et al.* (1981) it is useful to use Eq. (2.34) to re-express the amplitude $\widehat{\psi}_s$ in Eq. (2.25) in terms of the corresponding dimensionless island widths given by $\widehat{\psi}_s = sW_I^2/16q_s$. In the example used here, the rotational transform or q-profile is taken as

$$q(r) = 1.344\left[1 + \left(\frac{r}{0.56}\right)^{2\alpha}\right]^{1/\alpha} \quad \text{with} \quad \alpha = 3.24 \tag{2.35}$$

2.8 Magnetic Islands and the Unstable-Chaotic Trajectories of ...

and only the two terms $m = 3, n = 2$ and $m = 2, n = 1$ are retained in the perturbation. These terms are resonant at the $q(r_{s1}) = 3/2$ and $q(r_{s2}) = 2/1$ surfaces where $r_{s1} = 0.491$ and $r_{s2} = 0.650$. The local resonant Hamiltonian for the resonant trajectories is the effective pendulum given by

$$H^{\text{res}}(\Delta r, \theta) = \frac{s}{2q_s}(\Delta r)^2 - \widehat{\psi}_s \cos(m\theta)$$

describing phase oscillations and the separatrix for each resonant annulus. In the first case both island widths are taken as $W_I = 0.1$ which gives the set of isolated period-2 and period-3 islands in the surface of section plot in Fig. 2.4. In this small amplitude case the separation between the resonant surfaces $|r_{s2} - r_{s1}| = 0.159$ is greater than the island width W_I and the stochasticity is confined to narrow layers along the separatrices given by Eq. (2.33). The width of the stochastic layer can be computed from the Melnikov-Arnold integral method in Sec. 3.3. In the second case in Fig. 2.5 the island width is increased to $W_I = 0.4$ for both the 3/2 and

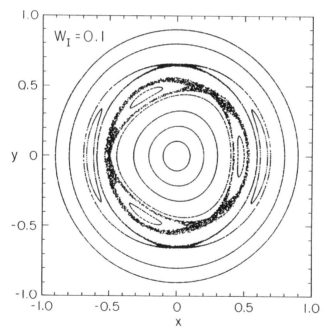

Fig. 2.4. A field line surface of section plot made by integrating the Hamiltonian field line equations (2.24) and (2.25) for the inner $m/\ell = 3/2 = q(r_1)$ and the outer $m/\ell = 2/1 = q(r_2)$ island chains from magnetic perturbations produced by resistive MHD instabilities in a tokamak. Here the amplitude of the perturbation is such as to make the magnetic island width $W_I = w/a = 0.1$.

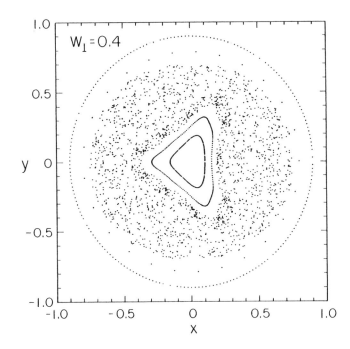

Fig. 2.5. Same surface of section plot as in Fig. 2.4 except that the amplitude of the magnetic perturbation is increased to make the island width $W_I = w/a = 0.4$. Now there are no more integrable field lines in the radial annulus between the 3/2 and 2/1 islands. Field lines started between these island chains ergodically fill the phase space volume $V(r, \theta, \phi) = 2\pi^2 R(r_2^2 - r_1^2)$ between the resonant surfaces.

2/1 perturbations. Now the island width is 2.5 greater than the separation of the resonant surfaces so that the annular region $r = [0.19, 0.75]$ is filled by chaotic magnetic field lines any one of which ergodically fills the unstable region in the $\lim |\phi| \to \infty$.

The strength of the instability of the trajectories of the magnetic field lines is made quantitative by calculating the Lyapunov exponent giving the average rate of divergence of the neighboring trajectories $\mathbf{r}_1(\phi)$, $\mathbf{r}_2(\phi)$. The divergence rate λ must be calculated by pulling or cutting back along the separation $\delta \mathbf{r} = \mathbf{r}_1 - \mathbf{r}_2$ before the separation vector $|\delta \mathbf{r}|$ becomes large enough for the nonlinear terms δr^2 to be important. A second method is to integrate the flow and the tangent flow simultaneously. The pull-back method is given by Benettini et al. (1976) and described in detail by Lichtenberg and Lieberman (1991). Mathematically the Benettini et al. (1976) method requires taking both the long-time limit $\phi \to \infty$ and the limit $\mathbf{r}_1 \to \mathbf{r}_2$ or $\|\delta \mathbf{r}\| \to 0$. When the limit is well defined, the Lyapunov exponent λ_m for the

2.8 Magnetic Islands and the Unstable-Chaotic Trajectories of ...

magnetic field lines is defined as

$$\lambda_m = \lim_{\phi \to \infty} \lim_{\mathbf{r}_1 \to \mathbf{r}_2} \left[\frac{\ln \|\mathbf{r}_1(\phi) - \mathbf{r}_2(\phi)\|}{\phi} \right] \qquad (2.36)$$

where the trajectories $\mathbf{r}(\phi) = (r, \theta)$ are computed from Eqs. (2.21) and (2.22). Greene and Kim (1989) show that the choice of the metric (g_{ij}) used to define the norm $\|\delta \mathbf{r}\|$ makes an insignificant change in the value of λ_m. For this problem in real space it is natural to take the Euclidean norm, but for more general systems the choice of the metric is not always clear.

In Figs. 2.6 and 2.7 we show the calculation for λ_m for the integrable and stochastic regions in Figs. 2.4 and 2.5. In Fig. 2.6 the trajectories are in the region between $r_{\min} = 0.52$ and $r_{\max} = 0.6$ and the monotonically decreasing value of $\lambda(\phi)$ indicates that the limit in Eq. (2.36) is $\lambda_m = 0$. In Fig. 2.7 the limit in Eq. (2.36) is well defined and gives $\lambda_m = 0.058 \pm 0.004$ for the finite value of $\phi_{\max}/2\pi = 500$ turns around the major axis of the torus.

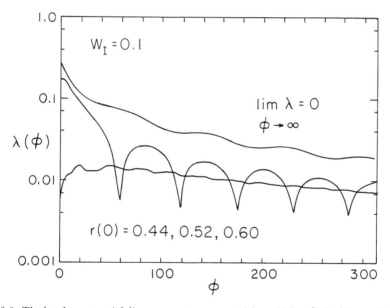

Fig. 2.6. The local exponential divergence rate computed from $\lambda(\phi) = [\ell n |\mathbf{r}_2(\phi) - \mathbf{r}_1(\phi)|] / \phi$ for two neighboring points $\mathbf{r}_1 \to \mathbf{r}_2$ in the case of the small islands ($W_I = 0.1$) shown in Fig. 2.4. Here the Lyapunov exponent $\lambda \equiv \lim_{\phi \to \infty} \lambda(\phi) = 0$.

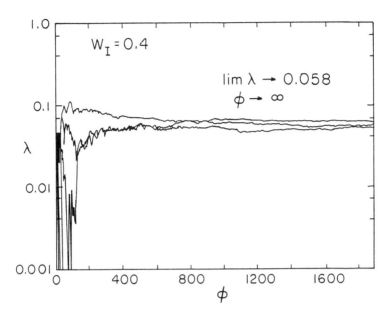

Fig. 2.7. The local exponential divergence exponent $\lambda(\phi)$ computed for two neighboring points $r_1 \to r_2$ in the case of the larger islands ($W_I = 0.4$) shown in Fig. 2.5. Here the Lyapunov exponent $\lambda \equiv \lim_{\phi \to \infty} \lambda(\phi) \simeq 0.06$. The three traces are for three different sets of initial data showing the convergence to the same Lyapunov exponent.

As becomes clear from this example the distinction between rational and irrational winding numbers and the definition (2.36) of the Lyapunov exponent are based on the mathematical idealization of infinite time orbits. In physical systems such infinite time limits are never achieved. For finite times the orbit traces out the sets of points seen clearly in Figs. 2.4 and 2.5 where it is the role of the physicist to discern whether to treat the orbit as a high order m/n-rational winding number or a finite time sample of an ergodic orbit with an irrational winding number q. Greene (1986, p. 429) devotes a lengthy paragraph to discussing the physical meaning of the distinction between the rational and irrational winding numbers. As a physicist one must make sure the lifetime of the contained orbit or the time between collisions, or other physical effects, is long enough to apply the infinite time concepts.

Soft X-ray emission diagnostics in tokamaks provide evidence for low order magnetic islands as shown in Fig. 2.4 rotating with the electron diamagnetic frequency due to the radial gradients of density and temperature producing phase shifts $-\omega t$ in Eq. (2.24) (see the related problem in Sec. 3.5.4). In the high ion temperature regimes ($T_i \sim 20-30\,\text{keV}, T_e \sim 10\,\text{keV}$) of the TFTR tokamak the $m/n = 3/2$ magnetic island (but not the 2/1 island) is commonly observed to grow and saturate at a width of about 5 cm. Chang *et al.* (1995) propose to explain the time evolution of $W(t)$ by growth from the ion pressure gradient driven bootstrap current.

2.9 Measure of the Resonant Rational Surfaces

From the example in Sec. 2.8 the question arises as to when the density of the resonant surfaces that occur at the rational values of rotational transform together with the island widths $W(m,r)$ in Eq. (2.34) will be sufficient to continuously cover a given radial region.

The same question arises in the particle dynamics in the phase space of action (I)-angle (θ) variables where for each rational value of $\Omega(I) = dH/dI = n/m$ there are islands of width $W(I,m)$.

To answer this question, and related questions on transport, it is necessary to consider the spacing of the rational surfaces. The spacing of the resonant toroidal surfaces is determined by the spacing of the rational numbers m/n in the profile of the functions for the rotation rate $1/q(r)$ or $\Omega(I)$. The answer to this question is found in the Euler-phi function $\varphi(n)$ that counts the number of irreducible rational fractions with denominators not exceeding the integer n.

2.9.1 The number of irreducible fractions

The number of irreducible simple fractions (those without common divisors of numerator and denominator) can be expressed in terms of the Euler-Totient function $\varphi(n)$, which is defined as the *number of integers not exceeding and relatively prime to* n. Thus, the number of proper irreducible fractions with denominators less than n is

$$\Phi(n) = \sum_{k=1}^{n} \varphi(k). \tag{2.37}$$

At the same time the number of all simple fractions without such limitations is

$$N(n) = \sum_{k=1}^{n} k = \frac{n(n+1)}{2}. \tag{2.38}$$

Using asymptotic formula for $\varphi(n)$ from the *Handbook of Mathematical Functions* (Abramowitz and Stegun, 1970 p. 826),

$$\frac{1}{n^2} \sum_{k=1}^{n} \varphi(k) = \frac{3}{\pi^2} + \mathcal{O}\left(\frac{\ln n}{n}\right), \tag{2.39}$$

we obtain the relative number of irreducible fractions as

$$G = \lim_{n \to \infty} \left[\frac{\Phi(n)}{N(n)}\right] = \frac{6}{\pi^2}. \tag{2.40}$$

Now consider the rotational profile $1/q(r)$, or the action-angle rotation rate $\Omega(I) = dH/dI$ in particle phase space. The resonant surfaces of periodicity-m and orbit length $nq(r)$ occur at r_{mn} (or I_{mn}) where $q(r_{mn}) = m/n$. The spacing between the resonant surfaces of fixed m and lengths nq and $(n+1)q$ are determined by

$$nq = m,$$
$$n\,\delta q + q\,\delta n = 0, \qquad (2.41)$$

which gives

$$\Rightarrow \delta n = -\frac{m\,\delta q}{q^2}. \qquad (2.42)$$

Here δ denotes the change of the quantity over the interval δr. Using $\delta q = q'(r)\delta r$, we obtain the density of rational surfaces with fixed m as

$$\frac{\delta n}{\delta r} = \left(\frac{m|q'|}{q^2}\right). \qquad (2.43)$$

This is not a derivative and the value has a meaning only in the average sense: by definition $\delta n \gg 1$ but $\delta r/r \ll 1$, which is consistent only for $m \gg 1$.

In the same limit we must take care to eliminate from consideration all fractions where numerator and denominator are multiples of the same integer. Indeed, the ratio of the number of these irreducible fractions to the number of all simple fractions with denominators less than k tends to the constant $G = 6/\pi^2 \approx 0.608$ as $k \to \infty$ as given by Eq. (2.40).

We can now define

$$f_m(r) = G\frac{\delta n}{\delta r} \approx m\frac{6|q'|}{\pi^2 \, q^2} \qquad (2.44)$$

to be the *mean density of states* with a fixed poloidal wavenumber m. With a good accuracy f_m is the radial density of significant resonances, and we can use it to calculate the measure (width) χ of the resonant domains up to order m_f by

$$\chi(r) = \sum_{m=1}^{m_f} f_m(r) \int_{-\infty}^{+\infty} \chi_{\rm sp}(r - r_\alpha, r_\alpha, \alpha)d(r - r_\alpha) \qquad (2.45)$$

where $\chi_{\rm sp}\,dr$ is the measure of the single helicity resonance. The subscript sp is for "spike" indicating $\chi_{\rm sp}(r)$ is sharply localized to the rational surface.

2.9.2 Measure of the resonant surfaces

Now we can calculate the measure of the islands of width $W(m,r)$ and period m. Using the *mean density of states* $f_m(r)$ the total width for islands from $m = 1$

2.9 Measure of the Resonant Rational Surfaces

up to order m_f is given by

$$\Delta x = \sum_{m=1}^{m_f} f_m(r) W_m(r). \tag{2.46}$$

The extent to which the radial region is covered by (2.46) depends on m-spectrum of W_m. Taking $W_m = W_1/m^p$ and replacing the m-summation with an integral we have for (2.46) that

$$\Delta x = \frac{f_1 W_1}{(p-2)} \left[1 - \frac{1}{(m_f)^{p-2}} \right] \quad \text{for } p > 2. \tag{2.47}$$

Thus, for $p > 2$ there is a threshold value of W_1, the strength of the perturbation, for the islands to cover a given radial region in the limit $m_f \to \infty$. For $p \leq 2$, however, the width of the covering from the higher order islands increases fast enough that *every* radial region is covered for any finite value of W_1. The critical spectrum is $W_m = W_1/m^2$ for which $\psi_m = \psi_1/m^4$ and the covered width increases with m_f as $f_1 W_1 \ell n(m_f)$.

The above covering calculation, or estimate of the measure of the islands, uses the mean density of states $f_m(r)$. In reality the spacing of the rationals is nonuniform with the low order rational surfaces $q = 1, 3/2, 2, \ldots$ having significant gaps around them. The effective gap width can be estimated as $\delta x_m = 1/m|dq/dr|$.

These gaps around the lower order rationals can lead to barrier regions with regard to the transport or diffusion of orbits across the confinement surfaces (Beklemishev and Horton, 1992). In Fig. 2.8 we show in the computation of Δx in Eq. (2.46) the exact spacing of the rationals compared with the mean density $f_m(r)$ of the rationals for a model q-profile. For the computation the model

$$q(r) = \frac{q_0}{(1+2r^2)} + 3.2\, r^2$$

which is monotonically increasing with r^2 is used to determine r_m/n. The island widths are taken as narrow Gaussians

$$W = W_0 \exp\left[-\frac{(r - r_{m,n})^2}{2\Delta^2}\right]$$

with $\Delta/a = 3 \times 10^{-3}$ and the spectral index is $p = 2$. The gaps around the rational surfaces at $q = 3/2, 2$ and $5/3$ are evident. Using the mean density of states $\rho_m(r)$ gives the smooth, monotonically increasing curve in Fig. 2.8.

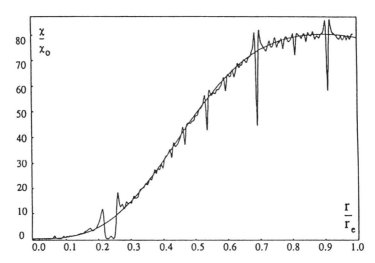

Fig. 2.8. The transport profile calculated as the superposition of identical local spikes for the model distribution of toroidal plasma current; $\Delta = 0.003a$. The smooth curve represents the same profile calculated with the continuous density of rational surfaces in Eq. (2.44).

The problem of the stochastic diffusion of the magnetic field lines is first considered in Rosenbluth et al. (1966) and the problem of the associated thermal transport in Rechester and Rosenbluth (1978). A recent analysis of the problem analyzed in Sec. 2.9 is given in Hatori et al. (1989).

2.10 Example 2.1 How a Swing Behaves

An example that requires the theoretical and computational methods presented in both Chapters 1 and 2 is the investigation by Greene (1986) entitled "How a Swing Behaves." This example is particularly useful since it allows a newcomer to nonlinear dynamics to put to test his intuitive understanding of real world physics in the esoteric domain of chaos and KAM surfaces in the phase space of the system. The first steps are to derive the equation of motion of the swing recognizing that the rider is a "pump" which periodically varies the length $L(t)$ of the ropes. Then one shows that in the linear approximation the stability analysis leads to the period doubling bifurcation given by the Mathieu equation as in Example 1.1. Now the problem is to determine invariant tori curves around the period-2 periodic orbits and the chaotic sea replacing the separatrices emanating from the hyperbolic fixed point at $\theta = \dot{\theta} = 0$.

Rather than giving here an edited version of this delightful work we ask the reader to find the article on pp. 427–447 in the generally available journal Physica D **18** (1986). This volume contains many useful articles in nonlinear science which were collected on the occasion of the Santa Barbara Conference to honor the sixtieth birthday of Martin Kruskal.

2.11 Invariant Tori and the KAM Theorem

The general behavior of two-degree-of-freedom $N = 2$ Hamiltonian systems is the same as the description of the magnetic field lines forming nested toroidal magnetic surfaces in $\mathbb{R}^3$ separated by resonant chains of islands with stochastic layers replacing the separatrices. The only aspect unusual about the toroidal magnetic field system is that the phase space is the real x, y, z configuration space and that the ergodic surfaces, the magnetic islands and the chaotic annular volumes can be measured by using injected small pitch-angle electrons to trace out the magnetic field structure down to the scale of the electron gyroradius. In the general Hamiltonian system the phase space is an abstract $\mathbb{R}^{2N}$ vector space with no possibility of actually measuring its structures and properties other than with computer simulations.

Now that the properties of a particular system (the toroidal magnetic field) are clear we state a few of the mathematical-physics results found in the theory of general Hamiltonian systems.

For N-degrees-of-freedom the Hamiltonian system is integrable when the system can be put in the form where

$$H = H_0(I_1, I_2, \ldots I_N) \tag{2.48}$$

independent of the angles $\phi_1, \phi_2, \ldots, \phi_N$. The flow in the $\mathbb{R}^{2N}$ phase space is given by

$$I_i = \text{const}$$
$$\phi_i(t) = \omega_i t + \phi_i(0) \tag{2.49}$$

where the frequencies $\{\omega_i(I_j)\}$ describe the rotation on an N-dimensional toroidal surface T^N. Thus the unperturbed flow gives an $\mathbb{R}^N \times T^N$ set of nested toroidal surfaces parameterized by the $\{I_N\}$ radius-like coordinates in $\mathbb{R}^N$. For $N = 2$, the energy surface is $2N - 1 = 3$ and the two-dimensional torus divides the phase space exactly as the magnetic surfaces divide the laboratory space in the confinement vessel. For $N = 3$ the energy surface is $2N - 1 = 5$, and the three-dimensional nested tori no longer divide the five-dimensional constant energy space. The analogy is similar to lines in a three-dimensional space since the difference in the dimensionality is two.

The flow in Eqs. (2.49) is ergodic on the surfaces and called equivalently either quasi-periodic or conditionally-periodic (Arnold, 1978) when the frequencies are

incommensurate, and thus the resonance since condition (2.7) cannot be satisfied. When the condition
nondegenerate

$$\det \left| \frac{\partial \omega_i}{\partial I_j} \right| \neq 0 \qquad (2.50)$$

on the determinant of the gradient of the frequencies is satisfied, the ergodic flows on the neighboring surfaces $I_j \to I_j + dI_j$ twist relative to one another, perturbations εV of the Hamiltonian

$$H = H_0(\mathbf{I}) + \varepsilon V(\mathbf{I}, \boldsymbol{\phi}) \qquad (2.51)$$

leave the orbits qualitatively unchanged. This lack of a fundamental change in the orbits is the meaning of the phrase *invariant tori* in the KAM theorem.

On the other hand, for orbits sufficiently close to the resonant tori a small perturbation results in a qualitative change in the dynamics. Just as shown in Sec. 2.8 for the magnetic field structure, the perturbation gives a bifurcation to a chain of alternating stable and unstable fixed points with a finite annular volume of chaotic orbits resulting from trajectories that pass close to the unstable hyperbolic points.

Following Arnolds' (1978) Appendix 8 we may state the KAM theorem as follows:

KAM Theorem

> If an unperturbed system is nondegenerate, then for sufficiently small Hamiltonian perturbations, most non-resonant invariant tori do not vanish, but are only slightly deformed, so that in the phase space of the perturbed system there are invariant tori densely filled with conditionally-periodic orbits with the number of independent frequencies equal to the number of degrees-of-freedom.

The invariant tori form a majority of the phase space in the sense that the measure of the complement of their union is small when the perturbation is small. In this statement of the theorem we follow precisely the wording of Arnold (1978, p. 405). We recommend the reading of his discussion of the meaning of the conditions for the theorem (and its implication) contained in that section of his text on classical mechanics. Here for the sake of uniformity of our text we briefly discuss a few of these points.

How close does one need to be to the resonance condition for the new orbits to set in? The answer clearly depends on the *order* of the resonance

$$n_1 \omega_1 + n_2 \omega_2 + \cdots n_N \omega_N = 0. \qquad (2.52)$$

The larger the magnitude of $|\mathbf{n}|$ the weaker the resonance, as already seen in Chapter 1. In addition, the answer depends on the analyticity (Arnold, 1961, 1963) or the differentiability (Moser, 1962) of the perturbing structure $H_1(\mathbf{I}, \boldsymbol{\phi})$. The effect of the resonance also becomes weaker as the number of degrees-of-freedom N increases.

2.11 Invariant Tori and the KAM Theorem

The usual statement is that for analytic perturbations the invariant tori exist when the off-resonance condition

$$|\mathbf{n} \cdot \boldsymbol{\omega}| = |n_1\omega_1 + n_2\omega_2 + \cdots + n_N\omega_N| > \frac{C}{|\mathbf{n}|^{N+1}} \quad (2.53)$$

is satisfied. For the magnetic field problem this condition reduces to

$$|n_1\omega_1 + n_2\omega_2| = |m - \ell q(r_s)| > \frac{C}{m^3}$$

where resonant surface is defined by $d\theta/d\phi = 1/q(r_s) = \ell/m$. The rate of convergence of $1/m^3$ with high m is already encountered in the consideration of ergodicity of the annulus map in Eq. (2.14). Namely, the rate $1/m^3$ is the slowest power law for which the sum over all rational surfaces gives a finite annular volume since the density of the rational surfaces increases in proportion to m as shown by Eq. (2.44).

Arnold discusses these conditions in more detail on pp. 399-412. We do not attempt to summarize his discussion, but we make only a few more remarks. From the magnetic field structure problem it is clear that the first condition in the theorem, namely, "if an unperturbed system is nondegenerate" is required. This condition for the magnetic field problem (and the annular twist map in Sec. 2.2) is that the magnetic shear parameter $s = rq'/q \neq 0$. When $s = 0$ the whole procedure developed in Sec. 2.8 breaks down. For small, nonvanishing s, the magnetic island width W_I in Eq. (2.34) becomes comparable to the radius of the confinement zone, and neighboring low-order resonances strongly overlap. (In the self-consistent resistive MHD dynamics this results in a major disruption of the plasma confinement.) For small, but finite s the island width W_I is large, and this is essentially the situation shown in Fig. 2.2 for the $m = 2$ island in ATF caused by the magnetic perturbations to the field line Hamiltonian.

More detailed conditions for the KAM theorem are best left to the mathematical physicists. Here we add one remark in this regard. Conditions for theorems are often crafted to allow a strict proof. Physical systems rarely satisfy all the conditions of the theorem. So much judgment and (numerical) experimentation is required to interpret the mathematical theorems. As an example, we will encounter simple two-degree-of-freedom Hamiltonian in the next Chapter (Sec. 3.4) in which the condition of nondegeneracy fails. At first this may seem disappointing; however, in reality much of the information from the KAM theorem still applies. The degeneracy allows a wider set of islands bordered by a stochastic sea to open up in the phase space.

Chapter 3

Stochasticity Theory and Applications in Plasmas

Predicting the trajectories of charged particles in specified magnetic fields is a central problem in both laboratory and space plasma physics. The basis for the fusion project, which has developed into a world-wide scientific-technological project over the past thirty years, is to understand and predict the particle orbits well enough to confine the ions and electrons with magnetic fields for periods of time sufficient to allow the slow nuclear reactions to take place. In space plasmas it is important to understand the trajectories of the charged particles in the magnetospheric fields. Near the earth or other planets the magnetic fields are dipole-like, but away from the planet they are weak and strongly varying. Near the magnetopause and in the magnetotail region on the night-side of the planets the magnetic field can form thin loops, reversed fields and cusps.

The geomagnetic field in the noon-midnight meridian taken from the well-known Tsyganenko model is shown in Fig. 3.1. The magnetic field model is a parametric fit to thousands of satellite measurements throughout the Earth's magnetosphere. Figure 3.1(a) shows the solar wind distorted magnetospheric cavity and the magnetopause boundary layer in which large shielding currents flow. In Fig. 3.1(b) the interplanetary magnetic field (IMF) with a northward orientation is added showing the two null points A and B where the magnetic field vanishes. The null points are critical points for the entry of charged particles from the interplanetary plasma into the magnetosphere. In the neighborhood of the null points the local tangent bundles of the magnetic field lines form the "spoke and spindle" pattern shown as inserts in Fig. 3.1b. The A-type null point has field lines **B** that converge along the spokes and diverge along the spindles as in the northern, high latitude cusp while for the B-type null point **B** lines diverge along the spokes and converge along the spindles as shown at the southern null point. Lau and Finn (1991) treat the three dimensional structures that occur with field nulls in detail and give the historical references. In the present text we are primarily concerned with geometries with nonvanishing magnetic fields.

In space plasmas it is important to understand the transition from trajectories in the strong dipolar magnetic fields where the magnetic moment is well conserved to the trajectories in the weaker magnetic field regions, such as in the geomagnetic tail, where the magnetic moment invariant is broken due to the rapid change in direction of the magnetic field vector $\mathbf{B}(\mathbf{x})$ over the size of the Larmor radius.

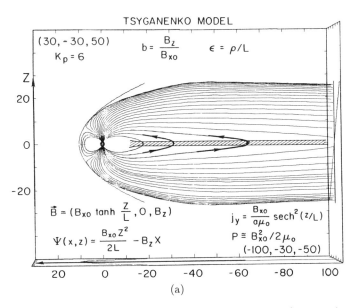

(a)

Fig. 3.1. The magnetic field lines forming the Earth's magnetosphere as given by the widely used magnetic field model Tsyganenko (1987) based on satellite data. The solar wind from the left stretches the dipole field out into the extended magnetotail to the right.

(a) The loops of magnetic field in the magnetic tail (on the right) may be described locally ($|y| \leq 15\,R_E, |z| < 20\,R_E$ and $x < -10\,R_E$) by the vector potential $A_y(x,z) = -\Psi(x,z)$ In the region $-65 \leq x/R_E \leq -15$ the parameters are in the range $B_{x0} = 10-30\,nT$, $B_z = 1-5\,nT$ and $L_z = 1-3\,R_E$ where $1nT = 10^{-9}$ Tesla and $R_E = 6370\,\text{km}$ is the reference distance taken from the mean radius of the Earth.

(b) The magnetosphere in the presence of a northward interplanetary magnetic field, called the IMF. The figure shows the two null points A and B where $\mathbf{B} = 0$: on the northern dayside cusp and on the southern cusp, which allow the entry of charged particles into the inner magnetosphere and the ionosphere. At the A and B null points the magnetic field lines follow the shape of "spokes and spindles" of a wagon wheel.

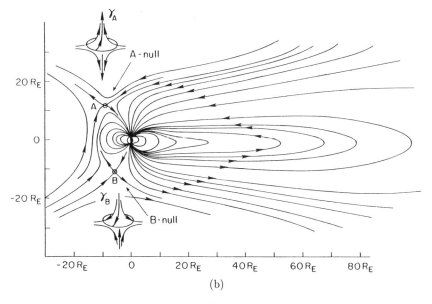

Fig. 3.1. (*Continued*)

In addition to the motion of the charged particles in the ideal case of a given, static magnetic field, both fusion and space plasma physicists and engineers must be able to predict the particle trajectories in the presence of time varying electromagnetic fields. The time varying fields arise either from the self-consistent dynamics of the long-range Coulomb interactions or the imposed electromagnetic fields such as radio frequency waves and laser beams launched from outside the plasma.

In Chapter 3 we give the principal results for the trajectories of the charged particles in increasingly complex magnetic and electric fields. The orbits show a rich variety of behavior and become chaotic in even very simple configurations. The chaos of the orbits in simple configurations is an unexpected, new result that has been really only appreciated in the past ten years, although early examples were known from the work of Soviet physicists such as Chirikov, Arnold, and Zaslavsky in the 1960s.

3.1 Trajectories in Straight, Nonuniform Magnetic Fields

As the first step in analyzing the motion of charged particles in general magnetic fields it is useful to consider first the motion in a straight, nonuniform magnetic field

$\mathbf{B}(\mathbf{x}) = B(x)\hat{\mathbf{e}}_z$. The translational symmetries in the y and z directions immediately yield the invariants p_y and p_z and the time independence of the system leads to the energy integral $E = mv^2/2 =$ constant. Thus, the 3 degrees-of-freedom problem with 3 invariants (p_y, p_z, E) is integrable reducing to a one-dimensional quadrature. The phase space portraits, however, in general contain a separatrix so that orbits sufficiently near the separatrix are unstable to structural changes in the system. It is in this sense that this simple geometry is not trivial and contains the seed for chaos. The seed is in terms of the homoclinic orbit, that connects the unstable hyperbolic fixed points. Orbits on either side of this critical orbit travel to completely separate regions of phase space. Thus, they form the so-called separatrix. For the neighboring problems (slightly perturbed Hamiltonian) with additional electric or magnetic field components the orbits neighboring the original homoclinic orbit become the chaotic orbits.

3.1.1 Integrable orbits and invariants of charged particle motion

The Hamiltonian $H(p,x)$ for the charged particle q, m in the nonuniform magnetic field, generated by the vector potential $A_y(x)$

$$B(x) = B_z(x) = \frac{dA_y}{dx}, \tag{3.1}$$

is given by

$$H = \frac{p_x^2}{2m} + \frac{p_z^2}{2m} + \frac{1}{2m}\left(p_y - \frac{q}{c}A_y(x)\right)^2 \tag{3.2}$$

which immediately gives the p_x, x motion for each constant value of $H = E$, p_y and p_z determined by the initial data. Thus, the component of the kinetic energy $mv_y^2/2$ in the symmetry direction expressed in terms of the constant of the motion p_y gives rise to a positive definite effective potential

$$V_{p_y}(x) = \frac{1}{2m}\left(p_y - \frac{q}{c}A_y(x)\right)^2$$

for the motion of the particles. Examples of the effective potential are shown in Fig. 3.2. The phase space curves in x, p_x corresponding to Fig. 3.2 are shown in Fig. 3.3.

3.1 Trajectories in Straight, Nonuniform Magnetic Fields 55

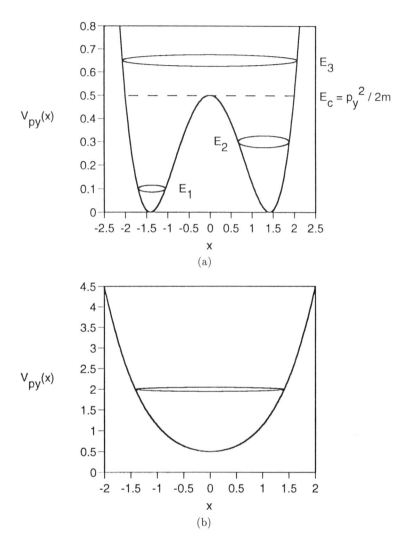

Fig. 3.2. The effective potential $V_{p_y}(x)$ controlling the motion of a charged particle in the presence of a nonuniform magnetic field. (a) the shape of the effective potential for the case of a reversed field $B(x)\mathbf{e}_z = B'x\,\mathbf{e}_z$ produced by a current sheet j_y. The orbits 1,2,3 are for the different kinetic energies $E_3 > E_2 > E_1$. The critical energy E_c divides the orbits which cross the reversal layer ($x = 0$) for $E > E_c$ and do not cross the reversal layer for ($E < E_c$). (b) The effective potential for the case of particles with $p_y B' < 0$.

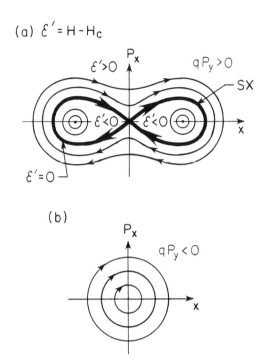

Fig. 3.3. The phase space diagrams for the Hamiltonian flow (or trajectories) corresponding to the cases shown in Fig. 3.2. (a) The case where $B(x) = xB'$ and $p_y B' = p_y j_y < 0$ with the phase space separatrix SX separating the $\varepsilon' = H - H_c > 0$ orbits crossing between the $B_z > 0$ and $B_z < 0$ regions. (b) The case $p_y B' = p_y j_y > 0$ where the low energy $\varepsilon' = H - H_c < 0$ confined to either the $B_z > 0$ or the $B_z < 0$ regions. The critical energy is $H_c = p_y^2/2m$.

The particle is trapped in a one-dimensional potential well $V_{p_y}(x) = (p_y - \frac{q}{c} A_y(x))^2/2m$ where the topology of the well changes with p_y. The potential is bounded by $V_{p_y} \leq H$ leading to the constraint $|A_y(x) - A_y(x_0)| < cmv/q$ where $v = (2H/m)^{1/2}$ and x_0 is defined by $p_y = q A_y(x_0)/c$. In a relatively strong magnetic field that does not vary strongly over the size of the particle orbit this constraint leads to a well-defined local gyroradius

Local gyroradius

$$\rho(X) = \frac{mc\, v_\perp}{qB(X)} \tag{3.3}$$

where we use $|A_y(x) - A_y(x_0)| = |x - x_0|B(X)$ with X in $[x, x_0]$ from the Cauchy mean-value theorem provided $B(x)$ is not too small. The bound $V_{p_y} \leq H$ forces the

3.1 Trajectories in Straight, Nonuniform Magnetic Fields

particle to oscillate in $V_{p_y}(x)$. The action for the oscillations

$$I_x = (mc/q)\mu = \oint p_x \, dx / 2\pi$$

is an adiabatic invariant with the value $\mu = \frac{1}{2} v_\perp^2 / B(x)$ so that $\rho(x) \propto \mu^{1/2} / B^{1/2}(x)$ provided $\varepsilon = \rho d\ell n \, B/dx \ll 1$. The parameter ε is called the finite gyro- or Larmor radius (FLR) parameters. Most, but not all, calculations in fusion systems assume $\varepsilon \ll 1$ and use the reduced set of equations known as the guiding-center equations.

Important cases occur both in the laboratory (the field reversed configuration called the FRC system) and space plasmas (the night-side magnetotail formed by the action of the solar wind producing the trapped plasma current sheet) where the local value of $\varepsilon > 1$. The generic situation for the occurrence of large orbits is when there is a strong, localized current sheet. Across a current sheet the magnetic field reverses direction so that $B(x) \simeq B'x$ where $B' = dB/dx$ is a measure of the strength of the current from Ampere's law $4\pi j_y(x)/c = -dB_z/dx$. Current sheets arise naturally in plasmas and are an effective means of containing high pressure ($p \sim B^2/8\pi$) plasma.

In addition, it is of fundamental interest to understand the transition from μ-conserving to μ-breaking motion. Studies of this problem by Chirikov (1973) and Taylor (1969) lead to the standard map analyzed in Chapter 4.

In the general case the trajectories are given by

$$\dot{x} = \frac{\partial H}{\partial p_x} = \frac{p_x}{m} \tag{3.4}$$

$$\dot{p}_x = -\frac{\partial H}{\partial x} = \frac{q}{mc}\left(p_y - \frac{q}{c} A_y(x)\right)\left(\frac{dA_y}{dx}\right) \tag{3.5}$$

with the fixed points at $p_x = 0$ and either

(i) $\quad v_y = \dfrac{p_y}{m} - \left(\dfrac{q}{mc}\right) A_y(x) = 0 \qquad$ (ii) $\qquad B(x) = \dfrac{dA_y}{dx} = 0. \tag{3.6}$

If the field $B(x)$ nowhere vanishes then only condition (i) gives a fixed point which is automatically a stable fixed point according to the bound after Eq. (3.3). The magnetic potential $V_{p_y}(x)$ is monotonically increasing away from the fixed point x_0 which is then the oscillation center given by the flux condition ($A_y(x)$ is the flux from $B_z(x)$ per unit length in y between x and a reference point)

$$A_y(x_0) = \frac{cp_y}{q}. \tag{3.7}$$

The fixed point x_0 is then the position of the local minima shown in Fig. 3.2(a) and the elliptic fixed points in the phase space curves in Fig. 3.3(b). The orbital excursion from x_0 is bounded by the constant $H = E$ due to the limits $-(2mE)^{1/2} \leq |p_y - qA/c| \leq (2mE)^{1/2}$ which with Eq. (3.7) imply that

$$\Delta x \leq \frac{2c(2mE)^{1/2}}{qB_{\min}}. \tag{3.8}$$

Clearly, the oscillations become strongly nonlinear when $\Delta x \gtrsim B/B' \equiv L_B$. The topology of the orbits $x(t)$, $y(t)$ are shown in Fig. 3.4 for the different cases (E, P_y) shown in Fig. 3.2 and Fig. 3.3 for the potential. The quantitative description of the orbits is given in Sec. 3.1.2 and 3.2.3.

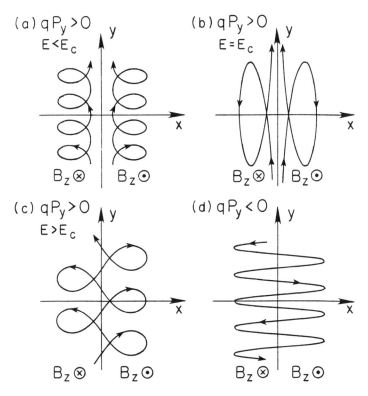

Fig. 3.4. The orbit trajectories $x(t)$, $y(t)$ for the different energy levels H and P_y values described in Figs. 3.2 and 3.3. (a) $qP_y > 0$ and $\varepsilon' = H - H_c < 0$ corresponding to small gyroradius particles trapped on either side of the reversal layer. (b) $qP_y > 0$ and $\varepsilon' = H - H_c = 0$ corresponding to the critical orbits along the separatrix in the phase space. These orbits are called the homoclinic. (c) $qP_y > 0$ and $\varepsilon' = H - H_c > 0$ corresponding to crossing orbits that flow in the direction to oppose the current. (d) $qP_y > 0$ and all H values corresponding to orbits drifting in the direction of the j_y current.

3.1.2 Nonlinear oscillator for weakly inhomogeneous magnetic field

In the limit that the kinetic energy is small or that the variation of $B(x)$ is sufficiently weak over the (energy dependent) scale of Δx in Eq. (3.8) we may expand

$$B(x) = B(x_0) + (x - x_0)B'(x_0) + \tfrac{1}{2}(x - x_0)^2 B''(x_0) + \cdots \quad (3.9)$$

to obtain the anharmonic oscillations of the particle in the bottom of the magnetic potential $V_{p_y}(x)$ in Eq. (3.2). With expansion (3.9) the Hamiltonian Eqs. (3.4) and (3.5) reduce to the form of Eq. (1.3) with

$$\ddot{x} = -\Omega^2 x \left[1 + \frac{3}{2}\frac{B'}{B}x + \left(\frac{1}{2}\left(\frac{B'}{B}\right)^2 + \frac{4}{3}\frac{B''}{B}\right)x^2\right]$$

where $\Omega = qB(x_0)/mc$.

Following the amplitude expansion given in Sec. 1.1 with $a = v_\perp/\Omega(x_0)$, the local gyroradius, gives the second and third order motions

$$x^{(1)}(t) = a\cos(\Omega t) \quad (3.10)$$

$$x^{(2)}(t) = -\frac{3}{4}\frac{B'}{B}a^2 + \frac{1}{4}\frac{B'}{B}a^2 \cos(2\Omega t) \quad (3.11)$$

$$x^{(3)}(t) = \frac{a^3}{32}\left(-\frac{3}{2}\left(\frac{B'}{B}\right)^2 + \frac{B''}{B}\right)\cos(3\Omega t) \quad (3.12)$$

and the shift of the oscillation frequency from the local gyrofrequency $\Omega = qB(x_0)/mc$ is

$$\omega = \Omega(1 + \kappa a^2) \quad (3.13)$$

with

$$\kappa = -\frac{3}{4}\left(\frac{B'}{B}\right)^2 + \frac{1}{2}\frac{B''}{B}. \quad (3.14)$$

For the linearly vanishing field $B(x) = B'x$ the gyrofrequency shifts toward zero when the oscillation amplitude a is comparable to $L_B = B/B'$. In a reversed field configuration ($B(x) = B'x$) the shift toward zero frequency connects to the vanishing rotation frequency $\Omega(H, P_y)$ as the critical separatrix region of the phase space is approached. In Fig. 3.2(a) this approach occurs for $E_2 \to E_c$. Also, see Fig. 3.5.

The second order effects in Eq. (3.11) describe the shift of the oscillation center toward the weaker field region and the generation of the 2Ω-harmonic oscillations in the orbits. Using the orbits in Eq. (3.10)–(3.12) to calculate the $v_y(x) =$

$(cp_y - qA_y(x))/c$ velocity and averaging over the cyclotron period we obtain a steady, energy dependent, drift velocity of the orbit given by
$\boxed{\nabla B \text{ drift velocity}}$

$$\bar{v}_y(x) = \frac{cmv_\perp^2}{2qB^2} \cdot \frac{dB}{dx} \quad (3.15)$$

where $v_\perp^2 = 2E/m - p_z^2/m^2$. In the nonuniform field the gyrocircles do not close on themselves due to the larger orbital radius (3.3) in the weaker magnetic field portion of the orbit.

3.1.3 Nonlinear orbits in the field reversed configuration or sheet pinch

Now in case (ii) in (3.6) where the magnetic field has a point where $B(x)\hat{e}_z$ reverses direction there is an additional fixed point at the reversal layer which is chosen as the $x = 0$ point so that locally $B(x) = B'x$. Not too far from the reversal layer the magnetic potential reduces according to

$$V_{p_y}(x) = \frac{1}{2m}\left(p_y - \frac{q}{c}A_y(x)\right)^2 \simeq \frac{1}{2m}\left(p_y - \frac{qB'x^2}{2c}\right)^2.$$

This potential has a bifurcation according to the sign of qp_yB' or for given qB', for the sign of the constant of integration p_y. The bifurcation of the potential is shown in Fig. 3.2 with respect to the change in sign of p_y. That is, the bifurcation parameter is p_y with $p_y = 0$ being the critical point. At the bifurcation the orbits change their character as shown in Fig. 3.3 for the phase space and in Fig. 3.4 for the real space trajectory. Due to the change of the direction of the magnetic field the orbits change their sense of rotation as the particle crosses the neutral sheet ($x = 0$).

The bifurcation condition can also be understood from the Taylor series expansion of the flux condition in Eq. (3.7) which gives the oscillation center at $X = cp_y/qB(X)$ for $B(x) > \Delta x B'(x)$ but for the opposite limit gives $X^2 = 2cp_y/qB' > 0$ or $X = 0$ depending on the sign of p_yqB'. For the maximum value of $p_y = (2mE)^{1/2}$ the value of $X = (2cp_y/qB')^{1/2}$ may be estimated as $(vL_B/\Omega)^{1/2}$ — the geometric mean of the gyroradius and the magnetic scale length $L_B = B/B'$.

Potentials of the form shown in Fig. 3.2 are called bi-stable potentials since there are two stable wells connected by a passage-way whose height may be altered with a perturbation to the system. In a bi-stable trajectory it is often sufficient to know if the system is in the right (R) or left (L) state. This allows the dynamics of the orbit to be truncated to form a coarse graining of the trajectory by the sequence RLLRLLL R... for each time period of importance. The transitions from R to L or L to R are most likely when the phase of the perturbation and the particle is

3.1 Trajectories in Straight, Nonuniform Magnetic Fields

such as to lower the potential barrier when the particle rotation is in the small $|x|$ crossing region.

The phase space portraits of the orbits given in Fig. 3.4 show how the bifurcation at $p_y = 0$ occurs with the stable elliptic fixed point $(x = 0, p_x = 0)$ for $p_y < 0$ changing into an unstable hyperbolic fixed point for $p_y > 0$ by splitting off two new elliptic fixed points at $(\pm x_0, 0)$ with $x_0 = (2p_y c/qB')^{1/2}$. Clearly there is a factor of 2 increase in the period of the orbit in going across the separatrix in Fig. 3.3.

To understand the bifurcation directly from the Lorentz force we consider the x-acceleration

$$\ddot{x} = \left(\frac{q}{m}\right) v_y(x) B(x). \tag{3.16}$$

The symmetry in y allows the direct integration of $\dot{v}_y = -(q/m)\dot{x} B(x)$ to give

$$v_y(x) = \frac{p_y}{m} - \frac{qB'}{2mc} x^2 = -\frac{qB'}{2mc}(x^2 - x_0^2) \tag{3.17}$$

where p_y appears as the constant of integration. Thus the acceleration equation is

$$\ddot{x} = \frac{1}{2}\left(\frac{qB'}{mc}\right)^2 x \left(x_0^2 - x^2\right) \tag{3.18}$$

allowing one to see the bifurcation with the change in sign of $x_0^2 \equiv 2p_y/qB'$ directly from the Lorentz force for the nonlinear x-acceleration.

The critical orbits that asymptotically leave and enter the hyperbolic fixed point as $t \to \pm\infty$ are called the homoclinic orbits when they intersect with the same (homo) slope (clinic) in the finite t portion of the critical curve. In Sec. 3.3 where a perturbation is applied we will find that these intersections of the orbits occur at a finite angle leading directly to the (chaotic) mixing of the trapped and passing orbits. In this case the orbits are called heteroclinic.

From the potential diagram in Fig. 3.2 the homoclinic or separatrix orbits occur for

$$E = E_c = \frac{p_y^2 + p_z^2}{2m} \tag{3.19}$$

or

$$2mE_\perp = p_y^2$$

where $E_\perp \equiv E - p_z^2/2m$ is the kinetic energy in motion perpendicular to the magnetic field. The homoclinic orbits are easily derived either by solving for p_x from Eq. (3.2) with $H = E_c$ or more simply by multiplying Eq. (3.18) by $\dot{x}$, integrating and choosing the constant of integration to include the unstable fixed point $(0,0)$ which gives

$$\frac{1}{2}\dot{x}^2 = \frac{1}{8}\left(\frac{qB'}{mc}\right)^2 (2x_0^2 - x^2)x^2. \tag{3.20}$$

The solutions of Eq. (3.20) are

Homoclinic orbits

$$x_{sx}(t) = \pm\sqrt{2}\, x_0 \operatorname{sech}\left(\frac{qB'x_0}{\sqrt{2}\,mc} t\right)$$

$$(p_x(t))_{sx} = \mp 2p_y \operatorname{sech}\left(\frac{qB'x_0}{\sqrt{2}\,mc} t\right) \tanh\left(\frac{qB'x_0}{\sqrt{2}\,mc} t\right) \quad (3.21)$$

where the subscript sx denotes the orbit along the separatrix which is called the homoclinic orbit. Homoclinic indicates that the unstable trajectory and stable trajectories out of the ×-point meet with the same continuous curve to form the smooth orbits given in Eqs. (3.21). The critical orbits (3.21) describe at $t \to -\infty$ the exponential escape from the neighborhood of the ×-point along the unstable manifold with $(x, p_x) \sim \exp(\Omega_0 t)$ with $\Omega_0 = qB'x_0/\sqrt{2}\,mc$ and for $t \to +\infty$ the exponential approach to the ×-point along the stable manifold with $(x, p_x) \sim \exp(-\Omega_0 t)$. This critical orbit is not periodic and takes an infinite amount of time to complete its motion.

These critical orbits (on the right and left sides of the $x = 0$ surface) and the neighboring trajectories are unstable to infinitesimal structural perturbations to the system and will be used to construct the perturbation theory for the width of the chaotic phase space region surrounding the separatrix using the Melnikov-Arnold theory. Structural perturbations that have been studied in detail and that are known to produce chaos in systems close to the integrable one considered in this section are (i) the addition of small constant magnetic field components B_x and B_y and (ii) the addition of time varying electric fields $E_x(t)$ or $E_y(t)$. Some of the theories for these systems are developed in Secs. 3.3 and 3.4. Before considering the chaos that perturbations produce, we introduce the action-angle variables (I, θ) and the constraints placed on the dynamics by the adiabatic invariants.

3.2 Adiabatic Invariants

Limits on the chaotic regions of the phase space follow from the existence of the adiabatic invariants. For the periodic orbits with either oscillations or rotations (Fig. 3.4) in the phase space we define the action I by the area enclosed by the unperturbed phase space trajectory for each value of (H, p_y, p_z) by

$$I = \frac{1}{2\pi} \oint p_x\, dx = \frac{1}{\pi} \int_{x_{\min}}^{x_{\max}} \left[2m\left(H - \frac{p_z^2}{2m} - \frac{1}{2m}\left(p_y - \frac{q}{c}A_y\right)^2 - q\Phi\right)\right]^{1/2} dx. \quad (3.22)$$

The action $I(H, p_y, p_z)$ is approximately constant when the perturbed orbit essentially closes on itself during a period of its motion. Clearly, when the orbit crosses a

3.2 Adiabatic Invariants

separatrix it does not even approximately close on itself and there is a finite jump in the value of I with $I_i \to I_f$. In fact, it is clear that due to the divergence in the period $T(H \to E_c) \to \infty$ of the orbits approaching the separatrix the condition for the constancy of I, that the rate of change of the perturbation be slow compared with the frequency $\Omega(H)$ of the orbit, always fails when the orbit is sufficiently near the separatrix. The Melnikov-Arnold theory gives a precise meaning to the statement of when the frequency of the perturbation is comparable with that of the orbital motion and gives the method developed in Sec. 3.3.2 to describe the dynamics near the separatrix. This motion near the separatrix is generic to Hamiltonian systems and thus results in the standard map explored in Chapter 4.

The transformation to action-angle variables I, θ can be carried out with a canonical transformation using the $F_2(P, q)$ type generating function given by

$$S(I, x) = \int_{x_{\min}}^{x} \left[2m \left(H_\perp(I, P_z) - V_{p_y}(x') - q\Phi(x') \right) \right]^{1/2} dx' \qquad (3.23)$$

where we introduce the shorthand notation $H_\perp = H(I) - p_z^2/2m$ and $V_{p_y}(x) = (p_y - q A_y(x)/c)^2/2m$ and use $x_{\min}$ as the smallest accessible ($p_x^2 > 0$) value of the x coordinate. The new angle variable follows from

$$\theta = \frac{\partial S(I, x)}{\partial I} = \frac{dH}{dI} \int_{x_{\min}}^{x} \frac{m \, dx'}{[2m(H_\perp - V_{p_y}(x') - q\Phi(x'))]^{1/2}} \qquad (3.24)$$

where the angular rotation frequency of the orbit is

$$\Omega(H, p_y, p_z) = \frac{dH}{dI}, \qquad (3.25)$$

and the second factor in Eq. (3.24) is the time along the orbit $t(x) = \int^x dx'/\dot{x}$ with $\dot{x} = p_x/m$. The second member of the $F_2(P, q) = S(P = I, q = x)$ canonical transformation gives

$$p_x = \frac{\partial S}{\partial x} = \left[2m(H_\perp - V_{p_y} - q\Phi) \right]^{1/2}.$$

Now we evaluate the action-angle variables in several limits to develop an understanding of the physical regimes and the associated adiabatic invariants.

3.2.1 Small gyroradius limit $(2m\, E_\perp \ll p_y^2)$

For the case where the magnetic field is slowly varying over the size of the orbit such that $\varepsilon = \rho/L \ll 1$ then the magnetic potential $V_{p_y}(x')$ locally varies with x' as $A_y(x + \delta x') = A_y(x) + B(x)\delta x'$. The large value of $A_y(x)$ forces x to remain close to the fixed point X defined by $p_y = (q/c)A_y(X)$ to keep $|v_y(x)| \leq (2mE_\perp)^{1/2}$. In this regime the action integral in Eq. (3.22) reduces with the substitution

$$x' = X + \frac{c(2mH_\perp)^{1/2}}{qB(X)} \cos \zeta \qquad (3.26)$$

to

<u>First adiabatic invariant</u>

$$I = \frac{2mH_\perp}{(2\pi)(qB(X)/c)} \int_0^{2\pi} (1-\cos^2\zeta)^{1/2} \sin\zeta\, d\zeta = \frac{mcH_\perp}{qB(X)} = \left(\frac{cm^2}{q}\right)\mu \quad (3.27)$$

where

<u>Magnetic moment</u>

$$\mu = v_\perp^2/2B$$

is the conventional normalization for the magnetic moment or the first adiabatic invariant. The action integral (3.24) gives $\theta = \zeta = \Omega(X)t$ with $\Omega = dH/dI = qB(X)/mc$ where X is a constant of the motion given by $p_y = (q/c)A_y(X)$. The Hamiltonian in the action-angle variables is thus

$$H = \frac{p_z^2}{2m} + m\mu B(x) \quad (3.28)$$

independent of ζ. Equation (3.28) follows from Eq. (3.27) and the definition of $H_{\perp 0}$.

The condition that the second term in Eq. (3.2.6) is small compared with the first term is equivalent to $2m\, E_\perp \ll p_y^2$ since $p_y = (q/c)A_y(X) \cong (q/c)B(x)X$ locally.

The motion governed by the reduced Hamiltonian in Eq. (3.28) describes the mirroring of particles from regions of high $|B|$. The action associated with $\oint p_z\, dz$ from the mirroring motion is called the second adiabatic invariant. The second adiabatic invariant is then

<u>Second adiabatic invariant</u>

$$J(E,\mu,\mathbf{x}_\perp,t) = \int_{s_1}^{s_2} \sqrt{2m(H/m - \mu B)}\, ds$$

where s is the coordinate along the field line.

3.2.2 Small gyroradius and a weak electrostatic potential

An important generalization of Eq. (3.27) for the study of plasmas in the presence of low frequency ($\omega \ll \Omega$) waves is the inclusion of an electrostatic potential $q\phi(x)$. In this case the action integral (3.22) for $q\phi \ll 2mH_\perp \ll p_y^2$ reduces to

$$I = \frac{1}{2\pi} \oint \left\{ \left[2mH_\perp - \left(p_y - \frac{q}{c}Bx'\right)^2\right]^{1/2} - \frac{q\phi(x')}{\left[2mH_\perp - \left(p_y - \frac{qBx'}{c}\right)^2\right]^{1/2}} \right\} dx'.$$

Now the same substitution (3.26) used in Eq. (3.27) together with the Fourier decomposition of $\phi(x') = \sum \phi_k \exp(ik_\perp x')$ (Taylor, 1969) allows the integral to be performed to obtain

$$I = \frac{cm}{q} \oint \left\{ \left[\frac{H_\perp}{B(X)} + \sum_k \frac{q\phi_k}{B(X)} \left(1 - J_0^2(k_\perp v_\perp/\Omega)\right) \right] \right\} d\zeta \qquad (3.29)$$

where $v_\perp = (2m\, H_\perp)^{1/2}$ and $\Omega = qB/mc$. Since plasmas often contain a spectrum of low frequency electrostatic fluctuations with $k_\perp v_\perp/\Omega \lesssim 1$ this formula is of considerable practical importance. For example, when the one-particle distribution function f is a function of $f(H, I_k)$ we have $f \to f(H_\perp, \mu) + (q\phi_k/B)(1 - J_0^2)\partial f/\partial \mu$ as the adiabatic change in the particle distribution function in the presence of a low frequency electrostatic wave.

3.2.3 Current sheet invariant

For general $B(x)$ the action integral I must be carried out numerically. In the important case where the field $B_z(x)$ reverses direction there is a generic behavior. In the neighborhood of the reversal $B(x) = B'x$ where B' measures the strength of the current density in the reversal layer with $j_y = -(c/4\pi)(dB_z/dx)$ at $x = 0$. In this region the action integral can be performed in terms of the elliptic function. For the low energy particles trapped away from the reversal layer the elliptic function I-integral reduces to the magnetic moment $H_\perp/B(X)$ derived in Eq. (3.27). For higher energy particles $2m E_\perp \gtrsim p_y^2$ the orbits cross the current sheet as shown in Figs. 3.3(b)–(d) and thus the particles visit regions where $\Omega(x) = qB(x)/mc \to 0$. For these orbits there is still a well-defined action (Sonnerup, 1971) called the current sheet action which is conserved except for disturbances that are faster than $\Omega(H)$. As shown in Sec. 3.2.1 and in Fig. 3.5, the presence of the separatrix yields a set of orbits with small $w = (H - H_c)/H_c$ such that the period $T(H) \to \infty$ for $|w| \to 0$. To understand the onset of the chaotic motions we need to derive the orbits and action in some detail. The surface of section studies of the perturbed systems in the next sections will show that only the phase space regions where neither the magnetic moment nor the current sheet action are invariant will show global chaotic motions. In regions where either form of $I(H_\perp, p_y)$ is conserved, the chaos is weak.

In the regime where $B(x) = B'x$ we factor the integrand of Eq. (3.15) as

$$I = \frac{1}{2\pi} \oint \left[\left((2m\, H_\perp)^{1/2} - p_y + \frac{qB'x^2}{2c} \right) \left((2m\, H_\perp)^{1/2} + p_y - \frac{qB'x^2}{2c} \right) \right]^{1/2} dx \qquad (3.30)$$

and define the critical excursion points $x_\pm$ by

$$\frac{qB'}{2c} x_\pm^2 \equiv (2m\, H_\perp)^{1/2} \pm p_y. \qquad (3.31)$$

With Eq. (3.31) the orbit quadrature becomes

$$\left(\frac{qB'}{2mc}\right) \int_0^t dt' = \int_{s_{\min}}^x \frac{dx'}{(2mH_\perp)^{1/2} \left[(x_+^2 - x'^2)(x_-^2 - x'^2)\right]^{1/2}} \qquad (3.32)$$

which is readily expressed in terms of the Jacobi elliptic functions $sn(u)$, $cn(u)$ and $dn(u)$ using the formulas in the table on p. 5596 in Abramovitz and Stegun. For the readers' convenience in Table 3.1 we repeat two of the formulas (17.4.44) and (17.4.52) from A and S.

Table 3.1. Jacobi Elliptic Functions for Orbits in a Current Sheet

$F(\varphi \backslash \alpha)$	Equivalent Inverse Jacobian Elliptic Function	φ	t Substitution
17.4.44			
$\int_x^a \dfrac{a\,dx'}{[(a^2 - x'^2)(x'^2 - b^2)]^{1/2}} = \mathrm{dn}^{-1}\left(\dfrac{x}{a}\,\Big|\,\dfrac{a^2-b^2}{a^2}\right)$		$\sin^2 \varphi = \dfrac{a^2 - x^2}{a^2 - b^2}$	$x = a\,\mathrm{dn}\,v$
17.4.52			
$\int_x^b \dfrac{(a^2+b^2)^{1/2}\,dx'}{[(x'^2+a^2)(b^2-x'^2)]^{1/2}} = \mathrm{cn}^{-1}\left(\dfrac{x}{b}\,\Big|\,\dfrac{b^2}{a^2+b^2}\right)$		$\cos \varphi = \dfrac{x}{b}$	$x = b\,\mathrm{cn}\,v$

There are three cases for the integrals according to the orbits indicated in Fig. 3.3. Let us write out the two cases for $p_y > 0$ where the separatrix divides the orbits into crossing and non-crossing orbits. For $p_y > 0$ and $2m\,H_\perp > p_y^2$ the orbits cross the neutral sheet $(B = 0)$ so that $x_{\min} = 0$ and Eq. (3.32) yields

$\boxed{\text{Crossing orbits}}$

$$x(t) = x_+ \, cn(u, m) \tag{3.33}$$

with

$$m = \frac{x_+^2}{x_+^2 + x_-^2} = \frac{(2mH_\perp)^{1/2} + p_y}{2(2m\,H_\perp)^{1/2}} \tag{3.34}$$

and

$$u = t\left(\frac{qB'}{2mc}\right)(x_+^2 + x_-^2)^{1/2} = \Omega(H_\perp, p_y)t$$

where the angular frequency is given by

$$\Omega(H_\perp, p_y) := 2\left(\frac{qB'}{mc}\right)^{1/2}(4m\,H_\perp)^{1/2}\left(\frac{\pi}{2K(m)}\right). \tag{3.35}$$

3.2 Adiabatic Invariants

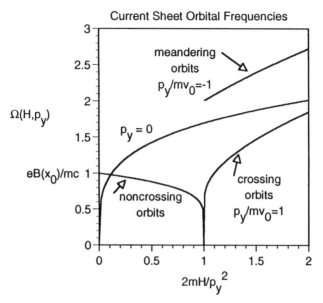

Fig. 3.5. The dispersion of the angular rotation frequency $\Omega = dH/dI$ of the orbits in the reversed magnetic field as described in Fig. 3.4. The curve for $p_y = 0$ correspondence to the case where the potential $V_{p_y}(x)$ is at the bifurcation point. For $p_y j_y < 0$ the origin is an ×-point while for $p_y j_y > 0$ the origin ($p_x = 0, x = 0$) is an 0-point. For $p_y = 0$ the frequency is $\Omega = H^{1/4}$ while for $p_y > 0$ the frequency starts at Ω_0 for $H = H_c = p_y^2/2m$ and increases as $\Omega = \Omega_0 H^{1/4}$. For $p_y j_y < 0$ the low energy particles make cyclotron orbits with $\Omega = eB(x_0)/mc$ with x_0 determined from p_y by $mv_y = p_y - (q/c)A_y(x_0) = 0$. As H increases toward H_c the period diverges to infinity due to the X-point at $(p_x = 0, x = 0)$. For $H > H_c$ the frequency rapidly recovers from zero and increases at $H/H_c \gg 1$ at the rate $H^{1/4}$.

Unfortunately, the standard symbol for the index of the elliptic function is also m, as we have used for the mass of the particle. We must leave it as a test of the readers' understanding of Eqs. (3.33)–(3.35) to determine which m is the elliptic function index which is the mass of the particle.

The orbits are of the type indicated in Fig. 3.4(c) crossing the neutral sheet, and for high energy $(2m H_\perp)^{1/2} \gg p_y$ become $x(t) = x_+ cn(\Omega t, 1/2)$ corresponding to the cubic force law with

$$H_\perp = \frac{1}{2} p_x^2 + \frac{1}{8} \left(\frac{qB'}{mc} \right)^2 x^4.$$

Orbits of this type for the pure x^4-oscillator are thoroughly studied in the review article of Chirikov (1979) beginning in his Sec. 2.3. The $\Omega(H)$, the action and the

Fourier series for $x(t)$ are analyzed there in detail. The principal physical feature is that the frequencies are strongly dispersive with $\Omega(H) = dH/dI \propto H_\perp^{1/4}$ while the harmonic components $n\Omega(H)$ are weak for large n.

The angular rotation frequency $\Omega(H, p_y) = \Omega(2mH_\perp/p_y^2)$ for these trajectories is given by Eq. (3.35) in Fig. 3.5. The frequency vanishes at the critical point $p_y = (2mH)^{1/2}$ where the homoclinic orbit given in Eq. (3.21) occurs.

Lowering the energy to $(2mH_\perp)^{1/2} \to p_y > 0$, the critical energy $H_\perp = E_c$, the oscillations become highly nonlinear with

$$x_+ cn(u,m) \to \operatorname{sech} u - (m_1/4)(\sinh u \cosh u - u)\tanh u \operatorname{sech} u$$

where $m_1 = 1 - m \to 0$. These orbits are approaching the critical orbits in Eqs. (3.21) and the period is approaching infinity.

For still lower energies $(2m H_\perp)^{1/2} < p_y$ the allowed orbits oscillate between x_- and x_+ and $-x_+$ to $-x_-$ and thus do not cross the $B = 0$ region. In this regime the orbits, as shown in Fig. 3.4(a), approach the usual gyro-orbits in a weak field with adiabatic invariant μ and a grad-B guiding-center drift. For $(2mH_\perp)^{1/2} \leq p_y$ the quadrature in Eq. (3.39) is re-expressed with elliptic function as

Non-crossing orbits

$$x^2(t) = x_+^2 - (x_+^2 - x_-^2)sn^2(v,m) = x_+^2\left[1 - m\, sn^2(v,m)\right] \tag{3.36}$$

with

$$m = \frac{x_+^2 - x_-^2}{x_+^2} = \frac{2(2mH_\perp)^{1/2}}{(2mH_\perp)^{1/2} + p_y} \tag{3.37}$$

and

$$v = \frac{1}{2}\Omega(H_\perp, p_y)t$$

$$= t\left(\frac{qB'}{2mc}\right)^{1/2}\left((2mH_\perp)^{1/2} + p_y\right)^{1/2}\left(\frac{\pi}{2K(m)}\right). \tag{3.38}$$

In the limit $m \to 1$ the orbit becomes $x = \pm\sqrt{2}\,x_0 \operatorname{sech}(\Omega_0 t)$ using $sn(v, m \to 1) \to \tanh v$ and $x_0 = (2p_y/qB')^{1/2}$. Now for lower energies the $sn(v,m) \to \sin v = \sin(\Omega_0 t/2)$ and the reduction of Eq. (3.36) gives

Cyclotron orbit limit

$$x(t) = x_0 + \frac{(2m\, H_\perp)^{1/2}\cos(\Omega_0 t)}{qB(x_0)} \tag{3.39}$$

where $\Omega_0 = qB(x_0)/mc$ is the local cyclotron frequency. Thus, at low energies and $p_y > 0$ the orbits are reduced to the usual cyclotron orbits at the value of x derived from the fixed point of Eq. (3.5) — related to the value p_y by the flux condition in Eq. (3.7).

From Eqs. (3.35) and (3.38) we compute the nonlinear orbital frequencies for the three types of orbits shown in Figs. 3.3 and 3.4. In the low energy limit the downward shift from the local cyclotron frequency is given by

$$\Omega = \left(\frac{qB(x_0)}{mc}\right)\left(1 - \frac{1}{16}\left(\frac{2m\,H_\perp}{p_y^2}\right)\right) \tag{3.40}$$

for $2m\,H_\perp \ll p_y^2$. Near the separatrix the orbital period diverges with $w = (H - H_c)/H_c \to 0$ as

$$\Omega(H) = \frac{\pi}{\sqrt{2}}\left(\frac{qB(x_0)}{mc}\right)\frac{1}{\ell n(64/|w|)} \tag{3.41}$$

and for $2m\,H_\perp \gg p_y^2$ the frequency becomes

$$\Omega(H) \simeq \left(\frac{qB(x_0)}{mc}\right)\left(\frac{2m\,H_\perp}{p_y^2}\right)^{1/4}, \tag{3.42}$$

increasing with the square root of the particles speed.

3.3 Onset of Chaos from a Transverse Electric Field

In both the laboratory and magnetospheric configurations with reversed magnetic fields the compression and heating of the plasma takes place from the application of a time dependent external inductive electric field $E_y(t)$ transverse to the magnetic field. The case of a transverse electrostatic field is also important in the theory of plasma heating and is treated by Karney (1983) and Fasoli et al. (1993). The transverse electric field arises from the changing magnetic flux $A_y(x,t)$ as the flux surfaces are compressed or expanded. In the MHD limit where the gyroradius ρ is small $\varepsilon = \rho/L \ll 1$ the charged particles are trapped in the bottom of the magnetic potential wells shown in Fig. 3.2a. The motion of the magnetic wells shown in Fig. 3.2 is given by the $\mathbf{E} \times \mathbf{B}$ velocity with $v_x = cE_y/B_z$. To show that the potential wells move with the $\mathbf{E} \times \mathbf{B}$ velocity, we take the time derivative of condition (3.7) giving the fixed point of the Hamiltonian flow at the bottom of the magnetic well defined by $\partial V_{p_y}/\partial x = 0$. Allowing $A_y \to A_y(x,t)$ the motion of the fixed point determined by Eq. (3.7) is given by

$$\partial_t A_y + \dot{x}\partial_x A_y = 0$$

which, with

$$B_z = \partial_x A_y$$

and

$$E_y = -\frac{1}{c}\partial_t A_y,$$

gives that $\dot{x}$ is the same as the MHD $\mathbf{E} \times \mathbf{B}$-velocity

$$v_x(x,t) = -\frac{\partial_t A_y}{\partial_x A_y} = \frac{cE_y}{B_z(x)} \tag{3.43}$$

for the velocity of the bottom of the moving magnetic potential well $V_{p_y}(x,t)$.

Now from the Hamiltonian we see that the particles trapped near the bottom of the $V_{p_y}(x,t)$ potential have no choice but to be convected with the $\mathbf{E} \times \mathbf{B}$ velocity. When the electric field is such that $j_y E_y > 0$ the motion of the particle is into the region of lower magnetic field where the gyroradius rapidly expands and begins to fill the trapped region defined by the separatrix in the p_x, x phase space in Fig. 3.3(a). It is at the stage where the particles reach the separatrix region that their orbits become chaotic and describe an irreversible heating when certain conditions are met. The $\mathbf{E} \times \mathbf{B}$-motion which compresses the plasmas is a reversible heating. Reversible heating is defined as follows.

If the electric field changes sign and returns to its initial value the net change in the kinetic energy vanishes showing that the local increase of the energy is reversible. The reversible or adiabatic change in energy is described by the conservation of $\mu = v_\perp^2/2B(x)$ and the change in the kinetic energy $\frac{1}{2}m(E_y/B_z)^2$ through the polarization current $\delta j_y = (mn/B_z)\frac{d}{dt}(E_y/B_z)$. Now, we consider in detail the mechanism for the onset of chaotic dynamics and the irreversible heating in the one-dimensional current sheet.

3.3.1 $1\frac{1}{2}$-D Hamiltonian

For a general time dependent perturbation to the flux function $A_y(x)$ the Hamiltonian flow occurs in a $d = 3$ phase space referred to as a $1\frac{1}{2}$-degree-of-freedom problem. The Hamiltonian $H(p_x, x, t)$ is given by Eq. (3.45) with $A_y \to A_y(x,t)$. The value of the Hamiltonian $H = E$ now changes along the trajectories due to the inductive electric field according to

$$\frac{dH}{dt} = \frac{\partial H}{\partial t} = -\frac{q}{mc}\frac{\partial A_y}{\partial t}\left(p_y - \frac{q}{c}A_y(x,t)\right) = qv_y E_y(x,t). \tag{3.44}$$

From which the change ΔH in the kinetic energy of a particle in time Δt follows as

$$\Delta H = \Delta E = q\int_{\Delta t} v_y E_y dt = q\int_{\Delta y} E_y dy.$$

In the case where there is a periodic change in the electric field $E_y(t+T) = E_y(t)$, the phase space structure is that of a two-dimensional toroidal surface $\mathcal{T}^2$ from the direct product of $\mathbb{R}^2(p_x, x)$ phase space with the periodic toroidal angle

3.3 Onset of Chaos from a Transverse Electric Field

$\phi = 2\pi t/T = \omega t$ giving the T^2 space. The Hamiltonian flow around the time torus gives an area preserving mapping of the $\mathbb{R}^2(p_x, x)$ plane at a given meridian (ϕ = const) onto itself. Thus, the *topology* of the space of the flow for the particle orbits is the same as that for the magnetic field lines in Sec. 2.6.

In the limit that E_y is sufficiently weak the KAM theory states that the integrable curves of the p_x, x phase space in Fig. 3.3 are preserved in the presence of the perturbation. In the new phase space structure with the modified integrable curves $\Delta H = 0$ over the period of the perturbed motion. The exception to the KAM theory result that $\Delta H = 0$ occurs near the boundaries of the resonance surfaces in the phase space. Resonances occur where the flow after n turns in ϕ returns to $\theta + 2\pi m$ to form the m/n-order resonance. Denoting the frequency of the external field by $\omega = 2\pi/T$ and recalling that Eq. (3.25) gives the frequency $\Omega(H, p_y)$ for the integrable motion, the rational resonances occur at

$$\frac{d\theta}{d\phi} = \frac{1}{\omega}\frac{dH}{dI} = \frac{\Omega}{\omega} = \frac{m}{n}. \tag{3.45}$$

Equation (3.45) defines a dense set of resonant values of H, p_y at which the phase space surface opens up into islands of period m in θ and period n in ϕ. Surrounding the islands is a stochastic layer formed from orbits lying near the m hyperbolic fixed between the m elliptic fixed points of the island chain.

3.3.2 Melnikov-Arnold integral

Let us consider the change of the energy of the particle during one cycle of the external field. For the current sheet model the time varying flux function is

$$A_y = \frac{B'x^2}{2} + \frac{E_y}{\omega}\left(\cos(\omega t + \alpha) - \cos(\alpha)\right)$$

where

$$E_y(t) = E_y \sin(\omega t + \alpha)$$

with α the phase of the electric field at $t = 0$. In this case the parallel kinetic energy $p_z^2/2$ is a constant of the motion and we introduce $H_\perp = H - p_z^2/2m$ and then drop the subscript $\perp$ in this section, throughout the following analysis. The instantaneous change in the kinetic energy is

$$\frac{dH}{dt} = \frac{qE_y}{m}\sin(\omega t + \alpha)\left[p_y - \frac{q}{c}A_y(x,t)\right] \tag{3.46}$$

and the change over a complete period is

$$\Delta H = \frac{q^2 E_y B'}{2m}\int_{-T/2}^{T/2} dt\, x^2(t)\sin(\omega t + \alpha). \tag{3.47}$$

Now, for the case where the motion $x(\theta) = x(\Omega t)$ is periodic with $x(\theta + 2\pi) = x(\theta)$ the value of $\Delta H = (q^2 E_y B'/2m) x_n^2 2\pi \delta(\omega - 2n\Omega)$, so that $\Delta H = 0$ except at the resonances $\omega = 2n\Omega(H)$.

For particles near the separatrix that separates the cyclotron-like orbits from the crossing orbits there is a broad spectrum of orbital harmonics in $x_n(\Omega t)$ going to a continuum in the limit $w = (H - H_c)/H_c \to 0$. The energy gained by the orbits near the separatrix can be calculated approximately by replacing the exact orbit with that along the separatrix given in Eq. (3.21). In this approximation, the jump in the kinetic energy ΔH in the divergent period of the orbit is given by

M-A integral

$$\Delta H = \frac{q^2 E_y B'}{2m} \sin\alpha \int_{-\infty}^{\infty} dt X_{sx}^2(t) \cos(\omega t) \tag{3.48}$$

from Eq. (3.47). The idea is that for small $\Delta H/H_c$ the integral in Eq. (3.47) can be calculated with the exact orbit approximated by $X_{sx}(t)$. Integrals of the type Eq. (3.48) are known as the Melnikov-Arnold (M-A) integrals and their properties have been extensively developed. For the case of the perturbed pendulum Chirikov (1979) develops their properties in detail in his Sec. 4.4 and his appendix on the M-A integral. We refer the reader to that appendix for the development of the properties of the M-A integral.

For the separatrix orbit in Eq. (3.21), the integral in Eq. (3.48) may be carried out exactly to find that

$$\Delta H = W \sin(\alpha) \tag{3.49}$$

where

$$W = \sqrt{2} q\, E_y x_0 \frac{\mu}{\sinh(\mu)} \tag{3.50}$$

with

$$\mu = \pi \frac{\omega}{\sqrt{2}\Omega_0}. \tag{3.51}$$

For $\mu \gg 1$ where the driving frequency is well above the local cyclotron frequency Ω_0 the change of the energy of the particle is exponentially small $W \propto \mu \exp(-\mu)$. In this case the dominant contribution to the ΔH integral comes from a jump in H at the time of crossing the maximum excursion in x when $p_x = 0$ and $x = \sqrt{2}x_0$. For driving frequencies ω below the local cyclotron frequency Ω_0 we have $\mu \ll 1$ and the increase of H per cycle is $\sqrt{2}x_0\, qE_y \sin\alpha$. This increase will take the particle out of the separatrix layer when W/H_c is finite.

3.3.3 The Whisker map and the standard map

In the limit where $W/H_c \ll 1$ the dynamics in the $d = 3$ phase space near the separatrix is given by the jump $\Delta H = W \sin\alpha$ accumulated over each passage around the time torus $t_n = 2\pi n/\omega$ with $n = 0, 1, 2, 3, \ldots$.

3.3 Onset of Chaos from a Transverse Electric Field

To calculate the advance of the phase α of the particle relative to the electric field in each passage around the time torus we must take into account that in formulating the ΔH integral in Eq. (3.48) the $t = 0$ point is specified as the point where $p_x = 0$ and $x = x_{\max}$ to take advantage of the t-symmetries $(x(-t) = x(t)$, $p_x(-t) = -p_x(t))$ in the orbit. The phase α is that of the external electric field at the time the orbit reaches the position $(p_x = 0, x_{\max})$.

The time between the successive maxima of $x(t)$ is given by the nonlinear period $2\pi/\Omega$ derived in Sec. 3.2.3. Near the separatrix the period becomes long as shown by the narrow dip in Fig. 3.5 and described by the logarithmic divergence in Eq. (3.41). For small $w = (H - H_c)/H_c$ the time between such maxima of $x(\Omega t)$ is given by

$$\frac{2\pi}{\Omega(H)} \simeq \sqrt{2}\Omega_0^{-1} \ln\left(\frac{64}{|w|}\right) \tag{3.52}$$

derived from the small w expansion of the $\Omega(H, p_y)$ formulas in Eqs. (3.35) or (3.38). Thus, the advance of the phase of the particle relative to the wave between successive passages is $2\pi\omega/\Omega(H)$ which, with approximation (3.52), gives

$$\alpha_{n+1} = \alpha_n + \frac{\sqrt{2}\omega}{\Omega_0} \ln\left(\frac{64}{|w|}\right). \tag{3.53}$$

From this equation we see that the phase change can be large even for small ΔH. Now taking into account the change ΔH and $\Delta \alpha$ in each successive passage around the time torus we derive the separatrix map

Whisker map

$$w_{n+1} = w_n + W\sin(\alpha_n)$$

$$\alpha_{n+1} = \alpha_n + \frac{\sqrt{2}\omega}{\Omega_0} \ln\left(\frac{64}{|w_{n+1}|}\right). \tag{3.54}$$

where the choice is made to use w_{n+1} rather than w_n in evaluating the phase advance to guarantee the area preserving property of the map. The map in Eq. (3.54) is called the "whisker map" and is of the same form as that derived for the perturbed pendulum by Chirikov (1979) although the physics of the problem is different. In an example, we compare the separatrix map with the Hamiltonian flow. When ΔH ($\leq W$) is not too large there is good agreement between the results. Now we consider a further reduction that brings the map (3.54) to the famous "standard map" that is generic to a wide class of Hamiltonian problems in the limit of being close to the transition to chaos.

3.3.4 The standard map

When the parameter ω/Ω_0 is large so that the change in energy is small it is possible to linearize the separatrix map around the resonant values of w_r for which the phase advance $\alpha_{n+1} - \alpha_n$ is $2\pi n$ with integer n. From Eq. (3.12) the resonance values of $|w|$ are

$$|w_r| = 64\exp\left(-\frac{2\pi n}{\lambda}\right) \tag{3.55}$$

with $\lambda = \sqrt{2}\omega/\Omega_0$. Expanding the phase shift $\ln(64/|w|)$ about the resonance, while introducing the new variable

$$I = -\frac{\lambda}{w_r}(w - w_r), \tag{3.56}$$

and the parameter

$$K = -\frac{\lambda W}{w_r}, \tag{3.57}$$

the whisker map reduces to the standard map

Standard Map

$$\bar{I} = I + K\sin\theta$$
$$\bar{\theta} = \theta + \bar{I}. \tag{3.58}$$

There are numerous studies about the properties of the standard mapping (Greene, 1979), so that the motion in the approximation leading to Eq. (3.58) may be considered as completely known. In particular, the critical value for the break-down of the last KAM surface is $K_{\text{crit}} = 0.971635...$, and for $K > K_{\text{crit}}$ the diffusion coefficient $D(K)$ is known to be well approximated by certain analytic formulas given in Sec. 4.2.

The properties of the standard map, and its relativistic generalization, are developed in Chapter 4.

3.4 Onset of Chaos from a Normal Magnetic Field Component

A typical structural perturbation to the reversed magnetic field configuration of Sec. 3.2 is the addition of a normal magnetic field component $B_x \neq 0$. Often there is also a $B_y \neq 0$ component which we will consider in Sec. 3.7. The $B_x \neq 0$ component is said to be the "reconnection" field component since the reversed B_z fields on the two sides of the current sheet are now connected by the small B_x field and particles with a sufficiently small gyroradius $\rho = v_\perp/\Omega$ and small pitch angle $\alpha \cong v_\perp/v \ll 1$ can freely cross between the previously separated regions. The change in the magnetic field line topology is shown in Fig. 3.1 and is a common configuration in both laboratory and space plasmas.

Static magnetic field configurations of the type shown in Fig. 3.1 have a vector potential $A_y(x,z)$ with

$$\mathbf{B} = \nabla \times (A_y \hat{\mathbf{e}}_y) = \nabla A_y \times \hat{\mathbf{e}}_y. \tag{3.59}$$

3.4 Onset of Chaos from a Normal Magnetic Field Component

Thus the magnetic field lines are the isolines of the potential $A_y(x,z)$. We may also call the function $\psi = -A_y(x,z)$ the magnetic flux function (with the conventional minus sign for the definition of ψ) such that $\mathbf{B} = \hat{\mathbf{e}}_y \times \nabla\psi$ and $\nabla \times \mathbf{B} = \nabla^2\psi\hat{\mathbf{e}}_y = (4\pi j_y/c)\hat{\mathbf{e}}_y$.

The Hamiltonian for the particle is now a 2-degree-of-freedom problem with

$$H(P_x, P_z, x, z) = \frac{p_x^2}{2m} + \frac{p_z^2}{2m} + \frac{1}{2m}\left(P_y - \frac{q}{c}A_y(x,z)\right)^2 \quad (3.60)$$

where the effective potential is

$$V_{p_y}(x,z) = \frac{1}{2m}\left(P_y - \frac{q}{c}A_y(x,z)\right)^2. \quad (3.61)$$

The constancy of $H = E = \frac{1}{2}mv^2$ and P_y (the sum of the kinetic and magnetic momentum) again work to confine the particle motion to $0 < V_{p_y}(x,z) < E$ which constraints the motion in the direction ∇V_{p_y} but not perpendicular to ∇V_{p_y} which is the $\mathbf{B}$-direction. [The perpendicular direction is $\mathbf{e}_y \times \nabla V_{p_y} \propto \mathbf{B}$].

Generalizing Eq. (3.7) for the location of the oscillation center specified by the flux surface

$$p_y = \frac{q}{c}A_y(x_0, z_0), \quad (3.62)$$

we have the bound on the motion

$$|A_y(x,z) - A_y(x_0, z_0)| \leq \frac{c(2mE)^{1/2}}{q} \quad (3.63)$$

which locally gives

$$\Delta r_\perp < \frac{c(2mE)^{1/2}}{q \min|\nabla A_y|} = \frac{cmv}{q B_{\min}}. \quad (3.64)$$

Thus, the motion across the gradient of A_y is bounded by the maximum $\rho = cmv/q B_{\min}$. The motion across $\mathbf{B}$ in the symmetry direction ∇y is not bounded, however, since the nonlinear harmonics will give secular terms, as in Eqs. (3.10)–(3.15), producing a drift velocity in the symmetry direction ∇y.

3.4.1 2D nonlinear coupled oscillations

The local magnetic field given by

$$B_x = -\frac{\partial A_y}{\partial z} = \text{constant}$$

$$B_z = \frac{\partial A_y}{\partial x} = \frac{\partial B_z}{\partial x}x \equiv B'x \quad (3.65)$$

has the polynomial vector potential

$$A_y(x, z) = \frac{B'x^2}{2} - B_x z. \tag{3.66}$$

The curvature of the line $z = B'x^2/2B_x \equiv x^2/2R_c$ is also the minimum radius of curvature R_c of the tangent magnetic field vector $\mathbf{b}$. To see this we introduce the curvature vector $\boldsymbol{\kappa}$ derived from the tangent field $\mathbf{b}(\mathbf{x})$ by

Curvature vector

$$\boldsymbol{\kappa} = (\mathbf{b} \cdot \nabla)\mathbf{b} = \frac{B_x^2 B_z'(B_x \widehat{\mathbf{e}}_z - x B_z' \widehat{\mathbf{e}}_z)}{(B_x^2 + B'^2 x^2)^2}. \tag{3.67}$$

At $x = 0$ we have

$$\boldsymbol{\kappa} = \frac{\mathbf{e}_z}{R_c} = \frac{B_z'}{B_x} \widehat{\mathbf{e}}_z$$

and $|\boldsymbol{\kappa}(x)|$ decreases rapidly as $1/|x|^3$ away from the reversal layer. The constant magnitude $B(\mathbf{x})$ surfaces are given by

$$B = \left(B_x^2 + (B')^2 x^2\right)^{1/2} = B_x \left(1 + x^2/R_c^2\right)^{1/2}. \tag{3.68}$$

The gradient of $B(x)$ produce a nonlinear drift in the $\widehat{\mathbf{e}}_y$-direction as shown in Sec. 3.1.2 and given in Eq. (3.15). The ∇B-drift is in the opposite direction to the curvature drift $\mathbf{v}_{\text{curv}} = mc\,v_\parallel^2\, \mathbf{b} \times \boldsymbol{\kappa}/qB$ of the guiding center. The curvature drift is easily understood as the generalization of the $\mathbf{E} \times \mathbf{B}$ drift when we take into account the centrifugal force from the $v_\parallel$-motion along the magnetic field line $q\mathbf{E} \to q\mathbf{E} - mv_\parallel^2\, \mathbf{b} \cdot \nabla \mathbf{b}$.

The Hamiltonian for the charged particle in the parabolic magnetic field is

$$H(p_x, p_z, x, z) = \frac{p_x^2}{2m} + \frac{p_z^2}{2m} + \frac{1}{2m}\left(p_y - \frac{q}{c}\left(\frac{B'x^2}{2} - B_x z\right)\right)^2 \tag{3.69}$$

which is a 2-degree-of-freedom Hamiltonian parameterized by p_y. The reference surface defined by Eq. (3.62) is the parabolic magnetic surface with a crossing of the symmetry plane at z_0 defined by

$$p_y = \frac{q}{c}\left(\frac{B'x^2}{2} - B_x z\right) \equiv -\frac{q}{c} B_x z_0 \tag{3.70}$$

giving the reference surface $z = z_0 + x^2/2R_c$ with $R_c = B_x/B'$ — the radius of curvature at the tip of the parabola.

If we translate the origin of z to z_0 (defined by p_y) the Hamiltonian may be written as the nonlinear coupling of the two integrable Hamiltonians $H_x(p_x, x) = I_x\,\theta_x$ and $H_z(p_z, z) = I_z\,\theta_z$ where

$$H = \frac{p_x^2}{2m} + \frac{p_z^2}{2m} + \frac{q^2 B_x^2}{2mc^2}\left(z - \frac{x^2}{2R_c}\right)^2 = H_x(p_x, x) + H_z(p_z, z) + V x^2 z \tag{3.71}$$

3.4 Onset of Chaos from a Normal Magnetic Field Component

with the two integrable degrees-of-freedom

$$H_x(p_x, x) = \frac{p_x^2}{2m} + \frac{m}{8}\frac{\Omega^2}{R_c^2} x^4 \tag{3.72}$$

$$H_z(p_z, z) = \frac{p_z^2}{2m} + \frac{m}{2}\Omega^2 z^2 \tag{3.73}$$

coupled by $V(x, z)$ with

$$V(x, z) = -\frac{m\Omega^2}{2R_c} x^2 z \tag{3.74}$$

where $\Omega = q\,B_x/mc$ is the cyclotron frequency in the "reconnection" or normal field component B_x and $R_c = B_x/B'$ is the radius of curvature of the field line at the reversal layer.

3.4.2 Two degrees-of-freedom and degeneracy

While the problem of two coupled nonlinear oscillations is a difficult, nonanalytic problem, the KAM theory of Chapter 2 applies under the following three conditions:

(i) non-degenerate lowest order system $H = H_0(\{I\}) + \varepsilon V(\{I, \theta\})$

$$\det\left(\frac{\partial^2 H_0}{\partial I_i\, \partial I_j}\right) \neq 0 \tag{3.75}$$

(ii) weak $\varepsilon \ll 1$ perturbation

$$\varepsilon V(\{I, \theta\}) \to 0 \tag{3.76}$$

generally equivalent to sufficiently small amplitudes or low oscillator energies

(iii) irrational rotational transform or winding number

$$n\Omega_x(H) + m\Omega_z(H) \neq 0$$

$$q(H) = -\frac{\Omega_x(H)}{\Omega_z(H)} \neq \frac{m}{n} \tag{3.77}$$

The Hamiltonian (3.71) does not satisfy these conditions so we expect and find a rich dynamic behavior beyond the simple opening up of island structures given in Chapter 2 for the magnetic island problem. Note that $H_z(p_z, z)$ in Eq. (3.73) is a linear oscillator.

We follow the Zaslavsky designation of the problem as a problem with degeneracy when condition (3.75) is not satisfied (Sagdeev *et al.* pp. 125–128, 1988). The occurrence of degeneracy allows a global chaos to occur and is a more common feature in plasma problems than first realized. In the present problem the degeneracy occurs from the linearity of the $H_z(p_z, z)$ oscillator.

The degeneracy of the linear $H_z(p_z, z)$ oscillator with $\partial H_0/\partial I_z = \Omega = \text{const}$ can be removed by including a z-gradient of the B_x field. For weak dB_x/dz the amplitude dispersion given by Eq. (3.13) applies. In the case of the magnetotail and other current sheets where this problem was originally studied the spatial gradient of B_x is sufficiently weak that $\Omega = \text{const}$ is a good approximation.

The oscillation frequency of the Hamiltonian $H_x(p_x, x)$ in Eq. (3.72) is

$$\Omega_x = \left(\frac{\Omega(2m\, H_x)^{1/2}}{R_c}\right)^{1/2} \left(\frac{\pi}{2K(1/2)}\right) \tag{3.78}$$

and that for $H_z(p_z, z)$ is

$$\Omega_z = q\frac{B_x}{mc} = \Omega \ .$$

The frequency Ω_x is easily understood from the estimate that $\Omega_x = v_x/x_t$ where $v_x \simeq (2H_x/m)^{1/2}$ and x_t is the turning point where $H_x(p_x = 0, x_t) = H_x$ = the kinetic energy in the x-degree-of-freedom. The properties of the oscillator in Eq. (3.72) are given in Sec. 3.2.3 and in Sec. 2.3 of Chirikov (1979). [Here $K(1/2) = 1.85407$.]

3.4.3 Resonance conditions and the surface of section

A simple way to see the connection between the two-degree-of-freedom autonomous system given by Hamiltonian (3.69) and the chaos in Sec. 3.3 is to consider the problem of the transfer of energy between the two degrees-of-freedom. In particular, we may consider the case where the initial data puts most of the energy into the $H_z(p_z, z)$ oscillator which is linear and has the solution

$$z(t) \cong a\, \cos(\Omega t) \tag{3.79}$$

describing cyclotron orbits in the small region $|x| \ll R_c$. With the $z(t)$ motion substituted into Hamiltonian (3.69) we see that the form of the $H_x(p_x, x, z(t))$ motion becomes identical to that in Sec. 3.3 where the perturbation is that from a transverse electric field. The periodic time torus $\phi = \Omega t$ that occurs in the applied electric field problem is replaced with a more abstract toroidal surface described by the two coupled angle variables θ, ϕ associated with $H_x(p_x, x)$ and $H_z(p_z, z)$ components of the motion.

The resonance condition in the limit of weak coupling of the two degrees-of-freedom is given by

$$m\, \Omega_x(H) + n\, \Omega_z(H) = 0 \tag{3.80}$$

where $\Omega_\alpha = dH_0/dI_\alpha$ are the rotation frequencies in the uncoupled system and m, n are positive and negative integers. The condition (3.80) is a statement that the winding number $\nu(H) = \Omega_x(H)/\Omega_z(H)$ on the toroidal surface is rational. For

3.4 Onset of Chaos from a Normal Magnetic Field Component

rational winding numbers in nondegenerate systems the phase space opens up into a chain of islands defined by fixed points of order m/n as described in Chapter 2. Here we show, as an example, that for $m = -1$ and $n = 5$ there is a new kind of structure to the phase space for the degenerate system with $\Omega_z = $ const.

The winding number for the uncoupled oscillators is given by

$$\nu = \frac{\Omega_x}{\Omega_z} = \frac{1}{\Omega_z}\left(\frac{\Omega(2H_x/m)^{1/2}}{R_c}\right)^{1/2}\left(\frac{\pi}{2K(1/2)}\right) = 0.847\left(\frac{v}{\Omega_z R_c}\right)^{1/2} \quad (3.81)$$

where we use Eq. (3.78). Now we introduce the kappa-parameter κ_{BZ} defined by Büchner and Zelenyĭ (1986).

$$\varepsilon = \frac{\rho}{R_c} = \frac{mcv}{q\,B_{\min}\,R_c} \equiv \frac{1}{\kappa_{BZ}^2}. \quad (3.82)$$

Equations (3.81) and (3.82) show that the winding number is controlled by the maximum of the finite Larmor radius parameter ε. For $\varepsilon^{1/2} \equiv 1/\kappa_{BZ} < 1$ the winding number ν is less than one. In this case $\varepsilon \ll 1$ or $\kappa_{BZ} \gtrsim 2$ the magnetic moment $\mu = v_\perp^2/2B$ is a good adiabatic invariant. The resonance condition (3.80) is only satisfied for high m values with negligible zones of instability for $\varepsilon \ll 1$.

For regimes with $\kappa_{BZ} = 1/\varepsilon^{1/2} \lesssim 2$ the resonance conditions are satisfied for low m, n values, and the phase space is filled with overlapping resonances. As discussed above, the conditions for the KAM theorem, Eqs. (3.75)–(3.77), are not satisfied so we must examine the surfaces of section to understand the nature of the chaotic motion.

For the study of the surface of section and the Lyapunov exponents of the chaotic orbits it is convenient to introduce dimensionless variables appropriate to the current sheet. For a given value of $H = \frac{1}{2}mv^2$ we define the rescaled $(\mathbf{x}, t)$ variables by

$$\mathbf{x} \to \left(\frac{mcv}{qB'}\right)^{1/2}\mathbf{x} \quad \text{and} \quad t \to \left(\frac{mc}{vqB'}\right)^{1/2}t \quad (3.83)$$

so that

$$\frac{d\mathbf{x}}{dt} \to v\frac{d\mathbf{x}}{dt}. \quad (3.84)$$

The dimensionless Hamiltonian is then $H = mv^2 h$ with

$$h = \frac{p_x^2}{2} + \frac{p_z^2}{2} + \frac{1}{2}\left(\kappa z - \frac{x^2}{2}\right)^2 = \frac{1}{2} \quad (3.85)$$

with κ (hereafter the subscript is dropped on the Büchner and Zelenyĭ parameter) being the only parameter that controls the chaos in Eq. (3.85). The κ parameter is

$$\kappa = \left(\frac{R_c}{\rho}\right)^{1/2} = \frac{B_x}{B_0(\rho_0/L_B)^{1/2}} \quad (3.86)$$

where we have introduce the auxiliary parameters $L_B = B_0/B'_z(0)$, $\Omega_0 = qB_0/mc$ and $\rho_0 = mcv/qB_0$.

A closely related two-parameter (b_n, ε) form of the dimensionless Hamiltonian is given by Chen and Palmadesso (1986) for the global magnetic field model

$$B_z(x) = B_0 \tanh\left(\frac{x}{L}\right).$$

In this global model there are two parameters $b_n \equiv B_x/B_0$ and the FLR parameter ε in Eq. (3.82). In the Chen–Palmadesso works the space variables are normalized by $R_c = B_x L/B_0$ and the time by $t \to t/\Omega_x$ with $\Omega_x = qB_x/mc$ so that the value of the kinetic energy $\frac{1}{2}mv^2$, measured relative to $mL^2\,\Omega_0^2(B_x/B_0)^4$, becomes the stochasticity parameter. Defining the Chen–Palmadesso energy parameter

$$\widehat{H}_{CP} = \frac{mv^2}{2(mR_c^2\,\Omega^2)} \tag{3.87}$$

with $R_c = LB_x/B_0$, we have the relationship

Chaos parameter

$$\kappa_{BZ} = \frac{1}{(2\widehat{H}_{CP})^{1/4}} = \left(\frac{R_c}{\rho_{\max}}\right)^{1/2}. \tag{3.88}$$

The Chen-Palmadesso (CP) Hamiltonian is

$$\widehat{H}_{CP} = \tfrac{1}{2}\dot{x}^2 + \tfrac{1}{2}\dot{z}^2 + \tfrac{1}{2}\left(b_n^{-2}\ell n \cosh b_n x - z\right)^2.$$

Widely used CP reference parameter values are $\widehat{H}_{CP} = 500$ and $b_n = 0.1$ which corresponds to

$$\kappa_{BZ} = 0.178 \quad \text{and} \quad \frac{\rho_{\max}}{R_c} = 31.6. \tag{3.89}$$

For motions confined well inside the current sheet $|x| \ll L$ the Chen-Palmadesso problem reduces to the BZ-Hamiltonian (3.85) with κ_{BZ} and $\widehat{H}_{CP}$ related by Eq. (3.88). For motion outside the current sheet $|x| \gg L$ the Chen-Palmadesso problem has $\mathbf{B} = B_x\,\mathbf{e}_x \pm B_0\,\mathbf{e}_z \simeq$ constant, and the orbits are integrable in this asymptotic region.

The surface of section is constructed by finding the crossings of the $x = 0$ plane by testing each time step for $x(t_n)x(t_{n-1}) < 0$ and interpolating to find the values of (z, p_z) at t_* where $x(t_*) = 0$. The results for the parameter value given in Eq. (3.89) are shown in Fig. 3.6.

The structure of the phase space is revealed by the surface of section in Fig. 3.6 where three types of orbits can be identified by the regions labeled A, B, and C. The orbit types are as follows. The intersection of the $H_{CP} = 500$ energy surface with the surface of section is a circle of radius $(2H)^{1/2} = 31.6$.

3.4 Onset of Chaos from a Normal Magnetic Field Component

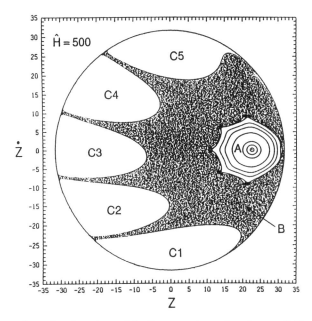

Fig. 3.6. The surface of section at $x = 0$ for the z, p_z phase plane for $\kappa = 0.18$ or $\widehat{H}_{cp} = 500$. The intersection of the energy surface $H_z = p_z^2/2 + z^2/2 = 500$ with the surface of section determines the circular boundary. The three types of orbits are: (A) integrable orbits with $H_x \gg H_z$, (B) stochastic orbits, and (C) the transient orbits. The transient orbits enter in the region C_1 and make successive x, p_x oscillations through C_2, C_3, C_4 and leave from region C_5 to follow the magnetic field line away from the field reversal zone. (Adapted from Chen and Palmadesso (1986).)

3.4.3.1 Integrable (ring) orbits with $H_x \gg H_z$

When most of the energy is in the $p_x - x$ oscillations there are invariant curves in the phase space as shown in Region A. The rotation rate near the stable fixed point is approximately qB_x/mc. The spatial configuration of an integrable orbit is that of oscillations on a cylindrical surface with its axis along $B_x \hat{e}_x$ and radius $r = mcv/qB_x$.

3.4.3.2 Stochastic orbits with $H_x \sim H_z$

In the region marked B the energy in the H_x and H_z oscillations are comparable and the motions are unstable in that two neighboring trajectories diverge exponentially in time.

In this regime the effective potential for the p_x, x-oscillator is rapidly changing from the stable to the unstable configurations shown in Fig. 3.2 — due to the periodic oscillations of the effective p_y for the p_x, x-oscillator for which

$$p_y \to p_y^{\text{eff}}(t) = z(t). \tag{3.90}$$

At the reversal layer where $v_y(x=0) = z$ the effective potential has an unstable fixed point for $z > 0$ and a stable fixed point for $z < 0$. The energy in $H_z(p_z, z)$ oscillator forces the system to make repeated separatrix crossings in the p_x, x-phase space giving rise to the chaotic scattering of the orbit as it passes close to the ×-point of the unperturbed Hamiltonian.

3.4.3.3 Transient unbounded orbits with $H_x \ll H_z$

In this regime marked C the orbits have small pitch angles in the exterior region so that they pass through the reversal layer and make long excursions into the strong left ($x < 0$) and right ($x > 0$) exterior magnetic field regions.

While transiting through the reversal layer the particle is rotated in the y-z-plane by the B_x magnetic field at the angular frequency $\Omega = qB_x/mc$. Chen and Palmadesso (1986) show that when there are an integer number n of (x, p_x) oscillations in the half period rotation time π/Ω that the particles enter and exit the white regions labeled C_1 and $C_2, \ldots, C_n$ with only a small fraction of the particles entering into the stochastic domain B. The condition for n-oscillations is given by $\Omega_x \Delta t = (n + 1/2)\pi$ with $\Delta t = \pi/\Omega$. This resonance condition defines a sequence of resonant energies E_n since $\Omega_x \cong (v\Omega_0/L)^{1/2}$ by Eq. (3.78). In terms of $\widehat{H}$ the resonance condition (Burkhart and Chen, 1991) is

$$\widehat{H}^{1/4} = n + 1/2. \tag{3.91}$$

For $\widehat{H} = 500$ used in Fig. 3.6 this gives $n = 5$ and explains the five white regions labeled by $C_1, C_2, \ldots, C_5$. For $n = 1, 3, 5, \ldots$ the particle exits on the opposite side from its entry while for $n = 2, 4, 6, \ldots$ the particle is reflected by the current sheet. Off the resonance energies H_n defined by Eq. (3.91) there is a wider angle of pitch angles for which the transient particles will enter the stochastic region B through C_1. Thus, in this degenerate Hamiltonian system the $m = -1$, $n = 5$ resonance shows a different kind of phase space structure than given in Chapter 2.

These types of transient orbits were discovered by Speiser (1965, 1967) and are known as Speiser orbits. The Speiser orbits play in important role in the transport processes and the heating rate of the plasma trapped in the geomagnetic tail on the night side of the earth (Lyons and Speiser (1982, 1985) and Horton and Tajima (1991)).

Chen and Palmadesso (1986) have emphasized that the particle distribution function f_α for the three regions $\alpha = A, B, C$ do not mix in a collisionless, fluctuation free, plasma and thus can have different values. They use this property to predict that the value of f at the resonant energies E_n is much less than the value of f off the resonance. Thus $f(E)$ should show peaks and valleys spaced at E_n.

3.4 Onset of Chaos from a Normal Magnetic Field Component

The time that particles spend in the stochastic orbits making repeated crossings of the neutral sheet increases at the resonant energies defined by Eq. (3.91). In Fig. 3.7 from Burkhart and Chen (1991) the residence time for stochastic particles τ_s is shown as a function of $\widehat{H}^{1/4}$. The stochastic particle trapping time τ_s is computed by taking the difference between times between the first and last crossing of the reversal layer ($x = 0$) for particles launched toward the reversal layer at $x = 4L$ and taken from a Maxwellian velocity distribution at each $\widehat{H}$ energy value. The stochastic trapping time is given in units of $(qB_x/mc)^{-1}$.

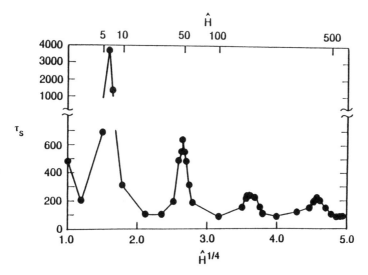

Fig. 3.7. The energy dependence of the residence time $\tau_s = \tau_s(H)$ for the stochastic particles trapped in the field reversed region. The number of the resonance N is given approximately by $N = \widehat{H}^{1/4} - 0.6$ and corresponds to $\omega_x/\omega_{cz} = N + 0.6$ for the number of $x = 0$ crossings of the reversal layer in the time $\Delta t = \pi/\omega_{cz}$. (From Burkhart and Chen (1991)).

Subsequently, study by Chen-Burkhart-Huang (1990) of the energetic ion distribution function obtained from the ISEE-3 satellite in the geomagnetic tail confirmed the bumpy structures of $F(E)$ associated with Fig. 3.7. The spacing of the resonances in energy E_n is such that $E_{n+1}^{1/4} - E_n^{1/4} = $ const proportional to $L^{1/2}$. The measured resonances conformed well to this prediction and suggest a method of determining the current sheet thickness from the resonances of $f(E)$ and the measured values of B_x and B_z.

Chen, Burkhart, and Mitchell (1990) have investigated the boundary between region B and region C orbits. By repeatedly blowing up smaller and smaller regions of the B-C boundary they show that the boundary is fractal in nature. This boundary curve is the large ε analog of the loss cone boundary in small ε (gyroradius-to-scale length) theory. The boundary is determined numerically following the definition of a loss-cone boundary in a μ-conserving theory. Sufficiently far from the current sheet $|x| \gtrsim 3L$ where $B = B_{\max} = (B_x^2 + B_0^2)^{1/2}$ the particles are launched toward the current sheet with a nearly vanishing parallel velocity; i.e. almost 90° pitch angle. Those with smaller pitch angles pierce the $x = 0$ plane inside region C. While those corresponding to punctures just outside region C will have insufficient H_z energy to reach the large $|x|$ region. Thus, they are reflected back into the reversal layer remaining in the stochastic B orbits for time τ_s. The current sheet Hamiltonian is thus a chaotic scattering system with integrable asymptotic states. Final state variables, such as pitch angle, have a high degree of sensitivity to the initial state variables. A comprehensive review of this nonlinear dynamical system is given in Chen (1992).

3.5 E × B Motion in Two Low-Frequency Waves and the Diffusion Approximation

The drift waves are low frequency electrostatic waves in which the cross-field particle motion is given by the $\mathbf{E} \times \mathbf{B}$ drift of the guiding center. The $\mathbf{E} \times \mathbf{B}$ motion is Hamiltonian $H(p, q, t) \to c\,\Phi(x, y, t)/B$ with the canonical momentum p being the radial coordinate $p \to x$ in the direction of the plasma inhomogeneity and the conjugate coordinate $q \to y$ being the spatial coordinate in the direction of symmetry mutually perpendicular to the inhomogeneity and the confining magnetic field. The isolating radial surfaces across which particles do not pass are the invariant KAM tori in the three-dimensional phase space of the Hamiltonian for the drift wave system. The $\mathbf{E} \times \mathbf{B}$ problem is that of transport in a $1\text{-}\frac{1}{2}$ D Hamiltonian system in which each wave acts over global regions of the p-q phase space. The onset of stochasticity then produces what is called a stochastic web covering the p-q phase space. Here we establish the conditions under which these isolating invariant tori break down allowing the radial transport of particles.

The Hamiltonian structure of the $\mathbf{E} \times \mathbf{B}$ drift equations is exploited to describe the onset of stochasticity for test particles in drift waves. In contrast to a longitudinal plasma wave (Example 1.3), a drift wave acts over the entire single particle phase space (Horton, 1981, 1985). This feature precludes a direct application of the Chirikov trapping velocity overlap criterion for predicting the onset of chaos. For two drift waves a generalized Chirikov overlap criterion is derived. Conditions on the drift wave spectrum for global stochasticity and the validity of the diffusion approximation for describing the motion are derived.

The drift wave instabilities produce convective transport of the plasma across the magnetic field that is described by anomalous diffusion coefficients and thermal conductivities. That anomalous transport occurs is well known from laboratory

3.5 **E** × **B** *Motion in Two Low Frequency Waves and* ...

experiments and computer simulations. Due to the simplicity of the theory, the anomalous particle flux from the drift waves is calculated with quasilinear theory based on short correlation time $\tau_c = 1/\Delta\omega$ due to the broad spectrum of small amplitude oscillations giving rise to a random $\mathbf{E} \times \mathbf{B}$ convection of the particles across the magnetic field. In contrast, a single drift wave only produces convection of the plasma with no net transport. The addition of a small secondary drift wave, however, dramatically changes the motion by producing a stochastic motion, at first localized along the boundaries of the single wave cells, which allows particles to cross from one cell boundary to another. This creates the stochastic web. The separatrix crossings now give rise to a net plasma transport. In this section the conditions for the drift wave system to change from a localized convection of plasma to a system described by diffusion are derived.

3.5.1 Test particle motion in drift waves

We consider a nonuniform, magnetized plasma with a fixed density gradient $dN/dr = -N/r_n$ and constant radial electric field $E_r = -d\Phi/dr$ in the slab approximation. The nonuniformity of the equilibrium density and potential gives rise to the electron diamagnetic drift velocity $v_{de} = cT_e/eBr_n$ and the $\mathbf{E} \times \mathbf{B}$ drift velocity $v_E = -cE_r/B$ in the symmetry direction $\hat{\mathbf{y}}$ perpendicular to the direction $\hat{\mathbf{x}}$ of the equilibrium gradient and the direction $B\hat{\mathbf{z}}$ of the magnetic field. [An alternative notation in common use for r_n is $L_n = r_n$ called the density gradient scale length.]

The waves are electrostatic with $\mathbf{E} = -\nabla\Phi(x,y,t)$ with a single wave $\mathbf{k}$ of the form

$$\Phi_{\mathbf{k}} = A_{\mathbf{k}} \sin(k_x x) \cos(k_y y - \omega_{\mathbf{k}} t).$$

In the presence of N drift waves the electrostatic potential is

$$\Phi(x,y,t) = -E_r x + \sum_1^N A_{\mathbf{k}} \sin(k_x x) \cos(k_y y - \omega_{\mathbf{k}} t + \beta_{\mathbf{k}}). \tag{3.92}$$

The frequency ω is determined by the wavenumber and the dispersion of the waves by $k_\perp \rho_s$ where $\rho_s = c(m_i T_e)^{1/2}/eB$ describes the dispersion of the waves due to the polarization of the plasma.

The considerations given here apply to a variety of drift waves and convective cell modes. For self-consistent plasma fluctuations it is sufficient to realize that each $\mathbf{k}$ has a given $\omega_{\mathbf{k}}$ with the phase velocity of order the diamagnetic drift speed or smaller. For ion temperature gradient driven drift waves there is a special wavenumber k_0 with $\tilde{\omega}_{k_0} = 0$ at $k_0 \rho_s = [(1-2\varepsilon_n)/(1+\eta_i)]^{1/2}$ where $\varepsilon_n = L_n/R$ and $\eta_i = \partial_r \ln T_i / \partial_r \ln n_i$ are two key parameters for the toroidal (major radius R) ion temperature gradient wave dynamics (Horton-Wakatani-Wootton, 1994).

Many hydrodynamic systems may also be described by the 2D Kelvin-Helmholtz instability, Rayleigh-Taylor and certain Benard convection systems. The breakdown of the approximation occurs when compression-expansion in the third dimension becomes significant $\nabla_\perp \cdot \mathbf{v}_\perp = -\partial v_z/\partial z \neq 0$. For certain estimates we will assume

$$\mathbf{k} = (k_x, k_y) \quad \text{related to } \omega \text{ through} \quad \omega_\mathbf{k} - k_y v_E = k_y v_{de}/(1 + k_\perp^2 \rho_s^2)$$

which is the linear dispersion relation for drift waves. Here the scale radius $\rho_s = c(m_i T_e)^{1/2}/eB$ determines the dispersion of the phase velocities of the waves.

In the hydrodynamic problem it is often possible to take $\omega_k = 0$ as in Benard convection, whereas in plasmas there is typically an intrinsic dispersive wave $\omega_k \neq 0$. The special case of stationary $\omega_k = 0$ convective cells or rolls has been analyzed analytically and numerically by Rosenbluth et al. (1987).

We consider the motion of a test particle with trajectory $\mathbf{r}(t) = (x(t), y(t))$ moving with the $\mathbf{E} \times \mathbf{B}$ drift velocity in the plasma. We need not specify in detail the type of particle. The test particle may equivalently be taken to be a fluid element with the Lagrangian fluid displacement $\boldsymbol{\xi}(t) = (x(t), y(t))$ moving with velocity $\mathbf{v}_E = c\mathbf{E} \times \mathbf{B}/B^2$. Thus, the fluid element may be called a fluid particle. For the ions and electrons the approximation of a pure $\mathbf{E} \times \mathbf{B}$ drift trajectory applies most directly to flute modes with $\omega \gg k_\| v_e \gg k_\| v_i$. For drift waves the motion of the thermal ions satisfy the relationship $v_i/L \ll \omega \ll v_e/L$ where $v_j = (T_j/m_j)^{1/2}$ is the average particle velocity and L is the effective length of the system along the magnetic field line. The electron motion is coupled to the parallel mode structure and is easily made stochastic by overlapping of $\omega = k_\| v_\|$ resonances in the parallel motion $\dot{z} = p_\|/m$ and $\dot{p}_\| = eE_\|$. The stochastic electron motion is given in Sec. 3.7. For trapped electrons, however, the effective $k_\| v_e \ll \omega$ and their $\boldsymbol{\xi}_\perp(t)$ motion is essentially incompressible. At short perpendicular wavelengths the electron circulation frequency Ω_E in its $\mathbf{E} \times \mathbf{B}$ motion can be resonant with the bounce frequency ω_{be} in the shallow magnetic well along the magnetic field lines.

Impurity ions of carbon, oxygen and heavy metal, common examples being iron, titanium, molybdenum, are trace elements that are passively convected by the $\mathbf{E} \times \mathbf{B}$ drift (Horton and Rowan, 1994).

The motion of the test particle given by $d\mathbf{r}/dt = \mathbf{v}_E$ is an incompressible two-dimensional flow given by

$$\frac{dx}{dt} = -\frac{c}{B}\frac{\partial \Phi(x,y,t)}{\partial y} \tag{3.93}$$

$$\frac{dy}{dt} = \frac{c}{B}\frac{\partial \Phi(x,y,t)}{\partial x}. \tag{3.94}$$

The flow $\dot{x}$, $\dot{y}$ is thus a one-and-a-half dimensional Hamiltonian system with the electrostatic potential $\Phi(x, y, t)$ being the Hamiltonian. The appropriate choice for the canonical momentum is the x coordinate and the conjugate coordinate is y. In equilibrium the coordinate y is cyclical (ignorable) and the particles are confined to $x(t) = x_0$ by the constant of the motion $\dot{x} = -\partial_y \Phi = 0$. In the presence of the waves

$\Phi = \Phi(x, y, t)$ in general has a complicated structure as shown by the contours of constant $\Phi(x, y, t)$ in Figs. 3.8 and 3.9.

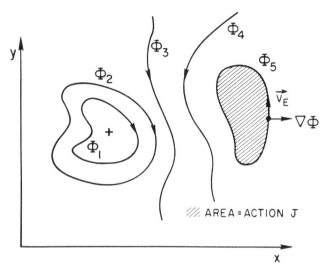

Fig. 3.8. Example of a set of electrostatic potential level lines (Φ = const) in the plane $x - y$ perpendicular to the magnetic field. The $\mathbf{E} \times \mathbf{B}$ flow is along the level lines and the area shaded within the $\Phi = \Phi_5$ contour defines the action integral $J(\Phi_5)$.

Although steep gradients in the Hamiltonian may arise, the physical condition that the energy density W of the waves be a finite fraction of the thermal energy density nT limits the gradients through the condition $\langle E^2 \rangle / 8\pi = \frac{1}{8\pi V} \int (\nabla \Phi)^2 dx dy dz < \infty$. In terms of the amplitude spectrum $A_\mathbf{k}$ in Eq. (3.92) the condition of finite $\langle E^2 \rangle$ requires $A_\mathbf{k} \leq C/|\mathbf{k}|^{2+\varepsilon}$ with $\varepsilon > 0$ as $|\mathbf{k}| \to \infty$.

3.5.2 Action-angle variables for $\mathbf{E} \times \mathbf{B}$ motion

For motions in general potentials it is useful to introduce action-angle variables J, θ. With time frozen the motion given by Eqs. (3.93)–(3.94) takes place along level contours of $\Phi(x, y)$ with the velocity $v_E = [(\partial_x \Phi(x, y))^2 + (\partial_y \Phi(x, y))^2]^{1/2}$. The action J is defined for a given Φ by

$$J(\Phi) = \frac{1}{2\pi} \oint x(y, \Phi) dy = \frac{\text{area of } \Phi \text{ contour}}{2\pi}. \quad (3.95)$$

For closed contours the integral (3.95) becomes

$$J(\Phi) = \frac{1}{2\pi} \int_{y_1}^{y_2} \left[x^+(y', \Phi) - x^-(y', \Phi) \right] dy'$$

where $y_{1,2}$ are the turning points where $\partial_x \Phi = 0$ and $x^\pm(y, \Phi)$ are the two branches of the inverse of $\Phi(x, y, t) = \Phi$. For open contours the action is

$$J(\Phi) = \frac{1}{2\pi} \int_0^{L_y} x(y', \Phi) dy'$$

where L_y is the periodic length of the system. The action for a closed contour is shown in Fig. 3.8.

The generating function $S(y, J)$ for the canonical transformation to action-angle variables (J, θ) is

$$S(y, J) = \int_0^y x(y', \Phi) dy' \qquad (3.96)$$

with $\Phi = \Phi(J)$ being the inverse of Eq. (3.95). The generating function (3.96) gives $x = (\partial S/\partial y)_J$ and

$$\theta = \frac{\partial S(y, J)}{\partial J} = \frac{\partial \Phi}{\partial J} \int_0^y \frac{\partial x(y', \Phi)}{\partial \Phi} dy'. \qquad (3.97)$$

Upon using Eq. (3.94) for $\dot{y}$ and defining the time

$$t(y) = \int_0^\ell \frac{d\ell'}{v_E} = \int_0^y \frac{dy'}{\dot{y}}$$

measured along the trajectory, the angle (3.97) may be written as

$$\theta = \frac{\partial \Phi}{\partial J} \int_0^y \frac{dy'}{\dot{y}(y', \Phi)} = \Omega_E(\Phi) t(y, \Phi) \qquad (3.98)$$

where the $\mathbf{E} \times \mathbf{B}$ circulation or eddy turn-over frequency is

$$\Omega_E = \frac{\partial \Phi}{\partial J}. \qquad (3.99)$$

The equations of motion in action-angle variables are

$$\frac{d\theta}{dt} = \frac{\partial \Phi}{\partial J} \quad \text{and} \quad \frac{dJ}{dt} = -\frac{\partial \Phi}{\partial \theta}. \qquad (3.100)$$

The contours of the potential $\varphi(x, y, t)$ change shape and reconnect on the period of the correlation time $\tau_c = 1/\Delta\omega$. When $\Omega_E \gg \Delta\omega$ the guiding center particles follow the contours of constant J until a separatrix of φ sweeps past them. During the separatrix crossing the particles begin convection in a different potential cell thus making a step $1/\Delta k_\perp$ given by the size of the potential cell. The separatrix crossing process is studied by Kleva and Drake (1984).

(ii) Integrable Curves for a Single Wave and the Trapping Condition

For a single drift wave in the expansion (3.92) the equations of motion

$$\dot{x} = k_y A_k \sin(k_x x) \sin(k_y y - \omega t + \beta_k) \tag{3.101}$$

$$\dot{y} = v_E + k_x A_k \cos(k_x x) \cos(k_y y - \omega t + \beta_k) \tag{3.102}$$

are integrable. The integral curves are found by making a canonical transformation to the wave frame x', y' with the generating function

$$F(y, x', t) = x'\left(y - \frac{\omega}{k_y} t\right) \tag{3.103}$$

giving $x = \partial F/\partial y = x'$ and $y' = \partial F/\partial x' = y - (\omega/k_y)t$. The Hamiltonian in the wave frame is $\Phi' = \Phi + \partial F/\partial t$ with

$$\Phi'(x', y') = \left(v_E - \frac{\omega}{k_y}\right) x' + A_k \sin(k_x x') \cos(k_y y'). \tag{3.104}$$

To study the integral curves we use phase coordinates $X = k_x x'$, $Y = k_y y'$ and measure time in units of the maximum $\mathbf{E} \times \mathbf{B}$ rotation frequency $\Omega_k^0 = k_x k_y A_k$ letting $\tau = \Omega_E^0 t$. The equations for the integral curves are

$$\frac{dX}{d\tau} = \sin X \sin Y \tag{3.105}$$

$$\frac{dY}{d\tau} = u + \cos X \cos Y \tag{3.106}$$

with the trapping parameter defined by

$$u = \frac{\omega - k_y v_E}{\Omega_E^0}. \tag{3.107}$$

For $u = 0$ all the orbits are closed and are given in terms of the elliptic functions by Horton (1981). For $0 < u < 1$ there are open (passing) and closed (trapped) orbits as shown in Fig. 3.9. For $u > 1$ all orbits are open or passing in the frame of the drift wave.

For the long wavelength drift wave dispersion relations $\tilde{\omega}_k = \omega - k_y v_E \cong \omega_{*e}$ the trapping condition $u \leq 1$ requires the single wave amplitude to satisfy

$$\frac{e\Phi_k}{T_e} > \frac{1}{|k_x r_n|} \sim \left(\frac{\lambda_x}{r_n}\right)$$

where $\lambda_x \ll r_n$ for the local wavenumber k_x to be well defined. For comparison with longitudinal trapping we note that the parallel trapping velocity

$$v_{\|,tr} = \left(\frac{e\Phi}{m}\right)^{1/2}$$

at this $\mathbf{E} \times \mathbf{B}$ trapping amplitude is

$$v_{\|,tr}^e = \left(\frac{e\Phi_k}{m_e}\right)^{1/2} = v_e \left(\frac{e\Phi}{T_e}\right)^{1/2} \cong v_e \left(\frac{\lambda_x}{r_n}\right)^{1/2}$$

for the electrons and negligible for the ions. The electron trapping velocity $v_{\|,tr}^e$ is small compared with $\omega/|k_\||| = (e\Phi/m_e)^{1/2}$ for small $\lambda_x/r_n \ll (\omega/k_\| v_e)^2$.

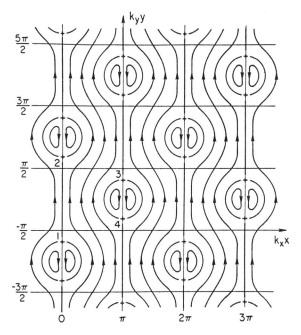

Fig. 3.9. Flow lines defined by the constant potential $\Phi' = (\omega_k/k_y)x'B/c + A_k \sin(k_x x')\cos(k_y y')$ in the frame of reference moving with the drift wave $y' = y - ut$ where $u = \omega/k_y$. The hyperbolic fixed marked by small +'s are stagnation points of the flow. The corresponding elliptic fixed points are in the interior of the trapped flow regions.

For $u < 1$ the unstable fixed points in a unit cell of phase space as designated by the labels 1, 2, 3, 4 in Fig. 3.9 are given by $(0, -\pi + \cos^{-1} u)$, $(0, \pi - \cos^{-1} u)$, $(\pi, \cos^{-1} u)$, $(\pi, -\cos^{-1} u)$ respectively, where $0 \leq \cos^{-1} u \leq \pi/2$. As $u \to 1$ the unstable pair y_3, y_4 and the two elliptic fixed points at $(\pi \pm \cos^{-1} u, 0)$ converge to $(\pi, 0)$, and the area of the trapped orbits vanishes as $J \propto (\cos^{-1} u)^2 = 2(1 - u)$.

The homoclinic orbits $y_{sx}(t)$ between $y_1 - y_2$ and $y_3 - y_4$ are calculated. The time variation of $y_{sx}(t)$ is given by $\Omega = \Omega_E^0(1 - u^2)^{1/2} = (\Omega_E^0{}^2 - \widetilde{\omega}_k^2)^{1/2}$ for $u < 1$. The dimensionless velocity along $y_1 - y_2$ is $\dot{Y} = u + \cos Y \leq 1 + u$ and along $y_3 - y_4$ is $\dot{Y} = u - \cos Y \leq 1 - u$. The homoclinic orbits $y_{sx}(t)$ along the long side $y_1 - y_2$ are given here by

$$\tan\left(\frac{Y_{sx}^{(t)}}{2}\right) = \begin{cases} \left(\dfrac{1+u}{1-u}\right)^{1/2} \tanh\left[\dfrac{\Omega_E^0 t(1-u^2)^{1/2}}{2}\right], u < 1 \\ \left(\dfrac{1+u}{u-1}\right)^{1/2} \tanh\left[\dfrac{\Omega_E^0 t(u^2-1)^{1/2}}{2}\right], u > 1 \end{cases} \quad (3.108)$$

and are required for the perturbation calculation in the next section. The harmonic motion around the elliptic fixed points $(\pi \pm \cos^{-1} u, 0)$ rotates with frequency $\Omega_b(u) = \Omega_E^0(1 - u^2)^{1/2}$.

3.5.3 The onset of stochasticity in two drift waves

In the presence of two drift waves $(A_{k_1}, \mathbf{k}_1, \omega_1; A_{\mathbf{k}_2}, \mathbf{k}_2, \omega_2)$ we again use the canonical transformation in Eq. (3.103) to the wave frame for the primary wave of the larger amplitude wave for which we choose the label 1. In this transformation the new $x' = k_{1x}x - r\pi$ where $r\pi$ is the rth radial node of the larger amplitude wave. Once we are in the wave frame we drop the primes on the new coordinates. The two wave Hamiltonian is

$$\Phi(x, y, t) = ux + \sin x \cos y + \varphi \sin(kx + \alpha) \cos[q(y - vt)] \quad (3.109)$$

where u is defined in Eq. (3.107) and

$$k = \frac{k_{2x}}{k_{1x}} \qquad q = \frac{k_{2y}}{k_{1y}} \qquad \varphi = \frac{A_2}{A_1} \qquad \alpha = \frac{r\pi k_{2x}}{k_{1x}} = r\pi k$$

and

$$v = \frac{\omega_2}{k_{2y}} - \frac{\omega_1}{k_{1y}}. \quad (3.110)$$

The test particle flow is given by

$$\begin{aligned} \dot{x} &= \sin x \sin y + q\varphi \sin(kx + \alpha) \sin[q(y - vt)] \\ \dot{y} &= u + \cos x \cos y + k\varphi \cos(kx + \alpha) \cos[q(y - vt)] \,. \end{aligned} \quad (3.111)$$

To determine the threshold for the onset of significant stochasticity we calculate the change in the Hamiltonian by perturbation theory along the separatrix of the single wave $(\varphi \to 0)$ flow from Eqs. (3.105)–(3.106).

The change in the Hamiltonian (electrostatic potential) $\Delta\Phi$ along the separatrix is given by the Melnikov-Arnold integral

$$\Delta\Phi = qv\varphi \sin(\alpha) \int_{-\infty}^{+\infty} \cos\left[q(y_{sx}(t) - vt)\right] dt \qquad (3.112)$$

where $Y_{sx}(t)$ is the trajectory along $1 \to 2$ or $3 \to 4$ in Fig. 3.9. Clearly, $\Delta\Phi = 0$ for $\omega'' = qv \to 0$ and is exponentially small for $\omega'' = qv \gg \Omega_E^0$. Analysis of the integral (3.112) leads to the results summarized here.

For small u the homoclinic orbits (3.108) are approximately

$$Y_{sx}(t) \cong -\frac{\pi}{2} + 2\tan^{-1}[\exp(\Omega_E^0 t)]$$

and the stationary phase condition $\dot{Y}_{sx}(t_s) = v$ is satisfied for real t_s when the frequency of the perturbation satisfies $\Omega_E^0 = qv$. This resonance condition gives the onset of strong stochasticity

$$\frac{ck_{2y}k_{1x}A_{k1}}{B} \gtrsim \left|\omega_{\mathbf{k}_2} - \frac{k_{2y}\omega_{\mathbf{k}_1}}{k_{1y}}\right| \qquad (3.113)$$

valid for $A_2 \ll A_1$. The stochasticity criterion (3.113) is easily satisfied for long wavelength drift waves, once the trapping condition ($u < 1$) or $\Omega_E > |\tilde{\omega}_k|$ is satisfied.

For comparable amplitude drift waves $A_1 \sim A_2$ symmetry of the two waves and numerical studies show that the threshold condition (3.113) generalizes to

$\boxed{\mathbf{E} \times \mathbf{B} \text{ resonance overlap}}$

$$\frac{ck_{1x}A_{k1}}{B} + \frac{ck_{2x}A_{k2}}{B} \gtrsim \left|\frac{\omega_{k2}}{k_{2y}} - \frac{\omega_{k1}}{k_{1y}}\right|. \qquad (3.114)$$

Condition (3.114) for the drift wave problem is the analog of the Chirikov overlap criterion

$\boxed{\text{Longitudinal wave overlap}}$

$$\left(\frac{e\Phi_1}{m}\right)^{1/2} + \left(\frac{e\Phi_2}{m}\right)^{1/2} > \left|\frac{\omega_1}{k_1} - \frac{\omega_2}{k_2}\right| \qquad (3.115)$$

for two longitudinal waves. The symmetric conditions (3.114) and (3.115) apply when the two wave amplitudes are comparable in magnitude.

When condition (3.114) is satisfied for drift waves stochasticity occurs in webs throughout the single particle phase space in contrast to the longitudinal wave problem. The longitudinal trapping begins at any finite φ_k whereas the $\mathbf{E} \times \mathbf{B}$ trapping requires a threshold amplitude $\Omega_E > |\tilde{\omega}_k|$.

3.5 **E** × **B** *Motion in Two Low Frequency Waves and ...* 93

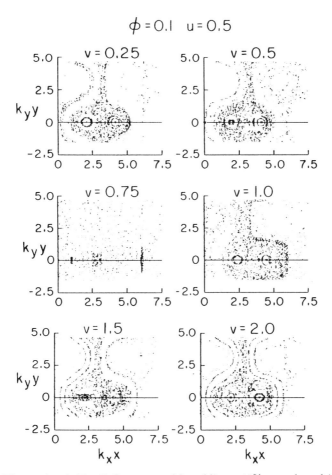

Fig. 3.10. The stochastic **E** × **B** flows created by adding a 10% secondary drift wave to the primary waves with cells as in Fig. 3.9. The amplitude of the primary wave is such that $\Omega_E/\omega_{k_y} = ck_x k_y \Phi/\omega_{k_y} B = 2.0$. The parameter varied is the speed v of the secondary wave relative to the primary wave. The frames show v increasing from $v = 0.25, 0.5, 0.75, 1.0, 1.5,$ and 2.0. The Melnikov-Arnold integral gives a resonance for $v \lesssim 1.0$ which is where the islands of integrable motion occupy the smallest fraction of the phase space.

For $u \lesssim 1$ the principal change is that the faster velocity along the 1-2 separatrix governs the breakdown of the invariant tori. The condition for a resonance along 1-2 becomes

$$\Omega_E(1+u) = \Omega_E + \tilde{\omega}_k \sim \omega'' = qv. \tag{3.116}$$

Away from the resonance conditions (3.113) or (3.116) the Melnikov-Arnold integral

(3.112) becomes exponentially small

$$\Delta \Phi \sim \exp[-qv/\Omega_E]. \tag{3.117}$$

The stochasticity then forms a thin web along the separatrix flows.

In Fig. 3.10 we show the variation of the stochastic region with varying v for fixed $\varphi = A_2/A_1 = 0.1$ and $u = 0.5$. The maximum velocity along the homoclinic orbit $1-2$ is $\dot{y}_{\max} = 1.5$. For $v = 0.25$ to 0.5 the invariant tori along 1-2 are still preserved. For $v = 3/4$ all the invariant tori appear to be destroyed. At $v = 1$ only small islands of stability about the elliptic fixed points remain. For $v \geq \dot{y}_{\max} = 1.5$ invariant tori along the 1-2 have returned.

(iv) Destabilization of the Elliptic Fixed Points

We would like to establish conditions under which the primary elliptic fixed points become unstable. To consider this problem analytically we construct from the N wave parameter set $\{\mathbf{k}, \omega_\mathbf{k}, A_\mathbf{k}\}$ special Hamiltonians which have an analytic elliptic fixed point at (x^*, y^*). Linearizing the equations of motion about x^*, y^* leads to a Mathieu equation for the motion of $\delta x(t)$ and $\delta y(t)$ about the fixed point so that the stability analysis in Chapter 1 applies. The stability condition on the parameters in the Mathieu equation then lead to the condition for the bifurcation of fixed points into hyperbolic-elliptic point pairs. The analysis is lengthy so we briefly summarize the results here.

The idea here is to follow the procedure of Schmidt and Bialek (1982) who show that the critical parameters for destabilization of the fixed points give approximations for the condition of breaking the KAM surfaces. The stability analysis of the fixed point follows the method given in Sec. 1.3 and the example in Sec. 1.5.

For resonant perturbations

$$2\Omega_E = |\omega''| \tag{3.118}$$

the orbits are destabilized for an arbitrarily small amplitude perturbation. In the Mathieu equation with parameters (a, q) the resonance condition (3.118) is $a = 1$. For $2\Omega_E < |\omega''|$ where $a^{1/2} = 2\Omega_E/|\omega''| < 1$ the neighboring orbits are stable for $-q^2/2 < a < 1 - q - q^2/8$. A simple estimate of the stable domain follows from $q_{\text{crit}} \leq 1 - a$ and yields the stability condition

$$\sum_i k_{ix}^2 \, \varphi_i < \omega'' \left(\frac{\omega''}{2\Omega_E} - \frac{2\Omega_E}{\omega''} \right). \tag{3.119}$$

For perturbing waves φ_i larger than the condition (3.119) the orbits exponentiate away from the original stable cycle becoming part of the stochastic web.

3.5 **E × B** Motion in Two Low Frequency Waves and ...   95

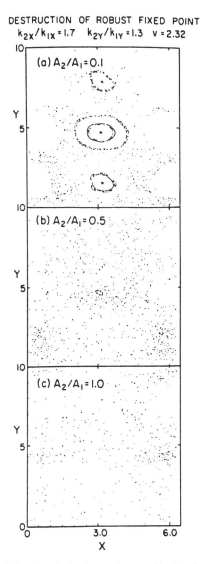

Fig. 3.11. Bifurcation of the islands for increasing amplitude of the secondary wave $A = \phi_2/\phi_1$ in the four wave example after Eq. (3.119) in Sec. 3.5.3. The fixed parameters are $k_{2x}/k_{1x} = 1.7$, $k_{2y}/k_{1y} = 1.3$ and $v = 2.32$. The stochastization condition is given by the reverse of inequality (3.119).

In Fig. 3.11 we show the results of increasing amplitude of the perturbing waves acting on a stable fixed point x^*, y^*. In this example the Hamiltonian is

$$\Phi(x, y, t) = ux + \sin x \cos y + A \left[\sin k_1(x - x_1^*) - \frac{k_1}{k_2} \sin k_2(x - x_1^*) \right] \cos[q(y - vt)]$$

with $A = A_2/A_1$. Instability of the elliptic fixed point is observed to occur near the condition given by Eq. (3.119).

Additional calculations for the onset of stochastic orbits and the associated diffusion may be made by deriving drift wave maps obtained by replacing the spectrum in (3.92) with a single **k**-vector with an infinite, discrete frequency model

$$\lim_{N \to \infty} \sum_{-N}^{+N} \cos(n\omega_0 t) = 2\pi \sum_{n=-\infty}^{+\infty} \delta(\omega_0 t - 2\pi n)$$

as a broad band frequency spectrum model characteristic of drift waves. The drift wave map analysis is complicated owing to the implicit form of the map. The map shows that incomplete global stochasticity occurs when the map amplitude parameter $K = ck_x k_y A_k / \omega_0 B \gtrsim 1$.

Even for simple systems the transitional region just beyond the critical resonance for the breakdown of the bounding invariant surfaces is difficult to analyze with regard to transport (Meiss, 1994). We do not attempt to analyze the leaky barrier (MacKay, Meiss, and Percival, 1984a,b) region where a slow, non-diffusive transport is expected. An example of the leaky barrier appears in the transitional parameter regime in Fig. 3.12 in the Sec. 3.5.6.

3.5.4 Example 3.1: $\mathbf{E} \times \mathbf{B}$ drift islands in the cylindrical plasma with sheared flow

From the particle point of view the favorable effect of the sheared rotation on transport and stability is related to the confining of the $\mathbf{E} \times \mathbf{B}$ excursions of the particles by the sheared angular rotation frequency $\Omega(r) = -cE_r(r)/rB$.

The linear plasma Lagrangian displacement $\xi(\mathbf{x}, t)$ in a plasma fluctuation $\delta \Phi_m \cos(m\theta - \omega t)$ has the radial component given by

$$\xi_m(r) = \frac{cm\delta\Phi_m}{rB(\tilde{\omega} - m\Omega(r))}$$

obtained from integrating

$$\frac{d\xi}{dt} = \frac{cE_\theta}{B} = -\frac{c}{Br} \frac{\partial \Phi (m\theta - \omega t)}{\partial \theta}$$

with r = constant. The r = constant approximation clearly fails for $|\tilde{\omega} \equiv \omega - m\Omega(r)| \to 0$ at $r = r_m$. Taking the minimum of $\tilde{\omega}$ as $\tilde{\omega} \equiv (d\tilde{\omega}/dr)\xi_m$ we find that the linear approximation fails within the trapping layer

$$|r - r_m| \leq \Delta r_t$$

where

$$\Delta r_t = \left(\frac{c\delta \Phi_m}{rB|d\Omega/dr|} \right)^{1/2}.$$

The plasma in the region $|r - r_m| \leq \Delta r_t$ becomes trapped in a vortical flow due to the $\mathbf{E} \times \mathbf{B}$ convection in the resonant layer. The situation is similar to the trapping of the magnetic field lines in the sheared magnetic field due to the mode rational surfaces at the resonant $k_\parallel(r) = 0$ from a magnetic perturbation in Chapter 2.6.

For given linear eigenfunctions it is straightforward to study the transport across the sheared layer by integrating the equations of motion

$$\frac{d\boldsymbol{\xi}}{dt} = \mathbf{v}_E(\mathbf{x} = \boldsymbol{\xi}, t) \tag{3.120}$$

for the test fluid cells. Here we analyze the fluid element trajectories close to the resonance. Finding all orbits for a given $\Phi(r, \theta, t)$ is equivalent to solving the continuity equation for the density $n(r, \theta, t)$ given the initial distribution $n(r, \theta, 0)$.

Introducing the action variable, $I = r^2/2$ the equations of motion $\dot{r} = v_{Er}$, $r\dot{\theta} = v_{E\theta}$ reduce to the $1 - \frac{1}{2} D$ Hamiltonian system with

$$\frac{dI}{dt} = -\frac{c}{B}\frac{\partial \Phi}{\partial \theta} = \frac{c}{B}\sum_m m|\Phi_m(r)|\sin\left[m\theta - \omega t - m\theta_0(r)\right] \tag{3.121}$$

$$\frac{d\theta}{dt} = \frac{c}{B}\frac{\partial \Phi}{\partial I} = \Omega(r) + \frac{c}{rB}\frac{\partial}{\partial r}\sum_m |\Phi_m(r)|\cos\left[m\theta - \omega t + m\theta_0(r)\right] \tag{3.122}$$

where $r = (2I)^{1/2}$. Note that for a variable B one may generalize to the toroidal magnetic flux $I = \int_0^r B_z(r')r'dr'$ for the action variable.

For a single mode the motion is integrable and reduces to the pendulum equation for $|\delta v_{E\theta}| \ll r\Omega$. In this approximation we transform to the frame rotating with the wave by letting $\psi = \theta - \omega t/m$ and obtain

$$\frac{d\psi}{dt} = \Omega - \frac{\omega}{m} = \left.\frac{d\Omega}{dr}\right|_{r_m} \Delta r(t) \tag{3.123}$$

$$\frac{d\delta r}{dt} = \frac{cm|\Phi_m|}{r_m B}\sin(m\psi - \beta_m) \tag{3.124}$$

where

$$\beta_m = \theta_0(r_m) \quad \text{and} \quad \Phi_m = \Phi_m(r_m).$$

For $d\Omega/dr < 0$ the motion (3.123) and (3.124) has m stable fixed elliptic points at $\psi = (2\pi p \beta_m)/m$ with $p = 0, 1, 2, \ldots, m-1$ between m unstable hyperbolic fixed points. The reduced Hamiltonian is

$$H_m^{\text{res}}(\Delta r, \psi) = \tfrac{1}{2} \left(\frac{d\Omega}{dr}\right)_m \Delta r^2 - \frac{c|\Phi_m|}{rB} \cos(m\psi - \beta_m)$$

$$= \tfrac{1}{2} \left(\frac{dr}{d\Omega}\right)_m \dot{\psi}^2 - \frac{cm|\Phi_m|}{Br_m} \sin(m\psi - \beta_m) \qquad (3.125)$$

where the effective mass of the fluid particle increases with shear as $m_{\text{eff}} = (d\Omega/dr)_m$. The equation for the separatrix between the trapped and circulating motion is

$$\Delta r_{sx}(\psi) = \pm 2 \left(\frac{c|\Phi_m|}{r_m B |d\Omega/dr|}\right)^{1/2} \sin\left(\frac{m\psi - \beta_m}{2}\right), \qquad (3.126)$$

and the characteristic frequency $\Omega_b^{n\ell}$ for convection around the vortex is

$$\Omega_b^{n\ell} = m\Delta r_{sx} \left|\frac{d\Omega}{dr}\right| = m \left(\frac{c|\Phi_m|}{r_m B} \left|\frac{d\Omega}{dr}\right|\right)^{1/2}. \qquad (3.127)$$

Global stochasticity occurs either from the presence of a second wave with $\Omega_b^{n\ell} \approx |\omega_1 - \omega_2|$ or the resonance $\Omega_b^{n\ell} \simeq n\pi v_\parallel/L$ due to the overlapping of the nonlinear resonances. Mathematically, the situation is similar to the enhanced transport due to the overlapping of magnetic islands produced by magnetic perturbations in a sheared magnetic field.

Even in the presence of a single mode the radial excursions $\Delta r = \pm \Delta r(\psi)$ with frequency $\Omega_b^{n\ell}$ of the $\mathbf{E} \times \mathbf{B}$ drifting guiding centers will produce an enhanced collisional transport from the broken symmetry of the equilibrium. The drift mode enhanced transport can easily exceed the neoclassical collisional transport for the wave amplitude above a critical value. Formulas (3.126)–(3.127) show that differential $\mathbf{E} \times \mathbf{B}$ rotation, which tends to stabilize the linear instabilities, also reduces the anomalous transport produced by the modes.

In the limit of solid body (sb) rotation the Hamiltonian becomes simply

$$H^{\text{sb}} = \Omega I + \frac{c}{B} \sum_{m,n} \Phi_{mn}(r) \cos(m\theta - \omega t - \beta_m^0) \qquad (3.128)$$

with $\Phi_{mn}(r)$ real and Ω and β_m^0 constants. The $\mathbf{E} \times \mathbf{B}$ transport for this system, which at small Φ_{mn} is bounded by the nodes of $\Phi_{mn}(r)$, becomes globally stochastic when a generalized resonance overlap criterion is satisfied. For the low-m modes the width of the radial convection increases from that with $\Delta r \sim a(\widetilde{\Phi}/\Delta\Phi_0)^{1/2}$ given by Eq. (3.126) for sheared rotation to the global value $\Delta r \sim a$ for solid body rotation.

3.5.5 The diffusion approximation

When the conditions for stochasticity given in Sec. 3.5.3 (e.g. Eq. (3.114)) are satisfied the phase space consists of large regions from stochastic orbits and smaller regions with chains of islands with stable orbits. The presence of small islands of stability in a sea of stochastic orbits is perhaps the generic form of the x-y phase space in drift wave turbulence. In this regime it is an important theoretical assumption to describe the motion by diffusion. The basic condition for diffusion is that the particle moving along its trajectory $\mathbf{r}(t)$ experience a short correlation time τ_c. The intrinsic stochasticity of the system gives rise to the short correlation time.

In the case of a broad spectrum of small amplitude waves the correlation time τ_c is given by $1/\Delta\omega$ from the band width rather than the intrinsic stochasticity. In the authors' view both experimental fluctuation levels in confined plasmas and numerical simulations show that τ_c is typically limited by the intrinsic stochasticity.

The definition of the diffusion tensor follows from the formal integration of the equations of motion

$$\mathbf{r}(t) = \int_0^t \mathbf{v}_E(\mathbf{r}(t_1), t_1) dt_1 \qquad (3.129)$$

along the unknown trajectory $\mathbf{r}(t)$.

Introducing the average $\langle \ \rangle$ over the initial conditions $\mathbf{r}(t=0) = \mathbf{r}_0$ in the stochastic region of phase space the diffusion tensor is defined through the limit

$$\left\langle \mathbf{r}^2(t) \right\rangle = \lim_{t/\tau_c \to \infty} \int_0^t dt_1 \int_0^t dt_2 \left\langle \mathbf{v}_E(t_1) \mathbf{v}_E(t_2) \right\rangle = 2t\,\boldsymbol{D}. \qquad (3.130)$$

The existence of the limit $\langle \mathbf{r}^2(t) \rangle \simeq 2\boldsymbol{D} t$ is expected to be valid for times long compared with the correlation time τ_c taken along the orbit. In terms of τ_c the integral definition of $\boldsymbol{D}$ is estimated by $D \simeq \langle v_E^2 \rangle \tau_c$. For a general potential $\Phi(x, y, t)$ we define the action J and $\mathbf{E} \times \mathbf{B}$ circulation frequency Ω_E through Eqs. (3.95) and (3.99). The dimensionless parameter determining the nonlinear regime of the system is

$$R = \Omega_E \tau_c \cong \left\langle v_E^2 \right\rangle^{1/2} \tau_c / \ell_c \qquad (3.131)$$

where τ_c is the correlation time and ℓ_c the spatial correlation scale of the potential. We assume here that $\varphi(x, y, t)$ is essentially isotropic in the x-y plane. The quasi-linear regime is defined by $R \ll 1$ where Eq. (3.130) gives $D = \langle v_E^2 \rangle \tau_c = (\ell_c^2/\tau_c) R^2$. The value of $R/2\pi$ determines the average number of rotations of the plasma in a correlation time.

In the strongly nonlinear regime $R \gg 1$ the trajectories make many rotations in one correlation time τ_c. In this regime the action integral J is a good adiabatic invariant. In this regime there is a diffusion which occurs in the action variable $\langle \Delta J^2 \rangle = 2 D_J t$ for time scales long compared with the correlation period τ_c. A fraction of the particles cross the moving separatrices in each correlation time. By crossing the separatrix they circulate in different convective cells moving across the distance $\ell_c(J)$ approximately independently of $\Phi_\mathbf{k}$, when averaged over Ω_E, in the time τ_c. The maximum rate of this diffusion is $D_m = \ell_c^2/\tau_c$.

From computer experiments with two drift waves with $\mathbf{k}_1 \times \mathbf{k}_2 \cdot \hat{\mathbf{z}} \neq 0$ and $k_y \Delta(\omega_k/k_y) \sim \omega_k$ we may expect to find a maximum of the diffusion coefficient for $R \gg 1$ with the maximum of D bounded by $D_{\max} = \Delta \omega \ell_c^2$ due to the adiabatic invariant $J(\Phi)$. Isichenko et al. (1992) have shown that this is not the case, however, due to particles that make long excursions in each correlation time along the "sea level" contours of $\Phi(x, y, t)$. These sea level contours are the generalization of the separatrices between the trapping zones of the maxima and minima of $\Phi(x, y, t)$. The theory for calculating the transport from these orbits is called percolation theory and is reviewed by Isichenko (1992). For the $\mathbf{E} \times \mathbf{B}$ drift problem the result is that the effective diffusion rate increases for $R_E \gg 1$ as

$$D = (\Delta \omega \ell_c^2)^{3/10} \left(\frac{c\tilde{\Phi}}{B} \right)^{7/10}$$

as given by Gruzinov et al. (1990). In addition to diffusion there are particles that have long flights that produce a "super diffusion" transport (Afanasiev et al., 1991). In a competition with long flights are long trapping times from the stickiness of islands. Recent experiments by Solomon et al. (1993, 1994) show this competition in geostrophic flows. The interpretation of the long flight transport is supported by the fractal transport theories of Zaslavsky et al. (1993).

To calculate the diffusion coefficient in terms of the Hamiltonian $\Phi(x, y, t)$ we introduce a statistical description of the test particle distribution function

$$n(\mathbf{x}, t) = \sum_{j=1}^{N_T} \delta\left(\mathbf{x} - \mathbf{r}_j(t, \mathbf{r}_0)\right)$$

for N_T identical test particles with different initial conditions $\mathbf{r}_0$. The initial coordinates are given by the normalized probability distribution $P(\mathbf{r}_0)$ with

$$\langle n(\mathbf{x}, t) \rangle = N_T \int d\mathbf{r}_0 P(\mathbf{r}_0) \delta\left(\mathbf{x} - \mathbf{r}_1(t, \mathbf{r}_0)\right)$$

with the initial value

$$\langle n(\mathbf{x}, t=0) \rangle = N_T P(\mathbf{x}).$$

3.5.6 Example 3.2: Onset of $\mathbf{E} \times \mathbf{B}$ diffusion from two equal amplitude waves

A simple example is shown in Fig. 3.12 where the initial distribution has $N_T = 50$ particles uniformly distributed in y at $X = k_{x1}x = \pi/4$ as shown in Fig. 3.12.

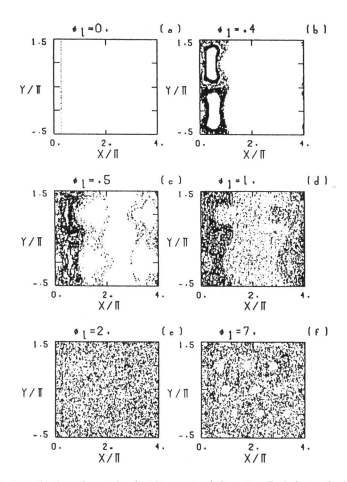

Fig. 3.12. Distribution of particles (guiding centers) from $\mathbf{E} \times \mathbf{B}$ drifts in the long-time limit with two equal amplitude drift waves with $k_{2x}/k_{1x} = 1$, $k_{2y}/k_{1y} = 2$ and $v = 1$. The critical condition from Eq. (3.113) is $\varphi_{\text{crit}} = 0.5$. The initial distribution is 50 particles at $k_{1x} x_0 = \pi/4$ uniformly distributed in $k_{2y} y$ as shown in frame (a) with $\varphi_1 = \varphi_2 = 0$ (b) shows the stochasticity around the convective cell boundaries at $\varphi_1 = 0.4$. (c) shows the critical point for breaking the last invariant curve at $\varphi_1 = 0.5$ (d) shows the nonuniform distribution just above the breaking of the invariant curves. (e) shows the nearly uniform distribution of particles for $\varphi_1 = 2$. (f) shows the emergence of nearly regular regions associated with the action integral in Eq. (3.95) for $\varphi_1 = 7$ where $R_E \gg 1$.

Figures 3.12 show the surface of sections for the two drift wave system

$$\varphi(x,y,t) = ux + \Phi_1[\sin x \cos y + \cos kx \cos q(y - vt)]$$

for increasing values of Φ_1 with the parameters $u = 0$, $k = v = 1$, and $q = 2$. The critical condition (3.114) for the onset of global stochasticity gives $\Phi_{1c} = kv/2 = 0.5$. At $\Phi_1 = 0.4$ in Fig. 3.12 the flow is stochastic around the boundaries of the convective cells. At $\Phi_1 = 0.5$ in Fig. 3.12 the last invariant surfaces are breaking up. At $\Phi_1 = 1.0$ in Fig. 3.12 there are no more invariant surfaces and no fixed points with islands visible on this scale. At $\Phi_1 = 2.0$ in Fig. 3.12 the system appears ergodic for the 200 iterations used in these figures. At $\Phi_1 = 7$ in Fig. 3.12 adiabatic islands produced by the rapid convection have emerged in the flow. (For irrational k, q values the adiabatic islands disappear). Figure 3.13 shows the diffusion coefficient D_x for this flow computed from Eq. (3.130) and averaged over the N_T test particles.

The phase space is three-dimensional with KAM surfaces given as 2D tori with toroidal angles $\zeta = qvt$ and θ of the action-angle variables given in Sec. 3.5.2.

For very large Φ_1 the white adiabatic regions of Fig. 3.12 expand to form a pattern of nearly regular convective cells and the diffusion is reduced. To analyze $\Phi_1 \to \infty$ let $t' = \Phi_1 t$, then $\varphi(x,y,t) \to \varphi'(x,y,t')$ with new parameters $\Phi_1' = 1$, $u' = u/\Phi_1$, $v' = v/\Phi_1$ and $k' = k$, $q' = q$. In the limit $\Phi_1 \to \infty$, $u' = v' = 0$ so that $\varphi' \to \varphi'(x,y)$ is integrable and the diffusion coefficient vanishes. Large values of Φ_1 are required, however, to reach this regime as seen from Figs. 3.11 and 3.12.

3.6 Renormalized E × B Diffusion Coefficients

The distribution of test particles satisfies the conservation law

$$\frac{\partial n(\mathbf{x},t)}{\partial t} = -\mathbf{v}_E \cdot \nabla n = -\frac{c}{B}\left(\frac{\partial \Phi}{\partial x}\frac{\partial n}{\partial y} - \frac{\partial \Phi}{\partial y}\frac{\partial n}{\partial x}\right) \quad (3.132)$$

since $\nabla \cdot \mathbf{v}_E = 0$. In the diffusion approximation the average distribution $\langle n \rangle$ satisfies

$$\partial_t \langle n \rangle = -\nabla \cdot \langle n\mathbf{v}_E \rangle = -\nabla \cdot \sum_{\mathbf{q}\Omega}\left\langle \mathbf{v}_E^*(\mathbf{q}\Omega)\delta n(\mathbf{q},\Omega)\right\rangle$$

$$\cong \boldsymbol{D}{:}\nabla^2 \langle n \rangle \quad (3.133)$$

where the $\mathbf{q}, \Omega$ components of Eq. (3.132) are

$$\delta n_{\mathbf{q}\Omega} = \frac{c}{B}\frac{\Phi_{\mathbf{q}\Omega}\hat{\mathbf{z}} \times \mathbf{q} \cdot \nabla \langle n \rangle}{\Omega - q_y v_E + i\nu_{\mathbf{q}}}. \quad (3.134)$$

The solution of Eq. (3.133) gives $\langle \mathbf{r}^2 \rangle = \int \mathbf{x}^2 \langle n \rangle \, d\mathbf{x} = 2\boldsymbol{D}t$.

3.6 Renormalized E × B Diffusion Coefficients

To calculate the correlation function $\langle n\mathbf{v}_E \rangle$ in Eq. (3.132) we find the response $\delta n_{\mathbf{q}\Omega}$ to a gradient $\nabla \langle n \rangle$ through the renormalized perturbation expansion of Eq. (3.133). Following the well-known calculation for the renormalized response function $g_{\mathbf{q}\Omega} = (\Omega - q_y v_E + i\nu_\mathbf{q})^{-1}$ we obtain

$$\nu_\mathbf{q} = -\operatorname{Im} \frac{c^2}{B^2} \sum_{\mathbf{k}_1\omega_1} \frac{(\mathbf{k}_1 \times \mathbf{q} \cdot \hat{\mathbf{z}})^2 \left\langle |\Phi_{\mathbf{k}_1\omega_1}|^2 \right\rangle}{\omega_1 - k_{1y} v_E + i\nu_{\mathbf{k}_1}} \quad (3.135)$$

$$\boldsymbol{D} = -\frac{c^2}{B^2} \operatorname{Im} \sum_{\mathbf{q}\Omega} \frac{(\hat{\mathbf{z}} \times \mathbf{q})^2 \left\langle |\Phi_{\mathbf{q}\Omega}|^2 \right\rangle}{\Omega - q_y v_E + i\nu_\mathbf{q}}. \quad (3.136)$$

From Eq. (3.134) the flux $\langle n\mathbf{v}_E \rangle$ is calculated as

$$\langle n\mathbf{v}_E \rangle = \frac{c}{B} \sum_\mathbf{q} i\hat{\mathbf{z}} \times \mathbf{q} \Phi_\mathbf{q} \delta n_q = -\boldsymbol{D} \cdot \nabla \langle n \rangle. \quad (3.137)$$

Equations (3.135) and (3.136) determine the diffusion tensor from the spectral components of the Hamiltonian

$$\Phi(x, y, t) = \sum_{\mathbf{k}\omega} \Phi_{\mathbf{k}\omega} \exp(i\mathbf{k} \cdot \mathbf{x} - i\omega t).$$

The theory leading to $\boldsymbol{D}$ is not exact, but is an infinite order perturbation expansion in R. The terms retained in the summations leading to (3.135) and (3.136) are the highest degree secular terms at each order in the power series in Φ_k. The details of the selection of the secular terms is given in Horton and Choi (1979). In the renormalized perturbation expansion the value of $R \sim ckq\Phi_k/B\nu_k$ is limited even for large Φ_k since the decorrelation rate ν_k increases with Φ_k according to Eqs. (3.135) and (3.136).

The renormalized equations for $\boldsymbol{D}$ contain two limiting cases. In the presence of many small amplitude waves the correlation time in Eq. (3.136) is determined by the dispersion $\Delta(\Omega - q_y v_E) > \nu_q$ rather than the nonlinearity. The diffusion becomes the quasilinear diffusion

$$\boldsymbol{D} \simeq \boldsymbol{D}^{q\ell} = \left\langle \mathbf{v}_E^2 \right\rangle \tau_c^{q\ell}$$

with

$$\tau_c^{q\ell} = \left\langle |\Omega - q_y v_E|^{-1} \right\rangle \sim 1/(\Delta\Omega - \Delta q_y v_E). \quad (3.138)$$

For larger amplitude waves, where R approaches unity, the nonlinear decorrelation $\nu_\mathbf{q} \gtrsim \Delta(\Omega - q_y v_E)$ determines the rate of decorrelation and $D < D^{q\ell}$ (see Fig. 3.13). Although not exact, the equation (3.135) for $\nu_\mathbf{q}$ can be approximated to give the reliable estimate

$$\nu_\mathbf{q} \simeq \left(\frac{c^2}{B^2} \sum_{\mathbf{k}_1\omega_1} (\mathbf{k}_1 \times \mathbf{q} \cdot \hat{\mathbf{z}})^2 \left\langle |\Phi_{\mathbf{k}_1\omega_1}|^2 \right\rangle \right)^{1/2} = \left\langle \Omega_E^2 \right\rangle^{1/2} \simeq \left\langle (\mathbf{q} \cdot \mathbf{v}_E)^2 \right\rangle^{1/2} \quad (3.139)$$

for the nonlinear decorrelation rate for $R \lesssim 1$. With the nonlinear decorrelation rate $\nu_q \sim \langle \Omega_E^2 \rangle^{1/2}$ the diffusion coefficient becomes

$$D = \frac{c^2}{B^2} \sum_{\mathbf{q}\Omega} \frac{(\hat{\mathbf{z}} \times \mathbf{q})^2 |\Phi_{\mathbf{q}\Omega}|^2}{\nu_{\mathbf{k}}} \simeq \langle v_E^2 \rangle^{1/2} / \langle k \rangle \quad (3.140)$$

with the maximum diffusion occurring for $R \sim 1$ where

$$D_{\max} \sim \frac{\ell_c^2}{\tau_c} \quad \text{at} \quad R \sim 1. \quad (3.141)$$

For the simple example in Figs. 3.12, the predictions of Eqs. (3.138) and (3.140) are that $\tau_c^{q\ell} = 1/qv = 0.5$ with $D_x^{q\ell} = c_1 \Phi_1^2$ for $\Phi_{1c} \ll \Phi_1 < \Phi_{1t}$ and $\tau_c = 1/\nu_q \simeq 3/kq\Phi_1$ with $D_x = c_2 \Phi_1$ for $\Phi_1 > \Phi_{1t}$. The transition occurs where $\nu_q \simeq kq\Phi_1/3 \sim qv$ or at $\Phi_{1t} \simeq 3$. The measured scaling of D_x with Φ_1 is shown in Fig. 3.13.

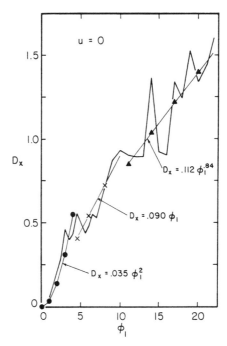

Fig. 3.13. Amplitude dependence of the $\mathbf{E} \times \mathbf{B}$ diffusion coefficient showing the three regimes: (i) $R_E \ll 1$ quasilinear diffusion with $D_x \propto \phi_1^2$ (ii) renormalized Lagrangian turbulence with $D_x \propto \phi_1$ and (iii) the percolation regime with $D_x \propto \phi_1^{7/10}$ from long excursions between the maxima and minima of the potential.

3.6 Renormalized **E** × **B** Diffusion Coefficients

For $R \gg 1$ we find the diffusion continues to increase as a fractional power of R due to the percolation of particles along the boundary layers (Isichenko et al., 1992).

The scale for anomalous transport produced by drift wave turbulence follows from $D_{\max}$ with the estimate of the cross-scale correlation distance as $\ell_{cx} \sim 1/\langle k_x^2 \rangle^{1/2} \sim \rho_s$ and the correlation time as $\tau_c = 1/\Delta\omega_k \simeq r_n/c_s$. The drift wave diffusion coefficient is then

drift wave diffusion

$$D_{dw} = \tau_c^{-1}/\langle k_x^2 \rangle \sim \left(\frac{\rho_s}{r_n}\right)\left(\frac{cT_e}{eB}\right)$$

times a function of order unity containing the other dimensionless parameters of the system.

In summary, like the motion of a particle in two longitudinal waves, the **E** × **B** flow of plasma is a $d = 3/2$ dimensional Hamiltonian system. In contrast to the Hamiltonian for motion in longitudinal waves where localized velocity resonances occur, each wave in the **E** × **B** system acts over the entire single particle phase space producing a two-dimensional array of trapping cells. In the presence of a second perturbing wave stochasticity of the motion sets in along the web of critical contours given by $\Omega(\Phi) = \partial\Phi/\partial J \sim ck_x k_y \Phi/B \sim 0$ which define the separatrices of the integrable single wave system.

For the two wave system the onset of stochasticity is derived from an analysis of the Melnikov-Arnold integral for the effect of perturbations along the critical flow contours defined by $\Omega_E(\Phi) = 0$ and leads to the condition in Eq. (3.114).

The onset of stochasticity guarantees the presence of anomalous transport but is not sufficient for the diffusion approximation to be valid. Partial or leaky barriers may be present as found by MacKay, Meiss, and Percival, (1984a,b) in the standard map discussed in Chapter 4. For the **E** × **B** flow different transport regimes are shown in Fig. 3.13 for a two-wave system.

In the globally stochastic regime the diffusion process takes on different forms depending on the strength of the convection $\tilde{\Omega}_E = \langle \Omega_E^2 \rangle^{1/2}$ and the correlation time τ_c through the parameter $R = \tilde{\Omega}_E \tau_c = \tilde{v}_E \tau_c/\ell_c$ where $\tilde{v}_E = \langle v_E^2 \rangle^{1/2}$ and ℓ_c is the spatial correlation distance of $\Phi(x,y,t)$. For $R < 1$ the diffusion occurs in the x, y coordinates and is given by renormalized turbulence theory. Summation of the secular contributions to perturbation theory gives the renormalized formulas (3.135) and (3.136) for the nonlinear correlation time and the diffusion coefficient valid for $R \lesssim 1$. For $R \lesssim 1$ the diffusion coefficient increases as $D \simeq \langle v_E^2 \rangle^{1/2} \ell_c$ reaching the maximum $D_{\max} = \ell_c^2/\tau_c$ at $R = 1$.

For $R \gg 1$ the convection is rapid compared with the rate of decorrelation. In this regime the motion occurs as shown in Fig. 3.12 with the level contours of $\Phi(x,y,t)$ changing slowly on the time scale τ_c. The adiabatic invariance of the action forces most of the diffusion to occur from the process of long excursions along the $|\Phi| \ll \Phi_{\min,\max}$ boundaries between the vortex cells. The result is a stochastic boundary layer between the vortex cells.

The motion studied in this chapter is that of test particles in a two-dimensional incompressible flow. The test particles may be actual charged particles of the collisionless plasma or the Lagrangian motion of fluid elements with a velocity $\mathbf{v} = (-\partial_y \psi, \partial_x \psi)$ given by the stream function $\psi(x, y, t)$. For drift waves with $T_i \ll T_e$ and $k_\perp \rho_i \ll 1$ the distinction between the fluid and ion motion is almost immaterial. For other systems, such as ideal MHD, the test particles must be taken as elements of a fluid field such as mass density. In this case the fluid particle does not have the same stream function Hamiltonian as the charged particles due to diamagnetic and finite Larmor radius currents.

With a change in point of view the same results are called Lagrangian turbulence theory. In this case the test particles are replaced with element of any physical quantity F convected $d_t F = \partial_t F + \mathbf{v} \cdot \nabla F = 0$ by a two-dimensional incompressible flow $\mathbf{v}(x, y, t)$. Immediate applications are the convection of the temperature $d_t T = 0$ (without thermal diffusivity) in temperature gradient driven instabilities, the convection of vorticity $d_t \zeta = 0$ in sheared flows (without viscosity), and the convection of magnetic flux $d_t \psi = 0$ without resistivity. In these examples, the anomalous transport and diffusion coefficient derived here produce an anomalous thermal conductivity, anomalous viscosity and anomalous resistivity, or magnetic diffusivity, respectively. In this context, the strong $R \sim 1$ rate of diffusion $D_m \simeq \ell_c^2/\tau_c$ derived here for the test particle motion is the same as the widely used mixing length diffusivity for the rate of anomalous transport. The mixing length level of saturation for drift wave turbulence $\lambda_x |dn/dx| \sim \delta n$ is, in fact, the $R = 1$ regime with $\ell_c = \lambda_x$ and $\tau_c = 1/\Delta\omega$.

The calculation of drift wave transport has been extensively carried out in both the quasilinear and the renormalized turbulence theory approximations. Studies of the measured amplitudes of drift waves in confinement experiments or the amplitudes observed in numerical simulation experiments indicate that the transport may often be in a higher amplitude regime $R \gtrsim 1$ where both approximations are inadequate. Thus, we are motivated to consider the exact nonlinear motion of the particles in finite amplitude drift waves.

The prediction of renormalized turbulence theory is

$$D_x = -\left(\frac{c}{B}\right)^2 \text{Im} \sum_{\mathbf{k}\omega} k_y^2 |\Phi_\mathbf{k}|^2 g_{\mathbf{k}\omega}(v_\parallel) \qquad (3.142)$$

where the propagator $g_{\mathbf{k}\omega}(v_\parallel)$ for the wave-particle correlation is the solution of

$$\left[\omega - k_\parallel v_\parallel - k_y u(x) - \left(\frac{c}{B}\right)^2 \sum_{\mathbf{k}_1} (\mathbf{k} \times \mathbf{k}_1)^2 |\Phi_{\mathbf{k}_1}|^2 g_{\mathbf{k}-\mathbf{k}_1}\right] g_\mathbf{k} = 1. \qquad (3.143)$$

For $\Delta k_\parallel v_\parallel < \Delta \omega = qv$ and small $\phi = \phi_1 = \phi_2$ the prediction above the threshold for global stochasticity is $D_x \simeq \langle v_x^2 \rangle \tau_c \simeq q^2\phi^2/4qv$ for $R = k^2 D_x \tau_c \sim k^2 \phi^2/4v^2 < 1$

and $D_x \simeq \langle v_x^2 \rangle^{1/2}/k \sim q\phi/2k$ for $k\phi > v$. In the second regime the condition $R = k^2 D_x g_k \lesssim 1$ is formally met for all ϕ since $g_k^{-1} \sim kq\phi$. We observe these scalings for D_x for $\phi \lesssim 5-10$; for larger ϕ the diffusion reaches a plateau. In the plateau regime where $R = k\langle v_x^2 \rangle^{1/2} g_k \gg 1$ the particles make many revolutions around the potential hills and valleys of size π/k_x, π/k_y while jumping randomly at a rate given essentially by $\Delta\omega$ independent of ϕ. (Of course, there is a weak dependence (increase) of $\Delta\omega$ on ϕ in this regime.) In this plateau regime a kind of universal diffusion $D_x \simeq \Delta\omega/k_x^2$ appears to apply.

There is a small fraction of the particles near sea level $\Phi(x, y, t) \simeq 0$ that transport faster than the diffusion of the bulk particles. These particles are sensitive to the equilibrium $\mathbf{E} \times \mathbf{B}$ shear flow and to magnetic shear and the ∇B-curvature drifts. The effect of the $\mathbf{E} \times \mathbf{B}$ shear flow in $u(x) = u_0 + u'x$ is to limit the radial excursion of the particles as given in Example 3.1.

3.7 E × B Motion in a Sheared Magnetic Shear

The characteristics of the quasilinear regime and the renormalized turbulence regimes are that the anomalous diffusion increases with the square of the amplitude and linearly with the amplitude, respectively. In the strongly nonlinear regime, however, the rate of increase can be much weaker than linear due to the short nonlinear correlation time from the intrinsic stochasticity.

Drift waves are low frequency waves in which the guiding center equations of motion for $x, y, z, v_\parallel$ are valid. The first order in m/e drift equations are

$$\frac{d\mathbf{r}}{dt} = v_\parallel \frac{\mathbf{B}}{B} + \frac{\mathbf{E} \times \mathbf{B}}{B^2} \tag{3.144}$$

$$\frac{dv_\parallel}{dt} = \frac{e}{m} E_\parallel \equiv \frac{e}{m} \widehat{\mathbf{b}} \cdot \mathbf{E} \tag{3.145}$$

where $\widehat{\mathbf{b}} = \mathbf{B}/B$. We consider first a uniform magnetic field $B\widehat{\mathbf{z}}$ and then a sheared magnetic field $B(\widehat{\mathbf{z}} + x\widehat{\mathbf{y}}/L_s)$ or $B[\widehat{\mathbf{z}}\cos(x/L_s) + \widehat{\mathbf{y}}\sin(x/L_s)]$. The ∇B-curvature drifts, polarization drifts, and mirror force $\mu \widehat{\mathbf{b}} \cdot \nabla B$ are neglected.

The drift waves are electrostatic with $\mathbf{E} = -\nabla \Phi$ where

$$\Phi(x, y, z, t) = \sum_i \Phi_i(x) \cos\left(k_{iy} y + k_{iz} z - \omega_i t + \beta_i\right). \tag{3.146}$$

For eigenmodes $\omega_i = \omega(\mathbf{k}_i)$, and the waves travel with different phase velocities.

In the simplest case of ions $\omega \gg k_\parallel v_i$ or trapped electrons where the effective $k_\parallel = 0$ the motion reduces to the $1\tfrac{1}{2}$-D Hamiltonian in canonical coordinates x, y

$$\frac{dx}{dt} = \frac{v_\parallel B_x}{B} - \frac{c}{B}\frac{\partial \Phi(x, y, t)}{\partial y} \qquad \frac{dy}{dt} = \frac{v_\parallel B_y}{B} + \frac{c}{B}\frac{\partial \Phi(x, y, t)}{\partial x}. \tag{3.147}$$

Since $v_\| = $ const with $E_\| = 0$ and the effective Hamiltonian becomes $H = (c\Phi_k - v_\| A_\|)/B$. We now consider the effect of magnetic shear which forces $E_\| = -ik_\| \Phi_k \neq 0$ due to $k_\|(x)$. In this case the x-motion in $k_\|(x)$ can limit the x-motion of the guiding center before the radial mode width limit from π/k_x analyzed in the previous sections occurs.

The first order in m/e drift Eqs. (3.144) and (3.145) do not preserve the Hamiltonian structure of the underlying Lorentz equations of motion as pointed out by Littlejohn (1981). Littlejohn (1983) has shown how the selective summation in m/e obtained by introducing the expansion in $1/B^*$ where $\mathbf{B}^* = \nabla \times \mathbf{A}^*$ with $\mathbf{A}^* = \mathbf{A} + mcv_\| \widehat{\mathbf{b}}/e$ restores the Hamiltonian structure to the guiding center drift equations.

To avoid effects arising from ∇B we take the helical, force-free magnetic field $\mathbf{A} = L_s \mathbf{B}(x) = L_s B_0 [\widehat{\mathbf{e}}_y \sin(x/L_s) + \widehat{\mathbf{e}}_z \cos(x/L_s)]$ with $4\pi \mathbf{j}/c = \mathbf{B}/L_s = \mathbf{A}/L_s^2$. The auxiliary field is then $\mathbf{B}^* = \beta \mathbf{B}$ where $\beta = 1 + mcv_\|/eBL_s$. In practice, the quantity $mcv_\|/eBL_s \sim \rho/L_s$ is usually small.

The magnetic shear $(1/L_s)$ can localize the motion of the particles in a single wave to a region Δx given by

$$\Delta x = (\rho_0 L_s)^{1/2} \tag{3.148}$$

with

$$\rho_0 = \frac{v_0}{\Omega} \qquad v_0 = \left(\frac{e\Phi}{m}\right)^{1/2} \tag{3.149}$$

where $\Omega = eB_0/mc$ and v_0 is the longitudinal trapping velocity. For small $k_x \Delta x \ll 1$ we neglect the x-dependence of $\Phi_i(x)$ in Eq. (3.147). We consider two different helicity modes with $d = \mathbf{k}_1 \times \mathbf{k}_2 \cdot \widehat{\mathbf{x}} = k_1 k_2 \sin\theta \neq 0$. We choose the origin $x = 0$ at the point where the pitch of the bigger wave $\mathbf{k}_1, \omega_1, \Phi_1$, matches the pitch of the magnetic field, i.e. where $k_{1\|} = 0$. For small θ the rational surface $k_{2\|} = 0$ where the pitch of the second wave matches the fields is at $x_2 = -\theta L_s$.

For $\theta \neq 0$, where θ is the angle between $\mathbf{k}_1$ and $\mathbf{k}_2$, we can make a transformation $\mathbf{r} = \mathbf{r}' + \mathbf{w}t$ that eliminates the time dependence of both the Φ_1 wave $\omega_1' = \omega_1 - \mathbf{k}_1 \cdot \mathbf{w} = 0$ and the Φ_2 wave $\omega_2' = \omega_2 - \mathbf{k}_2 \cdot \mathbf{w} = 0$. For $k_{1z} = 0$ the transformation is $w_y = \omega_1/k_{1y}$ and

$$w_z = \frac{(\omega_2 - k_{2y}\omega_1/k_{1y})}{k_{2z}} \simeq (\omega_2/k_{2y} - \omega_1/k_{1y})/\theta. \tag{3.150}$$

The principle of least action

$$\delta \int L(q, \dot{q}, t) dt = 0 \tag{3.151}$$

is a covariant statement of the equations of motion and in the wave frame the Lagrangian L' is

$$L' = \tfrac{1}{2} m v_\|^2 - e\Phi(\mathbf{r}') + \frac{e}{c}\left(\mathbf{w} + \frac{d\mathbf{r}'}{dt}\right) \cdot \mathbf{A}(x'). \tag{3.152}$$

Hereafter we work in the wave frame which moves with velocity **w** with respect to the laboratory frame, and we drop the primes on the wave frame coordinates. Thus we have an $N = 2$ degree-of-freedom Hamiltonian system.

The canonical momentum is

$$\mathbf{P} = mv_\| \widehat{\mathbf{b}} + \frac{e}{c}\mathbf{A}(x) = m(v_\| + \Omega L_s)\left[\widehat{\mathbf{y}} \sin\left(\frac{x}{L_s}\right) + \widehat{\mathbf{z}} \cos\left(\frac{x}{L_s}\right)\right] \quad (3.153)$$

and the Hamiltonian is

$$H = \frac{d\mathbf{r}}{dt} \cdot \mathbf{P} - L = \tfrac{1}{2} mv_\|^2 + e\Phi(y,z) - \frac{e}{c}\mathbf{w} \cdot \mathbf{A}(x) \quad (3.154)$$

with $H(x, y, z, v_\|)$ a constant of the motion. The dynamics may also be formulated in the canonical coordinates P_y, P_z, y, z, but one finds that it is cumbersome to do so because of the branches $\pm\left(P_y^2 + P_z^2\right)^{1/2}$ and the lack of a direct physical meaning of P_y and P_z.

3.8 Hamilton's Equations of Motion in Non-Canonical Coordinates

The dynamics and physical interpretation is clear, however, if we use the Hamiltonian equations in non-canonical coordinates. Such a procedure is an important one not often covered in the mechanics text book. The covariant formulation of Hamilton's equations uses the non-canonical coordinates (Robertson et al., 1987)

$$z^i = (v_\|, x, y, z). \quad (3.155)$$

Minimizing the action from Eq. (3.151) with L expressed in terms of H

$$\delta \int (\dot{z}^j P_j(z) - H(z)) dt = 0, \quad (3.156)$$

yields the equations of motion

$$\omega_{ij} \dot{z}^j = \frac{\partial H}{\partial z^i} \quad (3.157)$$

with the symplectic tensor ω_{ij} given by

$$\omega_{ij} = \frac{\partial P_j}{\partial z^i} - \frac{\partial P_i}{\partial z^j}. \quad (3.158)$$

The inverse of the symplectic tensor

$$\omega_{ij} J^{jk} = \delta_i^k \quad (3.159)$$

gives the Poisson tensor $J^{ij}(z)$ for Hamilton's equations

$$\frac{dz^i}{dt} = J^{ij}(z)\frac{\partial H}{\partial z^j}. \tag{3.160}$$

In the present problem, H given by Eq. (3.154), $P_j = (0,0,P_y,P_z)$ by Eq. (3.153), and $J^{ij}(z)$ is expressed in terms of $\cos(x/L_s)$, $\sin(x/L_s)$ and $\beta = 1 + mcv_\parallel/eBL$.

To see more clearly the structure of the equations of motion we scale the variables to those of the larger of the two waves, that is, the big wave Φ_1. We define the dimensionless set of noncanonical $\{z\}$ coordinates

$$u = \frac{v_\parallel}{v_0} \qquad \tilde{x} = \frac{x}{\Delta x} \qquad \psi_1 = k_{1y}y \qquad \psi_2 = k_{2y}y + k_{2z}z \tag{3.161}$$

using Eqs. (3.148) or (3.149) for $\Delta x, \rho_0$ and v_0. Now we define the dimensionless parameters

$$\phi = \Phi_2/\Phi_1 \quad , \quad \lambda = w_y L_s/v_0 \Delta x \quad \text{and} \quad \tilde{w}_z = w_z/v_0. \tag{3.162}$$

Note that the space-time scales here vary with the amplitude of wave one with $v_0 = v_e(e\Phi_1/T_e)^{1/2}$. The dimensionless phase velocity λ is an important parameter and varies as $(\omega_1/k_{y1} v_e)(L_s/\rho_e)^{1/2}(T_e/e\Phi_1)^{3/4}$. Restricting consideration to $|x| \ll \pi/k_x \ll L_s$ and small θ the Hamiltonian (3.163) reduces to

$$\widetilde{H} = \tfrac{1}{2}u^2 - \tilde{w}_z \frac{x^2}{2} + \lambda x - \cos\psi_1 - \phi\cos\psi_2. \tag{3.163}$$

The integrable single wave motion, $\phi = 0$ and $\tilde{w}_z = 0$, has (for $e = -|e|$)

$$\widetilde{P}_z = \frac{P_z}{mv_0} = u + \frac{x^2}{2} \quad \text{and} \quad \widetilde{H} = \frac{H}{\tfrac{1}{2}mv_0^2} = \text{constant} \tag{3.164}$$

and is shown in Fig. 3.14 in the $\tilde{x} - \psi_1$ plane for different values of λ and $\widetilde{P}_z$. Each frame $a, b, \ldots, \ell$ in Fig. 3.14 corresponds to the region labelled with the same letter in Fig. 3.15 of the $\lambda - P_z$ plane. Thus, we see that in the presence of magnetic shear even the motion in the single wave problem has a rich variety of topologies in the phase space.

From Fig. 3.14 we see that in the single wave particles moving parallel to the equilibrium current (adding to the current) are pinched into the $x = 0$ region giving a pendulum topology (negative P_z). Particles going opposite to the current are pushed out from $x = 0$ forming three parallel chains of X and 0 points along the lines at $x \cong 0$ and $x \cong \pm(2P_z)^{1/2}$. In dimensional units this separation is $\pm(2v_0 L_s/\Omega)^{1/2}$

3.8 Hamilton's Equations of Motion in Non-Canonical Coordinates

In the presence of the second wave the infinite period homoclinic orbits in Fig. 3.14 become stochastic. The direction of the second wave determined by the sign of $\tilde{w}_z$ has a strong constricting or pinching effect ($\tilde{w}_z < 0$) or expulsive effect ($\tilde{w}_z > 0$) on the orbits.

The problem of x-transport due to the second wave can be put in the language of tokamak physics by introducing the minor radius $\rho = \exp[\tilde{P}_z(x, u)]$, the poloidal angle ψ_1, and the toroidal angle ψ_2. For each H there is an effective tokamak with axisymmetric integrable curves $\rho = $ const for $\phi = \Phi_2/\Phi_1 = 0$. The presence of the second wave $\phi \neq 0$ breaks the toroidal symmetry and opens islands and stochastic bands, as shown in Fig. 3.16 drawn from an actual computation. Thus we see that the phase space torus defined here for each energy surface $H = $ const has the same behavior as the magnetic confinement torus analyzed in Secs. 2.5–2.10. The phase space tori have different separatrix webs according to the values of P_z and $\lambda = \omega/kv_0$. The separatrices' reconnection affects the type of KAM barrier and the value of the second wave amplitude required to break the barrier. A related map is discussed briefly in Sec. 4.9.

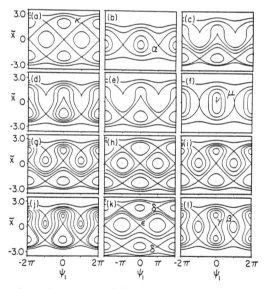

Fig. 3.14. The complex web structure of the phase space ($p = \tilde{x}$, $q = \psi_1$) for a guiding-center in a single wave as a function of the parameter domains shown in Fig. 3.15 of the particle canonical momentum $P_z = u + \tilde{x}^2/2$ and wave phase velocity parameter $\lambda = (\omega_1/k_y v_0)(L_s/\rho_0)^{1/2}$. The frame labels $a, b, c, \ldots \ell$ are for the (P_z, λ) values given in Fig. 3.15.

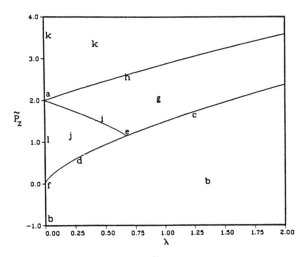

Fig. 3.15. Parameter space of $P_z = m_e v_0 \tilde{P}_z$ and λ giving the boundaries for different single particle guiding-center orbit topologies in the single wave-sheared magnetic field system in Sec. 3.8. Here $v_0 = (e\phi_1/m)^{1/2}$ and $\rho_0 = v_0/\omega_c = \rho_{th}(e\phi_1/T)^{1/2}$ with the phase velocity $\omega_1/k_y = \lambda v_0/(\rho_0/L_s)^{1/2}$.

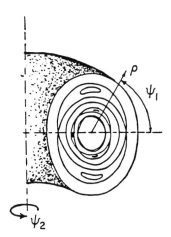

Fig. 3.16. Toroidal structure of the electron energy surfaces for the two-wave Hamiltonian of a guiding-center electron in a sheared magnetic field. The poloidal angle is the phase $\mathbf{k}_1 \cdot \mathbf{x}$ and the toroidal angle is the phase $\mathbf{k}_2 \cdot \mathbf{x}$. The minor radius is a measure of P_z which is proportional to the square of the spatial distance of the guiding center from the $k_\parallel = 0$ rational surface.

3.8 Hamilton's Equations of Motion in Non-Canonical Coordinates

From the formulas for the widths of the separatrices we have derived resonance overlap conditions for the various types of separatrices shown in Fig. 3.14. The resonance overlap criterion predicts global stochasticity for

$$S = \frac{1}{\theta}\left(\frac{v_0}{\Omega L_s}\right)^{1/2} \gtrsim 1 \tag{3.165}$$

where $\Phi_1 = \Phi_2$ and $v_0 = (e\Phi/m)^{1/2}$. The scaling of the overlap parameter with amplitude $S \propto \Phi^{1/4}$ predicts that the system is more fragile to the onset of stochasticity than the pendulum system where the dimensionless overlap parameter S scales $S = v_0/\Delta(\omega/k) \propto \Phi^{1/2}$. Numerical integrations confirm this prediction of a quick onset of chaos.

We conclude that the transport of particles in drift waves has a rich mathematical and statistical structure. For two coherent waves we find physically relevant regimes with strong diffusion-like transport. The scale of the anomalous diffusion $\langle x^2(t)\rangle = 2D_x t$ shown in the surface of section plot in Fig. 3.17 is

$$D_x^{\text{sto}} = (\Delta x)^2 k_y v_0 = k_y v_0 \rho_0^{3/2} L_s^{1/2} \tag{3.166}$$

for the electrons in the stochastic sea. We estimate the fraction n_e^{sto}/n_e of stochastic electrons by v_0/v_e, and then the net effective diffusion coefficient is given by

$$D_x = |k_y| v_e \rho_e^2 \left(\frac{L_s}{\rho_e}\right)^{1/2} \left(\frac{e\Phi}{T_e}\right)^2. \tag{3.167}$$

These results show that although magnetic shear confines particles in a single drift wave, the presence of a second wave with appropriate parameters can produce a diffusive radial transport. The stochasticity condition $S = 1$ in Eq. (3.165) gives the critical electrostatic potential

$$\frac{e\Phi_{\text{crit}}}{T_e} = \frac{L_s \theta^2}{\rho_e} \cong \frac{1}{k_y^2 \rho_e L_s}. \tag{3.168}$$

At the amplitude Φ_{crit} the diffusion rate from Eq. (3.167) is

$$D_x = \frac{cT_e}{eB}\left(\frac{L_s \rho_e}{r_n^2}\right)^{3/2} \tag{3.169}$$

where we use the lower limit on k_y as $k_y \rho_s \geq r_n/L_s$ from the neglect of ion-acoustic wave effects $\omega_{*e} > c_s/L_s$.

From the physical problems analyzed in this Chapter we see the generic behavior of physical systems with two degrees-of-freedom is to have a mixture of integrable orbits and chaotic orbits. The chaos starts as a stochastic web along the separatrices of some nearby integrable Hamiltonian system. Even at low levels of the

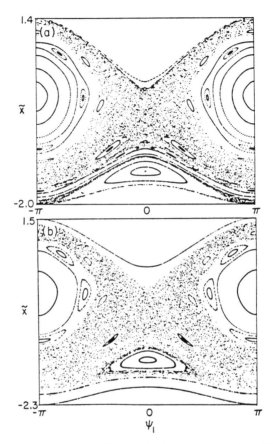

Fig. 3.17. Stochastization of the single particle phase space by the addition of a second wave with (a) $\phi_2/\phi_1 = 0.1$ and (b) $\phi_2/\phi_1 = 0.16$. The excursion distance $\tilde{x}$ is in units of $(\rho_0 L_s)^{1/2}$ and the circulation frequency is $k_y v_0 (\rho_0/L_s)^{1/2}$. The guiding-center diffusion is at the rate $D_x \cong (k_y v_0) \rho_0^2 (L_s/\rho_0)^{1/2}$ within the stochastic domain.

perturbation (second wave, normal component B_x of the magnetic field, etc.) the stochastic layers provide a mechanism for particles to move long distances in the phase space. At higher levels of the perturbation (or alternatively for generic two-degree-of-freedom Hamiltonians), there are broad regions of phase space with chaotic orbits that give rise to a diffusion-like transport in this region. The distribution becomes almost constant in this region with deviation from strictly constant phase space distribution having to do with the boundaries of the chaotic regions where there are long-time correlations due to the "stickyness" in orbit segments along the bordering island chains. The description of the transport including these features

3.8 Hamilton's Equations of Motion in Non-Canonical Coordinates

is the subject of the next chapter. Resolving the "stickiness" of the islands and the long-time correlations requires huge numbers of orbital periods ($\gtrsim 10^4$ to 10^6) so that typically maps representing the Hamiltonian flow are the practical tool for analyzing such details of the phase space structures and transport (Karney, 1979, 1983). For long-time Hamiltonian flows the use of symplectic integration is required (Forest and Ruth, 1990).

Chapter 4

Phase Space Structures in Hamiltonian Systems

In Chapter 4 we develop in some detail the phase space structures of the mixed regions of chaotic orbits and integrable orbits that act to define the generic Hamiltonian system. While Chapter 3 deals with phase space flows arising from Newton's laws taken with the Lorentz force acting on the charged particles, here the emphasis is changed to describe the mappings created by the intersection of the flows with a reference plane in phase space. Obtaining the maps on the surface of section for the flows is generally a difficult physics problem as shown in Chapter 3. With a given area preserving map the dynamics can be investigated in much more mathematical detail as made clear in the work of Henon (1983, for example). In this Chapter we use maps to determine the structure of the dynamics, the ergodicity, the transport and the diffusion approximation with more clarity and accuracy than is possible for the flows of Chapter 3.

Here again we must be selective. From the wide range of Hamiltonian (symplectic) maps we treat the standard map and its generalization, the relativistic standard map, in detail. The reader is referred to two books that have collections of the original articles concerning the theory of dynamical maps: *Hamiltonian Dynamical Systems*, MacKay and Meiss (1987) and *Chaos* by Hao (1984).

4.1 The Standard Map

The analysis of nonlinear motion in the vicinity of the separatrix leads to the standard map as a general description of the motion in many different physical systems. An example, typical of a class of such reductions to the standard map, is that given in Sec. 3.3.3. The essential conditions for a closed system to reduce to the standard map is that in some domain of the phase space

(i) One degree-of-freedom dominates to the extent that the coupling to the other degrees-of-freedom acts as a perturbation driving the dominant degree-of-freedom at a fixed frequency.

(ii) The phase advance per period for the dominant degree-of-freedom is linearly proportional to the increase of the momentum-like coordinate.

The corresponding physical conditions for the motion of a particular system to be represented by the standard map are quite varied. In some radio frequency (rf) heating examples (Stix, 1992, Ch. 16), the condition (i) of decoupling is naturally satisfied by the driver being an externally applied field. In the cases of 2D degrees-of-freedom, such as the magnetic moment invariance problem of Sec. 3.4, the approximation is that one motion, in this case the parallel transits between the high magnetic field mirroring points, can be taken as a given motion. Subsequently, the dynamical equations for the gyrophase $\zeta(t)$ and the perpendicular component of the velocity $v_\perp(t)$ are parametrically driven by the fixed parallel motion. Specifying the perpendicular velocity component is equivalent to giving the pitch angle $\alpha = \cos^{-1}(v_\parallel/v)$ since $v^2 = v_\perp^2 + v_\parallel^2 = \text{const}$. The map follows by recording the values of $(v_\perp, \zeta)$ at each crossing of a reference plane during the parallel $(v_\parallel)$ motion (Taylor, 1969).

The second condition on the linearity of the phase advance has to do with the strength of the variation of the dispersion of the nonlinear oscillator amplitude being described $\omega(h) = \omega_0 + \omega_1 h + \omega_2 h^2 + \cdots$ and the amplitude $(h_{\max})$ of the motion being considered. In the case of the whisker map the motion described by the standard map is obtained for motion sufficiently close to the critical energy H_{sx} that $h = (H - H_{sx})/H_{sx} \to 0$. In this limit the quadratic term $\omega_2 h^2$ is negligible compared with the linear term $\omega_1 h$. This linearization makes the phase space locally homogeneous.

In the case where we find that the motion from the standard map is characterized by diffusion D there will be two limits, $h \to 0$ and $t \to \infty$, that must be considered in the physical problem. In studies of the long-time behavior of the standard map the often hidden or implied assumption is that the limit $\omega_2 h/\omega_1 \to 0$ has already been taken. In Sec. 4.5 we will study the case where the nonlinear phase advance is retained before the $t \to \infty$ limit is taken. The nonlinear phase advance will restore integrable curves to the higher energy region of the phase space. In such cases where the nonlinear phase advance $\omega_2 h/\omega_1 \neq 0$ is important there is a much richer structure of fixed points and integrable curves than in the standard map. The nonlinear phase advance also severely restricts the domain where the phase dynamics is sufficiently uniformly random to allow the momentum-like coordinate to diffuse $\langle \Delta p^2 \rangle = 2Dt$. The nonlinear phase advance eliminates the so-called accelerator modes that require fixed periodic momentum jumps of size ℓ times the unit of momentum in the standard map.

4.1 The Standard Map

We illustrate the importance of the finite nonlinearity in the phase advance of the $d = 2$ area preserving map by exploring the example of the relativistic standard map (Nomura et al., 1992). Another example of a nonlinear map that reduces locally to the standard map is the tandem mirror map (Ichikawa et al., 1983) describing the radial excursions of ions bouncing between quadrupole magnetic mirrors separated by straight magnetic field lines.

Mathematically stated, the standard map S is the mapping of the two-dimensional plane $(d = 2)$ onto itself (X, P) given by

standard map

$$\begin{pmatrix} X_{n+1} \\ P_{n+1} \end{pmatrix} = S \begin{pmatrix} X_n \\ P_n \end{pmatrix} = \begin{pmatrix} X_n + P_{n+1} \\ P_n + A\sin(2\pi X_n) \end{pmatrix}. \quad (4.1)$$

For physical interpretation in terms of a linear harmonic oscillator of spring constant K about the principal fixed point, it is often useful to rewrite the amplitude constant $A = -K/2\pi$ so that in the limit of $K \to 0$ the motion about $(0,0)$ is given by $\dot{X} = P$ and $\dot{P} = -KX$ where for small fractional changes in X and P (for small K) we may take the continuum limit with $P_{n+1} - P_n \cong \dot{P}$ and $X_{n+1} - X_n \cong \dot{X}$.

The map (4.1) is an example of a class of $d = 2$ area preserving maps of the form

$$\begin{aligned} P_{n+1} &= P_n + F(X_n) \\ X_{n+1} &= X_n + G(P_{n+1}). \end{aligned} \quad (4.2)$$

In the standard map the momentum advance is the periodic function $F(X) = A\sin(2\pi X)$, or equivalently, $= -(K/2\pi)\sin(2\pi X)$, and the phase advance is linear $G(P) = P$. The condition for the existence of invariant surfaces (curves) is the twist condition that $G(P)$ be monotonic

twist condition

$$\frac{\partial G}{\partial P} > 0. \quad (4.3)$$

The twist condition for the standard map in Eq. (4.1) is especially simple since $\partial G/\partial P = +1$. This property guarantees that every vertical line $(y = P)$ maps into a curve after repeated applications of the map. This twist is the mechanism for the generation of the invariant curves in the long-time limit $(n \to \infty)$ of the map.

In the relativistic map the phase advance function $G(P)$ is

$$G(P) = \frac{P}{(1 + \beta^2 P^2)^{1/2}} \quad (4.4)$$

where the parameter β measures the strength of the relativistic mass increase. The dispersion of the $X_{n+1} - X_n$ advances is given by

$$\frac{\partial G}{\partial P} = \frac{1}{(1 + \beta^2 P^2)^{3/2}} \quad (4.5)$$

which at low momentum is $(1 - 3/2\,\beta^2 P^2)$ giving a perturbation to the shape of the critical curves in the standard map. At high momentum, however, $\partial G/\partial P$ tends to zero as $1/P^3$ with the result that a whole new ordered structure appears in the phase space. The weak dispersion of the high momentum phase advance, now controlled by the nearly fixed velocity $v \cong c =$ speed of light, introduces a lattice of stable and unstable fixed points bounded by a series of high momentum invariant (KAM) curves. We show how to determine the fixed points from the symmetries of space reflection and velocity inversion in Sec. 4.6.

4.2 Maps for the Motion of a Charged Particle in an Infinite Spectrum of Longitudinal Waves

The standard map is a generic or universal map for the local phase space in many dynamical systems as explained in Sec. 3.3. In Sec. 3.3.3 the standard map followed from the whisker map for the motion of a particle sufficiently near the separatrix of a driven system. Here we present another often cited example where the standard map describes the exact global phase space for the nonrelativistic motion of a particle in an idealized infinite wave spectrum of discrete constant amplitude waves at evenly spaced phase velocities. The special properties and limitations of the standard map become clearer if we start with the relativistic dynamics and proceed to derive the standard map as the limit in which $|\omega/k| \ll c$ — the speed of light. The motivation of this relativistic description of the dynamics is found a posterori, namely that when the Newtonian description is used, there are found to exist general sets of orbits called accelerator orbits in which the momentum increases linearly without bound with the iteration number or time. Even for the chaotic orbits the long-time diffusion, in the Newtonian dynamical description, will eventually bring the particles to relativistic speeds.

For the idealization of an infinite spectrum of equal amplitude waves with evenly spaced phase velocities the accelerating electric field is

$$E(x,t) = E_0 \sum_{n=-\infty}^{+\infty} \sin(kx - n\omega t) = E_0 \sin(kx) \sum_{-\infty}^{+\infty} \cos(n\omega t) \qquad (4.6)$$

where from the last form it is easy to see that due to the destructive interference the electric field components cancel except at the time sequence $t_n = 2\pi n/\omega$ where the waves are precisely in phase giving the impulsive forces of strength $(2\pi E_0/\omega) \sum_n \delta(t - t_n)$.

The relativistic motion of a charged particle is given by the Hamiltonian

$$H(x,p,t) = \left(p^2 c^2 + m_0^2 c^4\right)^{1/2} - \frac{eE_0}{k} \sum_{n=-\infty}^{+\infty} \cos(kx - n\omega t)$$

4.2 Maps for the Motion of a Charged Particle in ...

$$= (p^2c^2 + m_0^2c^4)^{1/2} - \frac{2\pi e E_0}{k\omega} \cos(kx) \sum_n \delta(t - n\tau)$$

where $\tau = 2\pi/\omega$. The dynamics of this $1\frac{1}{2}$ D Hamiltonian occurs in a $d = 3$ toroidal phase space where the toroidal angle around the principal axis is $\phi = \omega t$ and around the minor axis $\theta = kx$. All of the geometrical considerations of the magnetic field torus given in Chapter 2 now apply to this system. In addition the dynamics around the torus is analytically very simple being composed of free motion with $\dot{x} = \partial H/\partial p = p/\left(m^2 + p^2/m^2c^2\right)^{1/2} = \text{const}$ between the impulses at $t_n = n\tau$ ($\phi_n = 2\pi n$) and having a jumb in p at constant x during the infinitesimal time interval of the impulse of the torus onto itself. Thus, the motion of the particle can be completely described by keeping track of the position x_n and momentum p_n after each accelerating "kick" from the impulsive force. The motion is described by the $d = 2$ map of the cross-section

The appropriate dimensionless variables are clearly

$$X = \frac{k}{2\pi} x, \qquad P = \frac{k}{m_0\omega} p, \qquad T = \omega t. \tag{4.7}$$

With these dimensionless variables the amplitude of the electric field E_0 for each wave is transformed to the square of the ratio of the trapping velocity $v_{tr} = (e\phi_0/m_0)^{1/2}$ to the phase velocity ω/k (times $(2\pi)^2$). Thus, we define the two parameters of the relativistic standard map

$$K = \frac{4\pi^2 ek E_0}{m_0\omega^2} \qquad \text{and} \qquad \beta = \frac{\omega}{kc}. \tag{4.8}$$

With these dimensionless parameters the map is given by Eq. (4.2) with

$$F(X) = -\frac{K}{2\pi} \sin(2\pi X) \tag{4.9}$$

$$G(P) = \frac{P}{(1 + \beta^2 P^2)^{1/2}}. \tag{4.10}$$

For many studies it is appropriate to take the $\beta = 0$ limit of this map whereupon the dynamics is nonrelativistic (with $H = P^2/2m - eE_0(x,t)/k$) and the map is
standard map

$$P_{n+1} = P_n - \frac{K}{2\pi} \sin(2\pi X) \tag{4.11}$$

$$X_{n+1} = X_n + P_n. \tag{4.12}$$

Both maps (4.9)–(4.10), and (4.11)–(4.12) are area preserving twist maps. For typical parameters, both have a denumerable infinity of stable/unstable fixed points and regions of chaotic orbits. The important difference is that in the standard map the phase is invariant to $P \to P + \ell$ for integer ℓ so that we define a basic cell of unit size in both X and P for the standard map and pull all iterants back to this cell with $P = P$ modulo 1 and $X = X$ modulo 1. In the relativistic standard map the low $\beta^2 P^2$ region behaves similarly to the standard map, but for finite $\beta^2 P^2$ values the phase advance is strongly nonlinear. This nonlinearity of the phase advance introduces new invariant curves and a web of fixed points at high momentum.

Reviews and articles developing the properties of the standard map abound. For a reader to get started we suggest Chirikov (1979) and the collection of articles in MacKay and Meiss (1987).

4.3 Fixed Points, Accelerator Orbits and the Tangent Map

The key characteristic of a family of maps $\{K, \beta, \cdots\}$ is how they transform points, curves, and areas. In particular, those subset of points that are invariant are the fixed points — representing an equilibrium state — and those subsets that map into an invariant curve represent an integrable orbit. A closed invariant curve partitions the phase space into non-communicating or disconnected regions that can have completely different distributions of particles. This type of "trapping" of particles is a key feature of the confinement problem and the rf heating problem in fusion physics.

The critical values of map parameters $\{K, \beta, \ldots\}$ for an intrinsic change in the fixed points or the invariant tori are called bifurcation values. For the standard map the critical value of K for the breakup of the last invariant curve is

$$K_c = 0.971635406\ldots. \quad (4.13)$$

In the case of the standard map which is periodic in the momentum-like or action variable as well as the phase variable, there are fixed points orbits in the closed phase space $P \subset [-1/2, 1/2]$ with $P = P$ modulo 1 which in physical space have a steadily increasing momentum. The orbits associated with this type of fixed point are called ℓ-step accelerator orbits of period p since the momentum increases by the integer values of ℓ after p iterations of the map. We discuss these physically important orbits in Sec. 4.10.

The special equilibrium points (X_m, P_m) which do not change upon iteration are fixed points $S(X_m, P_m) = (X_m, P_m)$. Those points that return after the p-th iteration of the map

$$S^p(X_m, P_m) = (X_m, P_m) \quad (4.14)$$

are called period-p fixed points. As noted above in the special case maps that are periodic in P modulo 1 we may define the ℓ-step period-p fixed point by $P_{n+p} - P_p = \ell$.

4.3 Fixed Points, Accelerator Orbits and the Tangent Map

For the relativistic standard map in Eqs. (4.9) and (4.10) the fixed points are given for $m = 0, \pm 1, \pm 2, \ldots m_{\max}$ by $X_m = 0$ and $1/2$ and $P_m = m(1 - \beta^2 m^2)^{-1/2}$ giving the phase advance from the velocity in Eq. (4.10) the integer value m. Here $m_{\max} <$ integer part of β.

The mapping in the neighborhood of the fixed points is given by the tangent map ΔT which has a dependence on the fixed point (X_m, P_m) coordinates. For the period 1 fixed point the tangent map is just the linearization of S about (X_m, P_m). From Eq. (4.2) we get

$$\begin{pmatrix} \Delta P_{n+1} \\ \Delta X_{n+1} \end{pmatrix} = \Delta T \begin{pmatrix} \Delta P_n \\ \Delta X_n \end{pmatrix} \tag{4.15}$$

where ΔT is the tangent map
standard map

$$\Delta T = \begin{pmatrix} 1 & F'(X_n) \\ G'(P_n) & 1 + F'(X_n)G'(P_n) \end{pmatrix}. \tag{4.16}$$

associated with the map in Eq. (4.2). The stability of a point (X_n, P_n) is determined by the residue R. The residue R is given as

$$R \equiv \tfrac{1}{2} - \tfrac{1}{4} Tr(\Delta T) = -\tfrac{1}{4} F'(X_n)G'(P_n). \tag{4.17}$$

The fixed point is stable if $0 < R < 1$. For the relativistic standard map Eq. (4.17) gives rise to the stability condition

$$0 < K \cos(2\pi X_m) < 4(1 - \beta^2 m^2)^{-3/2}. \tag{4.18}$$

Therefore, we find that as far as $K > 0$, the fixed point $(X_m = 1/2, P_m = m(1 - \beta^2 m^2)^{-1/2})$ remains unstable, while the fixed point $(X_m = 1/2, P_m = m(1 - \beta^2 m^2)^{-1/2})$ is stable for K in the range

$$0 < K < 4(1 - \beta^2 m^2)^{-3/2} \tag{4.19}$$

For the stable orbits with $0 < R < 1$ the average rotation number ρ is given by

$$\rho \equiv \frac{1}{2\pi} \cos^{-1}(1 - 2R). \tag{4.20}$$

If the rotation number ρ becomes equal to p/q (p and q are primes), the Poincaré-Birkhoff period-q islands bifurcate out of the fixed point $(X_m, P_m)_s$. The onset of the period-q bifurcation takes place as K passes through the value

$$K(p/q) = 2(1 - \beta^2 m^2)^{-3/2} \left[1 - \cos\left(2\pi \frac{p}{q}\right)\right]. \tag{4.21}$$

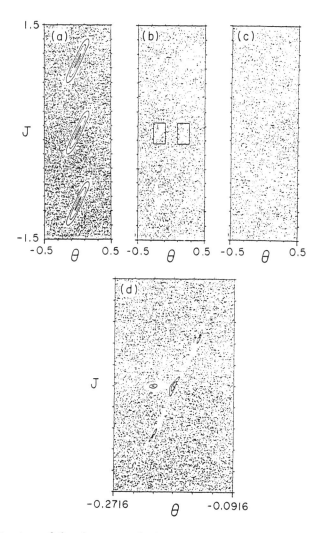

Fig. 4.1. Structure of the phase space for increasing values of the K parameter in the standard map. The points are generated by following 3×10^4 particles started around the hyperbolic fixed point at $(X, P) = (0.5, 0)$ (within a distance 0.01) for $n = 40$ iterations. (a) for $K = 3.8600$ there are large islands. (b) For $K = 6.9115$ the origin (0,0) has become a hyperbolic fixed point and there are period-3 accelerator mode tori within the two boxes. The island structure in the left box is shown in the enlargement of the left box in frame (d). (c) For $K = 10.053$ there are no visible invariant tori. Each island of (a) and (d) is encircled by a critical torus.

It is clear that at the $m = 0$ fixed point, i.e. the origin $(0,0)$, the onset condition $K > K(p/q)$ becomes identical with that of the standard map. Hence, the structure of the orbits around the origin is the same as that of the orbits in the standard map.

We show an example of the stability of the orbits and the period-3 bifurcation in Fig. 4.1 for the sequence $K_1 = 3.8600, K_2 = 6.9115$ and $K_3 = 10.053$ and $\beta = 0$. The stochastic sea is obtained by starting a cloud of 3×10^4 points at $(X = 1/2, P = 0)$ and iterating 40 times. For K_1 there are still integrable curves or invariant tori around the stable point $(0,0)$. For K_2 this point $(0,0)$ is hyperbolic, but period-3 accelerator mode tori are present in the boxed areas. The island structure of the period-3 accelerator modes and their tori are shown in the blowup in Fig. 4.1(d) of the left boxed region in Fig. 4.1(b). For K_3 there are no islands visible for this time scale.

For the relativistic standard map where $\beta = \omega/kc \neq 0$ there are no accelerator orbits due to the inhomogeneous momentum dependence of the phase advance in the dynamics. In the case of a small, but finite value of β, there are the large number of pairs of stable (elliptic) and unstable (hyperbolic) fixed points given by

$$(X_m, P_m)_s = \left(0, \ m(1 - \beta^2 m^2)^{-1/2}\right) \qquad (4.22)$$

$$(X_m, P_m)_u = \left(1/2, \ m(1 - \beta^2 m^2)^{-1/2}\right) \qquad (4.23)$$

for

$$m = 0, \pm 1, \pm 2, \ldots, \pm \text{Int}(1/\beta).$$

Hence, we find that the fixed points exist with unequal intervals within the range determined by the limit $|m| < \beta^{-1}$. For $\beta > 1$, the only fixed points (X, P) are $(0,0)$ and $(1/2,0)$. When $\beta = M^{-1}$, the last fixed point in the sequence, P_m goes to infinity, which is the resonance acceleration investigated by Chernikov, et al. (1989). They show that a chaotic channel is opened at $X = \pm 1/2$ for $\beta = M^{-1}$, and along the chaotic channel particles can be accelerated to arbitrary high energies.

4.4 Stochastic Motion and the Diffusion Coefficient

When the amplitude of the waves is such that the trapping velocity is greater than the phase velocity of the waves $(A = K/2\pi > 1)$ then the particles undergo a succession of trapping and detrappings in the nearest-neighbor waves. These chaotic orbits are characterized by the Lyapunov exponents describing the exponential divergence of neighboring trajectories. Chirikov (1979) describes the calculations for the Lyapunov exponents of the standard map and shows that the Lyapunov exponent λ is given by $\lambda = \langle \ell n(\lambda_+(\theta)) \rangle \simeq \ell n(A/2)$ for $A \gg 1$. Here $\lambda_+(\theta)$ and $\lambda_-(\theta)$

are the two local eigenvalues of the tangent map ΔT (4.15), and the average $\langle \ \rangle$ over the phase $\theta = 2\pi X$ is performed assuming a uniform distribution in time of the orbit in θ. Chirikov calls this divergence rate $h = \langle \ell n(\lambda_+(\theta)) \rangle$ the Kolmogorov-Sinai (KS) entropy. The term "entropy" is introduced since h measures the rate of mixing in the phase space.

The association of the time-average divergence rate with the thermodynamic concept of "entropy" is motivated by the coarse-grain averaging that takes place in using the long-time average from the chaotic orbit $Z(t) = (X(t), P(t))$ starting at Z_0. Properties of the Lyapunov exponent and the statistical mechanics of the chaotic orbits in the standard map are given in Grassberger et al. (1983, 1985). Horita et al. (1989, 1990) develop these statistical and thermodynamic-like properties of the standard map in some detail. Horita et al. show that the rate of convergence of the finite time t divergence rate

$$\Lambda_t(Z_0) = \frac{1}{t} \ell n |\lambda_+(Z_0)|$$

to Λ_∞ for $t \to \infty$ is slow due to the repeated sticking of the chaotic orbits to the critical invariant curves encircling the islands at the boundaries of the stochastic sea..

Horita et al. (1990) show that the probability distribution $f(\tau)$ for sticking to the critical tori for a number of iterations τ has the power law decay for large τ given by

$$f(\tau) \propto \tau^{-1-\beta} \qquad (1 < \beta < 2). \tag{4.24}$$

This power law decay for the sticking time implies that there are long-time correlations in the orbits (Karney, 1983). The long-time correlations give rise to the slow convergence of the finite time divergence rate Λ_t to Λ_∞ and a slow convergence of the diffusion coefficient. These issues are discussed in detail in the review of Mori et al. (1989) and briefly in the Sec. 4.4.1.

4.4.1 Probability distribution functions for the orbits

When neighboring orbits are separating in phase space at the exponential rate $\delta(t) = \delta_0 \exp(\Lambda^\infty t)$ where $\delta(t) = \|Z_1(t) - Z_2(t)\|$ and $Z = (X, P)$, it is usually not useful to study single particle trajectories except over periods of time $t \lesssim 1/\Lambda^\infty$. It is more profitable to introduce ensembles of trajectories described by the probability distribution function $f(x, p, t)$. The equation for $f(x, p, t)$ is called the Klimontovich equation and states the conservation of the number of trajectories in the phase space element $dx\, dp$. When the dynamics is an area preserving flow then the conservation of trajectories requires that $f(x, p, t)$ is constant along the trajectories

$$\frac{df}{dt} = \frac{\partial f}{\partial t} + \frac{p}{m} \frac{\partial f}{\partial x} + \frac{dp}{dt} \frac{\partial f}{\partial p} = 0. \tag{4.25}$$

4.4 Stochastic Motion and the Diffusion Coefficient

For N-trajectories or particles the formal solution of Eq. (4.25) is

$$f(x, p, t) = \sum_{i=1}^{N} \delta(x - x(x_i, p_i, t))\delta(p - p(x_i, p_i, t)) \qquad (4.26)$$

where x_i, p_i are the initial data of the ith particle or trajectory. Assuming distinct initial data (otherwise the trajectories are the same and the statistical weight of the trajectory is increased by a multiple) the label i gives the initial data (x_i, p_i). We define the ensemble average of a phase function $G(x, p)$ by

$$\langle G \rangle = \frac{1}{N} \sum_{i=1}^{N} G(x_i, p_i). \qquad (4.27)$$

When the system is ergodic then the phase space average is the same as the long-time average over a single trajectory (see Sec. 2.5) defined by

$$\overline{G}(x_0, p_0) = \lim_{T \to \infty} \frac{1}{T} \sum_{t=0}^{T-1} G(x(t), p(t)). \qquad (4.28)$$

In the chaotic sea of the standard map and other maps with chaotic domains the averages $\langle G \rangle$ and $\overline{G}$ are the same when sufficiently long orbits are taken. The convergence of the finite time orbit of length $t = n\tau$ to the infinite time limit (4.28) $\overline{G} = G^\infty$ can be slow and the fluctuations about G^∞ are of physical importance.

Let us consider the convergence of the finite time expansion rate $\Lambda_n(x_0, p_0)$ to the infinite time limit Λ^∞. It can be shown that the exponent for the fluctuations for the convergence $\overline{(\Lambda_n - \Lambda^\infty)^2} = n^{1-\beta}$ where β is the power law for the pdf of the sticking time $f(\tau) = \text{const}/\tau^{1+\beta}$ with $1 < \beta < 2$ given in Eq. (4.24). Horita et al. (1990) show that the probability $P(\Lambda, t)$ density for finding a given expansion rate Λ over a time interval $t \gg 1$ varies as

$$P(\Lambda, t) = \text{const} \exp(-t\psi_n(\Lambda)) \qquad (4.29)$$

where the characteristic function $\psi_n(\Lambda)$ is concave upward with a minimum at $\Lambda = \Lambda^\infty$ but has a small, perhaps vanishing value for all $0 \leq \Lambda < \Lambda^\infty$, in the $t \to \infty$. For $K = 6.9115$ they find that $\Lambda^\infty = 1.26$ and $\beta = 1.6$. The form of $\psi_n(\Lambda)$ for $K = 6.9115$ and $n = 100, 500$ and 1000 is given in Fig. 4.2. [Note that the exponent parameter β is defined by Eq. (4.24) and is not related to the relativistic β parameter of Sec. 4.1.]

128 Chapter 4 Phase Space Structures in Hamiltonian Systems

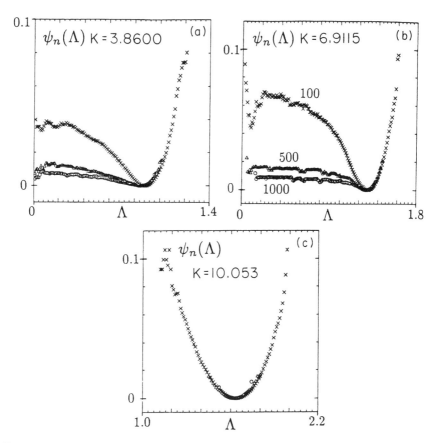

Fig. 4.2. The expansion-rate spectrum $\psi_n(\Lambda)$ for the probability density $P(\Lambda; n) = \exp(-n\psi(\Lambda))$ of finding the finite time $t = n$ Lyapunov divergence rate Λ. When $\psi(\Lambda)$ has a well-defined minimum $\psi \sim \frac{1}{2}\psi''(\Lambda - \Lambda_\infty)^2$ about Λ_∞ the Lyapunov exponent is well defined as Λ_∞ for $n \to \infty$. For the three K values in Fig. 4.1 we see from the shape of $\psi(\Lambda)$ that there is a "phase" transition. (a) For the lowest $K = 3.8600$ convergence for $n \to \infty$, is to $\psi_n(\Lambda) = 0$ for $0 < \Lambda \leq \Lambda_\infty$. (b) For $K = 6.9115$ the minimum of $\psi(\Lambda)$ is just beginning to form at $\Lambda_\infty \cong 1.26$. (c) For $K = 10.053$ there is a well-defined minimum of $\psi(\Lambda)$ at $\Lambda_\infty = 1.62$.

When velocity resonance overlapping occurs ($v_{tr} > \omega/k$ or $A > 1$) the momentum coordinate (for $\beta p \ll m_0 c$) diffuses with the diffusion coefficient or diffusivity D given by

$$D = \lim_{T \to \infty} \frac{1}{2T} \left\langle (P_i(T) - P_i(0))^2 \right\rangle \qquad (4.30)$$

4.4 Stochastic Motion and the Diffusion Coefficient

where $\langle \ \rangle$ is the ensemble average defined by Eq. (4.27). Computer calculations for the "running" diffusion coefficient

$$D(t) = \frac{1}{2t} \left\langle (P_i(t) - P_i(0))^2 \right\rangle \tag{4.31}$$

are easily performed on programmable calculators and personal computers. The student is encouraged to write programs for the maps discussed in this Chapter to look for the properties such as the islands, invariant curves, stochastic phase space regions and the running diffusion coefficient. Performing the simulations first hand will add both to the pleasure and depth of understanding nonlinear dynamics. Here we show the results of such calculations carried out on mainframe supercomputers.

A typical calculation for the diffusion coefficient as a function of A (for $\beta = \omega/kc = 0$) is to take the initial ensemble Z_0 given by the set of points

$$\{X, P, t = 0\} = \left\{ \frac{i}{N},\ P_0\ \Big|\ i = 1, 2, \ldots N \right\} \tag{4.32}$$

that is a uniform distribution in phase $kx = \theta = 2\pi X$ at a given momentum P_0. For $N = 10^5$ and $P_0 = 0$ the diffusion coefficient after $T = 300$ (300 iteration of the standard map) is shown in Fig. 4.3. In Fig. 4.3 the diffusivity defined by Eq. (4.31) with $T = 300$ is plotted after being normalized to the quasilinear value $D_Q = A^2/4$. The quasilinear D is calculated from Eq. (4.31) using $\langle (P_i(t) - P_i(0))^2 \rangle = A^2 \langle (\sin\theta_n + \sin\theta_{n-1} \cdots + \sin\theta_1)^2 \rangle = A^2 n/2$ where cross-terms $\langle \sin\theta_n \sin\theta_{n-1} \rangle \simeq 0$ due to the weak correlation between θ_n and θ_{n-1} at large A.

The solid line in Fig. 4.3 is the theoretical $D(A)$ derived by Meiss et al. (1983) taking into account correlations between successive θ_n values dropped in the quasilinear approximation. Figure 4.3 shows both regions of good convergence to the diffusion approximation $D(A)$ and regions near $A \gtrsim 1, 2, 3, \ldots$ where the behavior is not diffusive. Examination of the orbits in these shaded intervals shows that the momentum is increasing linear with time rather than as $(Dt)^{1/2}$ for diffusive orbits. Thus, the shaded regions are the regions of the accelerator modes.

For period 1 accelerator modes we have

$$A \sin(2\pi X^{(a)}) = \ell \qquad \ell = 1, 2, 3 \ldots.$$
$$P_{n+1} = P_{n+\ell}. \tag{4.33}$$

From the tangent map in Eqs. (4.15) and (4.16) we have determined the eigenvalues and the stability of period-1 step-ℓ accelerator orbits

$$X_n = X^{(a)} \tag{4.34}$$

$$P_n = \ell n + P^{(a)}. \tag{4.35}$$

The accelerator orbits are stable for

$$-4 < 2\pi A \cos(2\pi X^{(a)}) < 0 \tag{4.36}$$

130   Chapter 4   Phase Space Structures in Hamiltonian Systems

where Eq. (4.33) determines

$$2\pi X^{(a)} = \sin^{-1}(\ell/A), \quad (4.37)$$

and thus Eq. (4.36) yields the stable domain

$$|\ell| < A < |\ell|\left(1 + \left(\frac{2}{\pi\ell}\right)^2\right)^{1/2} \quad (4.38)$$

for the period-1 accelerator orbits. These intervals are the shaded zones above $A = \ell$ in Fig. 4.3. At the upper limit of the stability domain in Eq. (4.38) the period-1 accelerator orbit bifurcates into the period-2 accelerator mode.

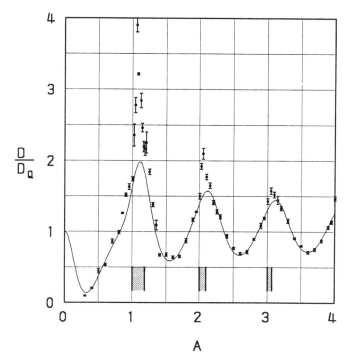

Fig. 4.3. The diffusion rate defined in Eq. (4.30) obtained from 10^5 particles started at $P = 0$ and uniformly distributed in x and advanced to $t = n = 300$. The diffusion D is normalized to the quasilinear value $D_{q\ell} = A^2/4$ that follows from the idealization of random, uncorrelated x_n values. By calculating the dominant correlations in the sequence x_n, x_{n-1}, x_{n-2} Meiss et al. (1989) obtained the formula (4.41) which is shown as the solid curve. The strongly correlated accelerator orbits occur just above $A = \ell = 1, 2, 3$ and violate the weak correlation approximation used to derive $D(A)$ and (4.41).

4.4.2 Evolution of the momentum distribution function

The mixed nature of the dynamics in the standard map that appears as the sum of a diffusive process given by $D(A)$ and the coherent acceleration of particles jumping by ℓ-steps in the unit of momentum every p time steps is clearly revealed by studying the time evolution of the momentum distribution function

$$F(P,t) = \int_0^1 dX\, f(X,P,t) = \frac{1}{N}\sum_{i=1}^{N}\delta(P - P_i(t)). \qquad (4.39)$$

In Figs. 4.4 and 4.5 we show at two times $t = 100$ and $t = 200$ the momentum distribution function $F(P,t)$ that evolves from the standard map taken with $N = 10^5$ particles with $P = P_0$ and uniformly distributed in X. In Fig. 4.4 where $A = 1.1$ and $P_0 = 0$, where there is a period-3 accelerator mode that causes intermittent transitions of the particle's motion between chaotic orbits and the accelerator orbits. In Fig. 4.5 where $P_0 = 1/2$ the period-3 accelerator mode is not present.

In Fig. 4.4 where $P_0 = 0$ the distributions at $t = 100$ and 200 show a diffusion component with

$$F(P,t) \simeq \frac{N}{(4\pi\, Dt)^{1/2}} \exp\left(-\frac{P^2}{4\, Dt}\right) \qquad (4.40)$$

with $D = 1.9$ which is shown as the solid line. In addition there is a substantial fraction of particles in a high momentum tail. These featues are also shown in Fig. 4.5 where we start the distribution with $F(X,P,0) = \delta(P - 1/4)$, that is $P_0 = 1/4$, rather than $P_0 = 0$. The period-3 accelerator orbits have $P_0 = 0$ so for the first case there are particles trapped in the period-3 step-3 invariant curves whereas in the second case this is not the case. Comparing Figs. 3 and 4 shows that this results in the presence or absence of a trapped, freely accelerating bunch of particles at the front $P = t$ of the momentum distribution. Of course, in a self-consistent field calculation the first case ($P_0 = 0$) with the positive slope, $\partial F/\partial P > 0$, at the front of the beam would be unstable to the emission of plasma waves.

4.4.3 Standard map at large values of A

There are systematic methods of calculating the correlations that give the deviation of the diffusivity from the quasilinear value. We do not present details here since they are given by Lichtenberg and Lieberman (1991). We note that procedure starts with writing Fourier integral representations for the probability densities

$$\delta(X_{n+1} - X_n - P_{n+1})\delta(P_{n+1} - P_n - A\sin(2\pi X_n))$$

which describe each step of the map. Once in Fourier integral form certain diagonal terms in the path integral over all possible trajectories are collected while off-diagonal term are dropped. We note that this Fourier path integral method was originated by Fenyman and used early on by Varma (1971, 1972) to derive formulas for the probability of a charged particle to escape from magnetic mirrors. The Varma (1972) works contain an interesting history of the theory and experiments of the fundamental problem of magnetic moment conservation. Work on this problem appears to be the main stimulus for the independent derivations of the standard map.

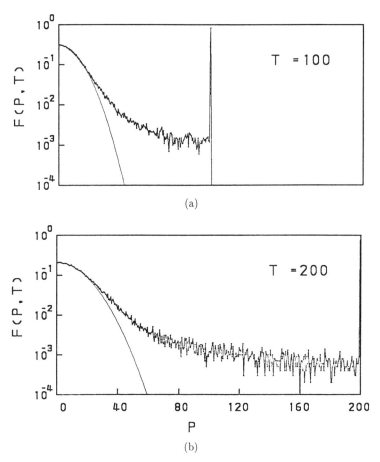

Fig. 4.4. Time evolution of the momentum distribution function $F(P,t)$ for $A = 1.10$. The distribution is computed from the initial distribution of 10^5 particles at $P = 0$ uniformly distributed in x. There a group of coherently accelerated particles show up as the spike at the leading front of F given by $P_{\max} = t$. (a) The momentum distribution at $t = 100$ and (b) the distribution at $t = 200$. The parabolic curve is the momentum distribution computed for Eq. (4.40) with $D = 1.9$.

4.4 Stochastic Motion and the Diffusion Coefficient 133

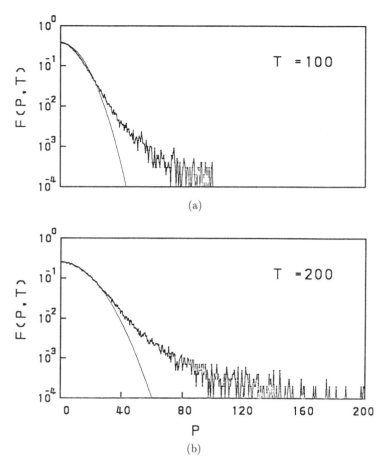

Fig. 4.5. The evolution of the momentum distribution function $F(P,t)$ for $A = 1.10$ as in Fig. 4.4 but starting the particles at $F(P, t = 0) = 10^5 \delta(P - 1/4)$. The P-distribution contains the front with $P_{\max} \simeq t$ but there is no trapped particle group at $P_{\max}$ as in Fig. 4.4. The fundamental (period-1 step-1) accelerator mode for $A = 1.10$ is at $X = 0.3183, P = 0.0$.

Using the Fenyman path integral method Meiss *et al.* (1983) sum certain diagonal contributions to all orders to obtain the diffusivity

$$D(A) = D_{QL} \left[\frac{1 - 2J_1^2(2\pi A) - J_2^2(2\pi A) + 2J_3^2(2\pi A)}{(1 + J_2(2\pi A))^2} \right] \quad (4.41)$$

where $J_n(X)$ is the nth order Bessel function. The solid line in Fig. 4.3 is the function

$D(A)/D_{QL}$ given in Eq. (4.41). The formula (4.41) is a remarkable prediction for the measured diffusivity. The only appreciable deviations are in the shaded regions given by Eq. (4.41) for the zone of stable period-1 accelerator modes and in the low A region $A < 0.4$ where the integrable curves emerge to block global diffusion. For $A \leq A_c$ the closed integrable curve appears across $P \approx 1/\gamma$ region rigorously bounding $(P_n - P_0)^2$ so that $D \equiv 0$.

4.4.4 Standard map at small values of A

The results for the simulations in the region $A < 1$ are given in Fig. 4.6 and also compared with Eq. (4.41). In this region there are no period 1 accelerator modes, but there are multi-period ($p > 1$) accelerator modes and they account for the enhanced D/D_Q shown in Fig. 4.6. The enhancements may be described as resonant structures in $D(A)$ for A passing through the regions where period p step ℓ modes exist according to

$$P_{n+p} - P_n = \ell \tag{4.42}$$

$$A \sum_{i=1}^{p} \sin(2\pi X_i) = \ell. \tag{4.43}$$

The phase space of the $p = 3$ accelerator mode for $A = 0.6144$ ($K = 3.860$) is shown in Fig. 4.1. The location $A, (X, P)$ and the nature (sharp or broad resonance) of these multi-period accelerator mode is given in Ichikawa et al. (1987).

4.4.5 Discussion of physics for the diffusion in the standard map

To understand physically how diffusion arises in the standard map we consider the motion of a particle in an infinite spectrum of waves with phase velocities evenly separated by the fundamental wave speed $v_{ph} = \omega/k$. In this case there are series of waves $E_0 \sin(kx - n\omega t)$ with essentially the same amplitude available to accelerate the particle in the vicinity of each of the velocities $v_n = n(\omega/k)$.

Consider the case of a charged particle in a spectrum of longitudinal waves. Now the behavior of the particle in the waves depends on its velocity relative to the neighboring waves $v_n < v < v_{n+1}$ and are the size of the trapping velocity $v_{tr} = (eE_0/mk)^{1/2} = (e\phi_0/m)^{1/2}$ compared with the separation of the phase velocities $v_{n+1} - v_n = \omega/k$. The details are made clear in Example 1.3. When v_{tr}/v_{ph} is small the particles that are trapped remain in the closest resonant wave defined by $\min|v - n\omega/k|$. On the other hand, when the amplitude of the trapping velocity is comparable to or larger than the separation of the phase velocities the particle easily accelerates from one resonant wave $n\omega/k$ to the neighboring resonant waves

4.4 Stochastic Motion and the Diffusion Coefficient

at $(n \pm 1)(\omega/k)$. Typically, the wave-particle phase $\varphi = kx - n\omega t$ at which the acceleration or deceleration to $(n \pm 1)(\omega/k)$ occurs is more or less evenly distributed over the full phase range of 2π. The net result over a number of these accelerations and decelerations is to produce a diffusion in the velocity of the particle described by $v^2 \to 2D(A)t$ for the (integer) time values such that $t \gg 1/k\Delta v$.

The analysis of the standard map in Sec. 4.3 and 4.4 shows there is an important, and generic, exception to the diffusion orbits that are called accelerator orbits. In the accelerator orbits the particles with certain initial data (X, P) are coherently accelerated to neighboring waves time after time. Thus, these orbits are called accelerator orbits, and the fundamental accelerator mode occurs only when the trapping velocity v_{tr} exceeds the phase velocity $v_{ph}(A > 1)$ and the velocity jumps with each interaction by v_{tr}. Higher order accelerator orbits can occur even at smaller wave amplitudes such as the fourth order accelerator orbit at $A = 3/(2\pi)$, where four iterations are required for the orbit to jump by v_{tr}. The analysis in Sec. 4.4 shows that diffusion behavior $D(A)$ in velocity and the accelerator modes dominate the dynamics in certain intervals of A.

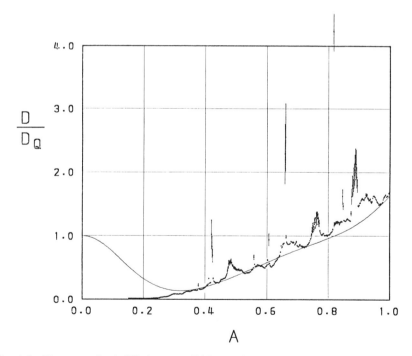

Fig. 4.6. The normalized diffusion rate D/D_Q in the region $A_c < A < 1$ for $N = 10^5$ particles uniformly distributed over X at $P_0 = 0.0$.

4.5 Regular Motion in the Relativistic Standard Map

In order to characterize the features of the dynamics predicted by the relativistic standard map, let us begin our analysis with the results of the numerical observations of the particle motions in the phase space. Since it is known that the standard map (4.11) and (4.12) exhibits the global stochasticity above the threshold value of $K_c = 0.97163$, we choose the nonlinear parameter $K = 1.30$, where the chaotic motion prevails over the regular motion. Equation (4.21) with $\beta = 0$ tells us that for $K = 1.30$ the period-6 and higher periodic Poincaré-Birkhoff (P.B.) islands bifurcate at the origin, but the period-5 P-B islands are not yet borne. In Fig. 4.7(a)–(d), we show traces for 5000 iterations ($t = 5000$) of the orbits of 50 particles, which are initially distributed uniformly over the range of $-0.5 < X \leq +0.5$ with $P = 0$. Increasing the value of β, we observe that the relativistic effect suppresses the deterministic diffusion drastically. In particular, we notice that at $\beta = 0.1\pi$, the KAM surface is formed between the $m = 2$ fixed point island and the $m = 1$ island, even though Eq. (4.22) allows the stable $m = 3$ and $m = 2$ fixed points. At the value of $\beta = 0.2\pi$, the KAM surface is squeezed down to the place between the $m = 2$ island and the $m = 1$ island.

Now, increasing the value of β above unity as shown in Fig. 4.8(a)–(d), we observe that when the relativistic effect dominates, the map exhibits a qualitatively different particle dynamics. At the larger value of β, the outermost KAM surface expands into the higher momentum region and coherent island structures prevail inside the region. In particular, one is able to count the period-20 islands in Fig. 4.8(c) for $\beta = 4\pi$, and as high as the period-54 in Fig. 4.8(d) for $\beta = 10\pi$. Nomura et al. (1992) show that it is straightforward to find an analytic approximation for the observed phase space structure. Firstly, we give the shape of the outermost KAM surface by assuming $|X_{n+1} - X_n| \ll X_n$ and $|P_{n+1} - P_n| \ll P_n$ for large n. Equation (4.2) can be reduced to

$$\frac{dP}{dX} = \frac{F(X)}{G(P)} \tag{4.44}$$

which, for the relativistic standard map (4.9) and (4.10), can be integrated to find that

$$\sqrt{1 + \beta^2 P^2} - \frac{K}{4\pi^2} \beta^2 \cos(2\pi X) = \text{const.} \tag{4.45}$$

Thus, for the initial condition of Fig. 4.8 given as $(X_0, P_0) = (1/2, 0)$, we get

$$P(X) = \frac{1}{\pi} \left[K \cos^2(\pi X) + \frac{K^2 \beta^2}{4\pi^2} \cos^4(\pi X) \right]^{1/2}. \tag{4.46}$$

4.5 Regular Motion in the Relativistic Standard Map   137

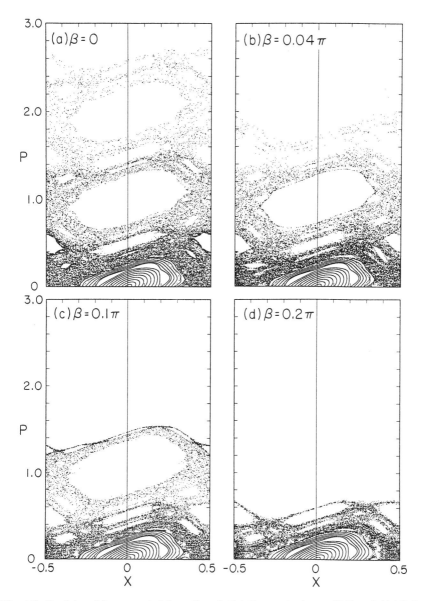

Fig. 4.7. Particle orbits computed from the relativistic standard map (4.9) and (4.10) for the stochastic parameter $K = 1.3$ and the relativistic parameter $\beta < 1$. Here (a) $\beta = 0$, (b) $\beta = 0.04\pi$, (c) $\beta = 0.1\pi$ and (d) $\beta = 0.2\pi$, respectively.

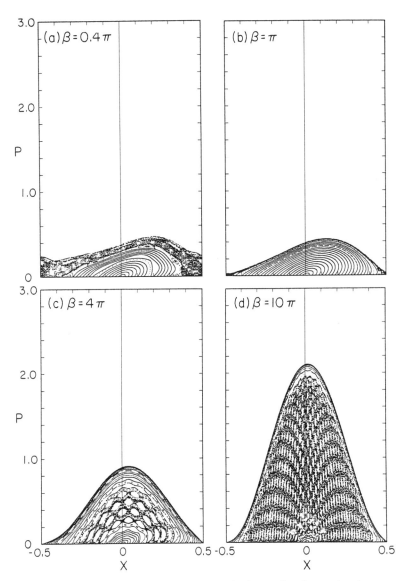

Fig. 4.8. Particle orbits of the relativistic standard map for the stochastic parameter $K = 1.3$ and the relativistic parameter $\beta > 1$. Here (a) $\beta = 0.4\pi$, (b) $\beta = \pi$, (c) $\beta = 4\pi$ and (d) $\beta = 10\pi$, respectively.

4.5 Regular Motion in the Relativistic Standard Map

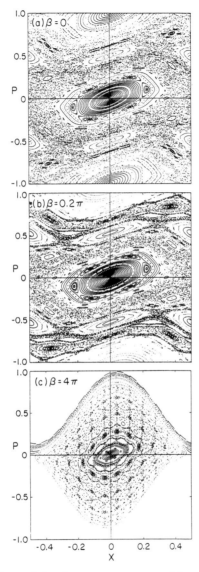

Fig. 4.9. The β-dependence of the phase space structure of the relativistic standard map with $K = 1.3$ for (a) $\beta = 0$, (b) $\beta = 0.2\pi$, and (c) $\beta = 4\pi$, respectively.

The maximum attainable momentum $P_{\max}$ is estimated by the condition $dP/dX = 0$ as

$$P_{\max} = \frac{1}{\pi}\left(K + \frac{K^2}{4\pi^2}\beta^2\right)^{1/2}. \qquad (4.47)$$

We confirmed that Eqs. (4.46) and (4.47) are in good agreement with the observed results given in Fig. 4.8.

Next, we consider the periodic islands. According to Eq. (4.21), we expect to have the period-6 islands around the origin for $K = 1.3$. In the high momentum region far out from the origin, however, we notice that the phase advancement $G(P,\beta)$ in Eq. (4.10) for one step of the iteration of map is approximately

$$\Delta X \simeq \frac{1}{\beta}\left(1 - \frac{1}{2}(\beta P)^{-3} + \cdots\right). \qquad (4.48)$$

Hence, the maximum periodicity N will be determined by the condition

$$N\,\Delta X \simeq 2, \qquad (4.49)$$

which gives $N \approx 2\beta$, namely $N = 24$ for $\beta = 4\pi$ and $N = 62$ for $\beta = 10\pi$. These maximum periodicity numbers are consistent with the observations.

Lastly, we show in Fig. 4.9 the details of the island structure for the values of $\beta = 0$, $\beta = 0.2\pi$ and $\beta = 4\pi$. Here, unlike Fig. 4.7 and Fig. 4.8, we put the particles at the places where the periodic islands exist. As for Fig. 4.9(a) and (b), we set the particles along the X-axis and also along the curves of $X = G(P)$ over the range of $0 < P < 1.0$. While for Fig. 4.9(c), the particles are distributed only along the curve of $2X = G(P)$ over the range of $0 < P < 1.0$, resulting in the empty period-8 islands around the origin. Figure 4.9(b) shows that the $m = 1$ fixed point is indeed stable, though it cannot be reached from the initial positions of $P = 0$. Furthermore, we call attention to the fact that the period-8 islands are stable on the X-axis ($P = 0$) at the value of $\beta = 0.2\pi$, while for the larger value of $\beta = 4\pi$, the period-8 orbits on the X-axis ($P = 0$) are unstable. This interchange of stable and unstable orbits in the phase space is analyzed in Sec. 4.8.

4.6 Symmetries of the Relativistic Standard Map

Although we have given a reliable estimate in Eqs. (4.49)–(4.50) for the multiplicity of the periodic orbits in terms of the constant phase advancement in the ultra-relativistic case, we wish to use the RSM as an example of how a theory for the overall structure of the regular motion of particles can be developed. For this purpose, the analysis of the symmetries of the mappings provides the key instrument. Greene *et al.* (1981) have discussed the global behavior of the area-preserving map

4.6 Symmetries of the Relativistic Standard Map

on the basis of its symmetries. Pina and Lara (1987) carried out an explicit analysis of the symmetries of the classical standard map. Introducing the symmetries with respect to the space inversion and to momentum inversion, Ichikawa et al. (1989) developed an extensive analysis of the regular motion of the classical standard map.

A map T is called reversible if there exists an involution I_0 which satisfies the relation

$$T \cdot I_0 \cdot T = I_0 \quad \text{and} \quad I_0 \cdot I_0 = I_d \tag{4.50}$$

where I_d stands for the unity matrix. This relation indicates that the reversible map can be expressed as the product of two involutions

$$T = I_1 \cdot I_0, \quad I_1 \cdot I_1 = I_d \tag{4.51}$$

with the definition of

$$I_1 = T \cdot I_0 \tag{4.52}$$

and the inverse transformation T^{-1} is given by

$$T^{-1} = I_0 \cdot I_1. \tag{4.53}$$

If we define I_j as the j-th iteration of the map T on the involution I_0, $I_j = T^j \cdot I_0$, we confirm that I_j is also an involution. The ensemble of I_j and T^k for arbitrary integers j and k forms a discrete infinite group with the relationships

$$I_j \cdot I_k = T^{j-k}, \quad T^j \cdot I_k = I_{j+k}, \quad I_j \cdot T^k = T^{j-k}. \tag{4.54}$$

It can be shown that the fixed points of the involution I_j form a curve Γ_j which is known as its symmetry line:

$$\Gamma_j: \quad \left\{ \mathbb{R} \middle| \ I_j \cdot \mathbb{R} = \mathbb{R} \right\}. \tag{4.55}$$

Therefore, the first equation of (4.54) states that the intersections of Γ_j and Γ_k determine the periodic points of T, of which period N divides $|j-k|$. From the second and the third relations of Eq. (4.55), we can deduce that the symmetry lines Γ_j are transformed by T^n into other symmetry lines according to the relation

$$\Gamma_{2N+j} = T^N \Gamma_j, \tag{4.56}$$

which enables us to construct the family of symmetry lines of arbitrary order.

For a generic form of the two-dimensional map as given in Eq. (4.2), if the transformation function of $F(X)$ is antisymmetric with respect to space inversion, $F(-X) = -F(X)$, the map is expressible as the composition of the following two involutions:

$$I_0: \quad P' = P + F(X), \quad X' = -X$$
$$I_1: \quad P' = P, \quad\quad\quad X' = -X + G(P)$$
(4.57)

The symmetry lines of these two fundamental involutions are given by

$$\Gamma_0: \quad X = 0 \quad\quad \Gamma_1: \quad 2X - G(P) = 0. \quad\quad (4.58)$$

Writing an expression for the symmetry line for I_j symbolically as $\Gamma_j(\mathbb{R}) = 0$, we have

$$\Gamma_j(\mathbb{R}) = \Gamma_{j-2}(T^{-1}\mathbb{R}) = 0$$
$$\Gamma_j(\mathbb{R}) = \Gamma_{j+2}(T\mathbb{R}) = 0$$
(4.59)

which provides us with the following recurrence formulas

$$\Gamma_j[X, P] = \Gamma_{j-2}[X - G(P), \; P - F\{X - G(P)\}]$$
$$\Gamma_j[X, P] = \Gamma_{j+2}[X + G\{P + F(X)\}, P + F(X)]$$
(4.60)

Explicitly, we have

$$\Gamma_2: \quad X - G(P) = 0$$

$$\Gamma_3: \quad 2X - 2G(P) - G[P - F\{X - G(P)\}] = 0 \quad\quad (4.61)$$

$$\Gamma_4: \quad X - 2G(P) - G[P - F\{X - G(P)\}] = 0$$

and

$$\Gamma_{-1}: \quad 2X + G[P + F(X)] = 0$$

$$\Gamma_{-2}: \quad X + G[P + F(X)] = 0 \quad\quad (4.62)$$

$$\Gamma_{-3}: \quad 2X + 2G[P + F(X)] + G[P + F(X)$$
$$\quad\quad + F\{X + G(P + F(X))\}] = 0.$$

In the previous section, we have made use of the Γ_2 symmetry line to construct Figs. 4.9(a) and (b), and the Γ_1 symmetry line to construct Fig. 4.9(c), so that we have been able to reproduce all aspects of the periodic island structure.

4.6 Symmetries of the Relativistic Standard Map

A factorization of the map into two involutions is not unique. Antisymmetry of the function $G(P)$ with respect to momentum inversion, $G(-P) = -G(P)$, gives rise to another involution decomposition $T = J_1 \cdot J_0$, with

$$J_0: \quad P' = -P, \; X' = X - G(P)$$

$$J_1: \quad P' = -P + F[X - G(P)] \tag{4.63}$$

$$X' = X - G(P) - G[P - F\{X - G(P)\}].$$

This factorization defines the momentum inversion symmetry as

$$\gamma_0: \quad P = 0$$
$$\gamma_1: \quad 2P - F[X - G(P)] = 0 \tag{4.64}$$

Since the same recurrence formula as Eqs. (4.63) are also valid for this momentum inversion symmetry, it is straightforward to write down the higher order symmetry lines as

$$\gamma_2: \quad P - F[X - G(P)] = 0$$

$$\gamma_3: \quad 2P - 2F[X - G(P)] = 0 \tag{4.65}$$

$$\gamma_4: \quad P - F[X - G(P)] - F[X - G(P)$$
$$-G\{P - F(X - G(P))\}] = 0$$

and

$$\gamma_{-1}: \quad 2P + F(X) = 0$$

$$\gamma_{-2}: \quad P + F(X) = 0$$

$$\gamma_{-3}: \quad 2P + 2F(X) + F[X + G\{P + F(X)\}] = 0. \tag{4.66}$$

Now, for the relativistic standard map, we show in Fig. 4.10 a family of the spatial inversion symmetry lines up to the 12-th order for the values of $K = 1.3$ and $\beta = 4\pi$. It is important to notice that in the region of $|P| \gtrsim \beta^{-1}$, the symmetry lines Γ_j become parallel with the constant phase separation of $(2\beta)^{-1}$. Thus we confirm from the symmetry analysis that the phase increment at each step of the mapping is indeed given by β^{-1} in accord with Eq. (4.49). Figure 4.11 illustrates a family of the momentum inversion symmetry lines up to the 12-th order. We observe that the higher order momentum inversion symmetry lines γ_j approach asymptotically

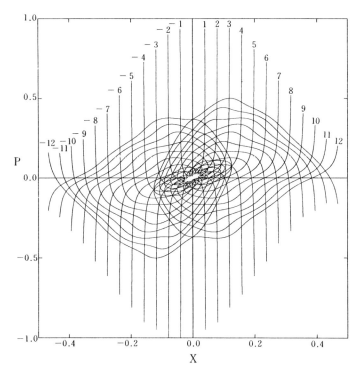

Fig. 4.10. Space inversion symmetry lines Γ_j of the relativistic standard map for $K = 1.3$ and $\beta = 4\pi$.

to the separatrix KAM surface. Thus, from the symmetry properties of the map, we are able to give a precise method for calculating the spacing of the islands and the limiting KAM surface given in Sec. 4.6.

To conclude the present section we illustrate, with Figs. 4.12(a) and (b), the results of a superposition of Fig. 4.9(c) with Fig. 4.10 and Fig. 4.11, respectively, that we can classify the even-number periodic islands into two groups, (i) $N = 2 \cdot (2\ell)$ and (ii) $N = 2 \cdot (2\ell + 1)$ with integers ℓ. For the first group, intersections of the odd-number symmetry lines Γ_{2j+1} determine the hyperbolic points and intersections of the even-number symmetry lines Γ_{2j} determine the elliptic points. For the second group, intersections of the odd-number symmetry lines Γ_{2j+1} determines the elliptic points and intersections of the even-number symmetry lines Γ_{2j} determine the hyperbolic points. As for the odd-number periodic islands, we remark that intersections of the symmetry lines Γ_j determine only the unstable odd-periodic islands. Contrary to the spatial inversion symmetry, Fig. 4.12(b) shows that intersections of the odd-number symmetry lines γ_{2j+1} determine all the elliptic even-periodic

4.7 Stability of the Periodic Orbits 145

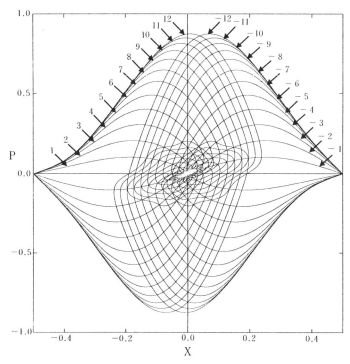

Fig. 4.11. Momentum inversion symmetry lines γ_j of the relativistic standard map for $K = 1.3$ and $\beta = 4\pi$.

orbits, while intersections of the even-number symmetry lines γ_{2j} determine all the hyperbolic even-periodic orbits. As for the odd-periodic islands, intersections of the symmetry lines γ_j determine all the stable odd-periodic islands.

4.7 Stability of the Periodic Orbits

In Sec. 4.6 we showed that the positions of the stable and unstable period-8 orbits are interchanged with increasing the value of β. Here this interchange of the stability of periodic orbits with respect to variation of β is investigated for several choices of the stochasticity parameter K. Figure 4.13 illustrates the features of this phenomenon. Keeping our attention on the periodic orbits on the X-axis ($P = 0$), we see that the period-4 orbit with $X > 0$ is stable up to a value around $\beta \approx 2.3$, then it turns unstable. At the same value the unstable period-3 orbit on the X-axis becomes stable. We also observe the occurrence of a stable period-6 orbit. Increasing β above 3.5, we see that the stable period-6 orbit is turning unstable

146   Chapter 4   Phase Space Structures in Hamiltonian Systems

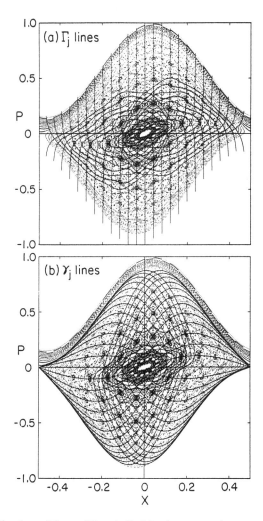

Fig. 4.12. Identification of the multi-periodic island structure in terms of families of the space inversion symmetry curves Γ_j in (a), and the momentum inversion symmetry curves γ_j in (b).

while the period-8 hyperbolic point changes into an elliptic point. Figure 4.14(a) and (b) show such an interchange of stability for the period-3 and period-4 orbits at $K = 3.3$ for the values of $\beta = 2.20$ and $\beta = 2.83$. In particular, we notice that Fig. 4.14(b) indicates the appearance of two sets of period-3 orbits very close to the origin. Increasing the stochasticity parameter K to 6.4717, we find a similar

4.7 Stability of the Periodic Orbits

phenomena shown in Figs. 4.15(a) and (b), where the stable fixed point at the origin has bifurcated into the period-2 orbit. We observe, however, that the stability interchange of the period-3 and period-4 orbits takes place somewhere in the interval $2.20 < \beta < 2.51$.

In order to calculate this interchange phenomena, we examine the local stability of the periodic orbits in some detail. The objective is to determine the critical values of β where the stability interchange occurs. Since a period-n orbit is a fixed point of the T^n map, its local stability is determined by the eigenvalue of a matrix L obtained by linearizing T^n about one of its fixed points. The stability of the fixed point (or orbit) is determined by the residue R at the fixed point, defined by

$$R \equiv \frac{1}{4}[2 - Tr(L)]. \quad (4.67)$$

The orbit is stable for $0 < R < 1$ and called elliptic. If $R < 0$, the orbit is directly unstable and called hyperbolic without reflection. For $R > 1$, the orbit is inversely unstable, and called hyperbolic with reflection.

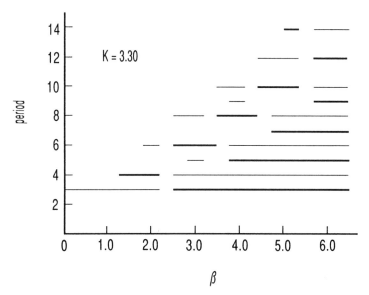

Fig. 4.13. Numerically observed feature of interchange of the stability of multi-periodic orbits with respect to the variation of β, for the value of $K = 3.30$. The heavy line indicates the region where the specified periodic orbit on the X-axis remains stable, while the thin line indicates an unstable region. Since the parameter β is varied by the amount of $\Delta\beta = 0.25$, the precise critical points are not identified here. The critical points are determined in the stability analysis shown in Figs. 4.16 and 4.17.

148 Chapter 4 Phase Space Structures in Hamiltonian Systems

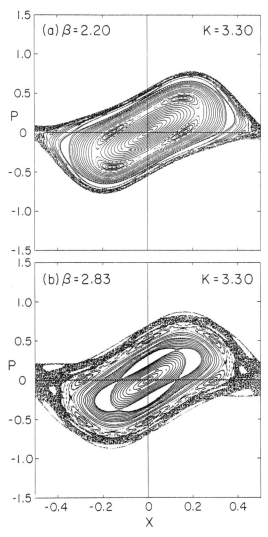

Fig. 4.14. Interchange of the stability of period-4 orbits between (a) $\beta = 2.20$ and (b) $\beta = 2.83$ at $K = 3.3$. Careful observation can identify the two sets of period-3 orbits for the case (b).

4.7 Stability of the Periodic Orbits 149

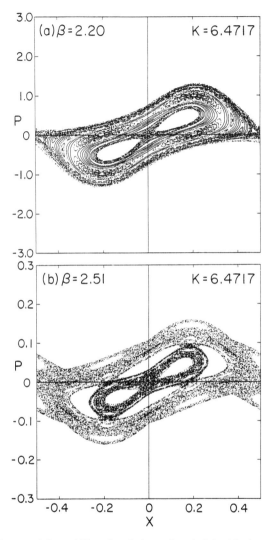

Fig. 4.15. Interchange of the stability of period-3 and period-4 orbits between (a) $\beta = 2.20$ and (b) $\beta = 2.51$ at $K = 6.4717$, where the fundamental fixed point at the origin bifurcates into the period-2 orbits.

150   Chapter 4   Phase Space Structures in Hamiltonian Systems

We find that it is tractable to develop the stability analysis for the period-3 and period-4 orbits that multifurcate from the origin. For period-3 orbits, there are two groups of orbits. The location of these orbits is determined by the symmetry analysis discussed in the preceding section. Firstly, a periodic orbit with one point on the X-axis can be determined by the intersection of the momentum inversion symmetry lines, γ_0 and γ_3 from Eqs. (4.64) and (4.65):

$$P_0 = 0, \quad 2F(X_0) + F[X_0 + G(F(X_0))] = 0. \tag{4.68}$$

After a lengthy but straightforward calculation we find that the residue of this orbit is

$$R^{(3)}(X_0, P_0 = 0) = -\frac{1}{4}[2F'(X_0)$$

$$+F'\{X_0 + G(F(X_0))\}\{1 + G'(F(X_0))F'(X_0)\}]$$

$$\times [2G'(F(X_0)) + G'(0)\{1 + G'(F(X_0))F'(X_0)\}]. \tag{4.69}$$

The other set of the period-3 orbits, with one point on the P-axis ($X_0 = 0$), is determined by the intersection of the space inversion symmetry lines, Γ_0 and Γ_3, from Eqs. (4.58) and (4.61),

$$X_0 = 0, \quad 2G(P_0) + G[P_0 + F(G(P_0))] = 0. \tag{4.70}$$

For this orbit, the residue is calculated as

$$R^{(3)}(X_0 = 0, P_0) = -\frac{1}{4}[2G'(P_0)$$

$$+G'\{P_0 + F(G(P_0))\}\{1 + F'(G(P_0))G'(P_0)\}]$$

$$\times [2F'(G(P_0)) + F'(0)\{1 + F'(G(P_0))G'(P_0)\}]. \tag{4.71}$$

Similarly, the location of the period-4 orbit, whose one point lies on the X-axis ($P = 0$), is specified by the intersection of the symmetry lines γ_0 and γ_4.

$$P_0 = 0, \quad F[X_0 + F(X_0 + G(X_0))] = 0 \tag{4.72}$$

from Eqs. (4.64) and (4.65). For the transformation function of $F(x) = -(K/2\pi)\sin(2\pi x)$, Eq. (4.72) gives rise to the following two solutions:

$$(a) \quad 2X_0^{(a)} + G(F(X_0^{(a)})) = 0, \tag{4.73}$$

and

$$(b) \quad G(F(X_0^{(b)})) = m + \frac{1}{2} \quad (m: \text{integer}). \tag{4.74}$$

4.7 Stability of the Periodic Orbits

When the nonlinear parameter K satisfies the condition

$$K > 2, \tag{4.75}$$

the a-branch period-4 orbits exist at $[X_0^{(a)}, P_0 = 0], 6[-X_0^{(a)}, F(X_0^{(a)})], [-X_0^{(a)}, P_0 = 0]$ and $[X_0^{(a)}, -F(X_0^{(a)})]$, which are mirror symmetric with respect to the origin. The residue of this period-4 orbit is calculated as

$$R^{(4)}(X_0^{(a)}, P_0 = 0) = -\frac{1}{4} F'(X_0^{(a)})$$

$$\times [2 + G'(F(X_0^{(a)}))][2 + G'(0)F'(X_0^{(a)})]$$

$$\times [2G'(X_0^{(a)})) + G'(0)\{2 + G'(F(X_0^{(a)}))F'(X_0^{(a)})\}]. \tag{4.76}$$

Equation (4.74) gives rise to the b-branch period-4 orbits at the positions of $[X_0^{(b)}, P_0 = 0]$, $[X_0^{(b)} + 1/2, F(X_0^{(b)})]$, $[X_0^{(b)} + 1/2, P_0 = 0]$ and $[X_0^{(b)}, -F(X_0^{(b)})]$, provided that the nonlinear parameter K satisfies the condition

$$K > 2\pi \left[(m+1/2)^{-2} - \beta^2 \right]^{-1/2}. \tag{4.77}$$

These orbits are mirror symmetric with respect to the point $\left(X_0^{(b)} + 1/4, P_0 = 0\right)$. Since the relativistic standard map is symmetric with respect to the origin, we have another set of the b-branch period-4 orbits at the positions of $[-X_0^{(b)}, P_0 = 0]$, $[X_0^{(b)} - 1/2, F(+X_0^{(b)})]$, $[-X_0^{(b)} - 1/2, P_0 = 0]$ and $[-X_0^{(b)} + F(X_0^{(b)})]$, respectively. The residue of the pairs of these period-4 orbits is calculated as

$$R^{(4)}(X_0^{(b)}, P_0 = 0) = -\frac{1}{4} G'(F(X_0^{(b)}))F'(X_0^{(b)})^2$$

$$\times [G'(F(X_0^{(b)}))G'(0)^2 F'(X_0^{(b)})^2$$

$$-4\{G'(0) + G'(F(X_0^{(b)}))\}]. \tag{4.78}$$

In addition to these sets of the period-4 orbit, we also have the period-4 orbit whose one point lies on the P-axis ($X = 0$). The location of this orbit is specified by the intersection of the space inversion symmetry lines, Γ_0 and Γ_4 (Eqs. (4.58) and (4.61))

$$X_0 = 0, \quad G(P_0) + G[P_0 + F(G(P_0))] = 0. \tag{4.79}$$

For the relativistic standard map $F(x)$, Eq. (4.79) reduces to

$$2P_0 + F(G(P_0)) = 0. \tag{4.80}$$

152   Chapter 4 Phase Space Structures in Hamiltonian Systems

The residue of this orbit is calculated as

$$R^{(4)}(X_0 = 0, P_0) - \frac{1}{4} G'(P_0)[2 + F'(G(P_0))][2 + F'(0)G'(P_0)]$$

$$\times [2F'(G(P_0)) + F'(0)\{2 + F'(G(P_0))G'(P_0)\}]. \qquad (4.81)$$

Having obtained analytic expressions for the residues of the period-3 orbits and the period-4 orbits, we can investigate the β-dependence of the residues by substituting values of the coordinates numerically determined for specified values of β. We show in Figs. 4.16 and 4.17 the β-dependence of the residue of the period-3 and the period-4 orbits at $K = 3.3$, respectively. We observe in Fig. 4.16 that the residues $R^{(3)}(X_0, P_0 = 0)$ and $R^{(3)}(X_0 = 0, P_0)$ of the period-3 orbits become zero at the value of $\beta = 2.40$, where the stability interchange takes place. As for the period-4 orbits, Fig. 4.17 indicates that all of them are unstable at $\beta = 0$. As β increases the b-branch turns stable and merges into the a-branch at $\beta = 0.612$. At the value of $\beta = 2.44$, the residues $R^{(4)}(X_0^{(a)}, P_0 = 0)$ and $R^{(4)}(X_0 = 0, P_0)$ of the period-4 orbits change their signs, leading to the interchange of their stability.

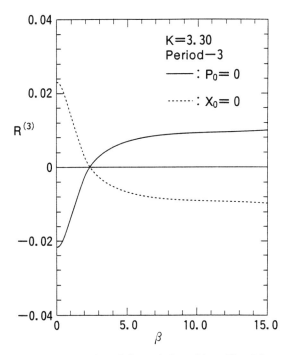

Fig. 4.16. Residue of the period-3 orbit at $K = 3.3$.

4.7 Stability of the Periodic Orbits

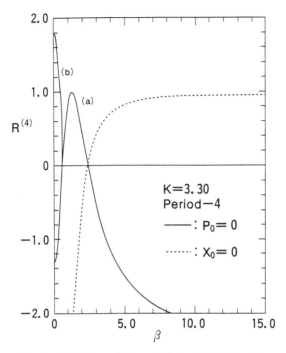

Fig. 4.17. Residue of the period-4 orbit at $K = 3.3$.

For the higher value of $K = 6.4717$, we show in Fig. 4.18 and 4.19 the β-dependence of the residue of the period-3 and the period-4 orbits. In contrast to the case of $K = 3.3$, the stable period-3 orbits survive only in a very narrow range of β. Consistent with the observed change in stability shown in Figs. 4.15(a) with $\beta = 2.20$ and (b) with $\beta = 2.51$, the stability interchange of the period-3 orbits takes place at $\beta = 2.34$. The residues of the period-4 orbits behave in a very complicated manner. At $\beta = 0$, every branch of the period-4 orbits is directly unstable. As β increases, the b-branch undergoes rapid change from stable to inversely unstable and back to stable, then merging into the a-branch at $\beta = 1.749$. The residues $R^{(4)}(X_0^{(a)}, P_0 = 0)$ and $R^{(4)}(X_0 = 0, P_0)$ at $K = 6.4717$ change their signs at the value of $\beta = 2.28$, which confirms the stability interchange phenomena found numerically in Fig. 4.15(a) and (b). Here, we notice that the residue $R^{(4)}(X_0 = 0, P_0)$ is increasing from zero to unity and then falls below unity again. This peculiar variation of the residue of the period-4 orbits will be discussed further in the next section.

154   Chapter 4   Phase Space Structures in Hamiltonian Systems

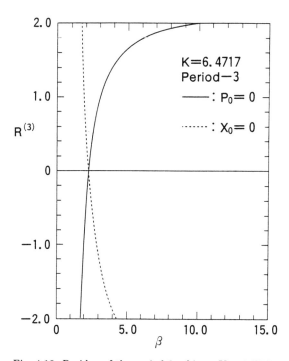

Fig. 4.18. Residue of the period-3 orbit at $K = 6.4717$.

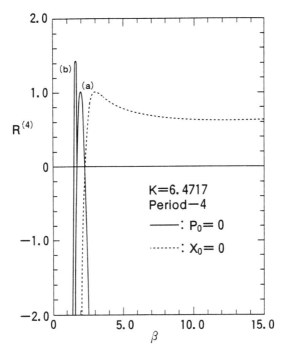

Fig. 4.19. Residue of the period-4 orbit at $K = 6.4717$.

4.8 Poincaré-Birkhoff Multifurcation for the Period-4 Orbit

In the preceding section, we showed that the residue of the period-4 orbit increases to unity and then falls again as β increases. We should be aware that stable orbits with the same value of residue but with different values of β exhibit distinct behaviors. Referring to Figs. 4.17 and 4.19, we observe that for the same value of $R^{(4)} = 0$ the merging of the b-branch into the a-branch takes place at a smaller value of β, and the stability interchange occurs at a larger value of β. In Fig. 4.20, a phase portrait around the period-4 orbits on the X-axis is shown for $K = 3.3$ and (a) $\beta = 0.933$ and (b) $\beta = 1.57$, respectively. For both cases, the residue $R^{(4)}$ is equal to ~ 0.75. The period-4 orbits on the X-axis $(X_0, P_0 = 0)$ and $(-X_0, P_0 = 0)$ are surrounded by the Poincaré-Birkhoff multifurcated period-3 orbits. In Fig. 4.20(a), we can recognize that here occurs the period-3 catastrophe, where the period-4 islands are squeezed into points by the period-3 orbits.

On the other hand, in Fig. 4.20(b), we see the period-3 multifurcation around the original period-4 orbits.

Now, at the value of $\beta = 1.26$, the residue $R^{(4)}(X_0, P_0 = 0)$ becomes unity, suggesting the occurrence of the Poincaré-Birkhoff multifurcation. Figure 4.21 shows that two pairs of the period-2 orbits are multifurcated at $\beta = 1.26$ for $K = 3.3$. Following Greene et al. (1981) let us examine this feature of the pair of period-2 orbits in some detail. Since the a-branch period-4 orbits are symmetric with respect to the origin, the points $(X_0, P_0 = 0)$ and $(-X_0, P_0 = 0)$ are regarded as the same point, and the Poincaré-Birkhoff period-4 sequence has a square root map. A study

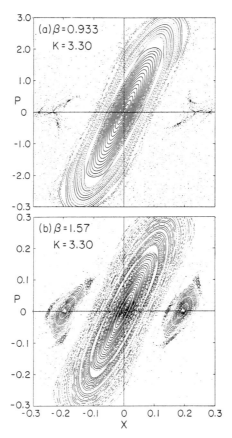

Fig. 4.20. Multifurcation of period-3 orbits out of the period-4 orbits at $K = 3.3$. (a) period-3 catastrophe at $\beta = 0.933$ and (b) two pairs of period-3 islands at $\beta = 1.57$.

4.8 Poincaré-Birkhoff Multifurcation for the Period-4 Orbit

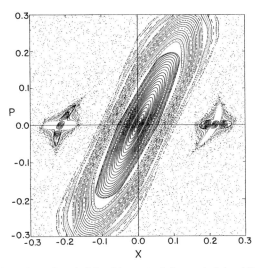

Fig. 4.21. Multifurcation of period-2 orbits out of the period-4 orbits at $K = 3.3$ and $\beta = 1.26$.

of the residue $R^{(2)}$ of the T^2 map gives

$$R^{(2)}(X_0, P_0 = 0) = R^{(2)}(-X_0, P_0 = 0)$$

$$= -\frac{1}{4} F'(X_0)[2G'(0) + 2G'(F(X_0))$$

$$+ G'(0)G'(F(X_0))F'(X_0)]. \tag{4.82}$$

Thus, we can confirm that the residue $R^{(4)}(X_0^{(a)}, P_0 = 0)$, in Eq. (4.76) may be factored into

$$R^{(4)} = 4R^{(2)}[1 - R^{(2)}] \tag{4.83}$$

which has a maximum of 1 at $R^{(2)} = 1/2$. We show in Fig. 4.22 the β-dependence of the residue $R^{(2)}$, in Eq. (4.83) and $R^{(4)}$, in Eq. (4.77) for $K = 3.3$. At the stable elliptic point, the characteristic multiplier λ

$$\lambda = 1 - 2R \pm 2[R(R-1)]^{1/2} \tag{4.84}$$

is expressed in terms of the rotation number ρ, previously defined in Eq. (3.4.19) as

$$\lambda = \exp(\pm i\, 2\pi\rho). \tag{4.85}$$

When the residue passes through the values

$$R(p/q) = \sin^2(\pi p/q) \qquad p, q: \text{ coprime integers,} \tag{4.86}$$

the Poincaré-Birkhoff islands of q times the original period are born. At $\beta = 0.612$, $R^{(4)} = 0$ and $R^{(2)} = 1$ so that this corresponds to $p/q = 1/2$ (period-doubling) bifurcation for the T^2 map. At $\beta = 0.933$, $R^{(2)} = R^{(4)} = 3/4$ and the value of p/q is $1/3$ for T^2 and $2/3$ for T^4, which is the case of the period-3 catastrophe observed in Fig. 4.15(a). On the other hand, at $\beta = 1.57$, the value $R^{(4)}$ is $3/4$ but $R^{(2)} = 1/4$, which correspond to the 2/6 multifurcation for T and the 1/6 multifurcation for T^2. Therefore, two sets of the period-3 islands turn up in Fig. 4.15(b) rather than the period-3 catastrophe. Lastly, at the value of $\beta = 1.26$, $R^{(4)}$ approaches unity and $R^{(2)}$ passes through $1/2$, which indicates that the 2/4 multifurcation for T^4 and the 1/4 for T^2 map. These 4-cycles are observed as two pairs of the period-2 orbits in Fig. 4.21.

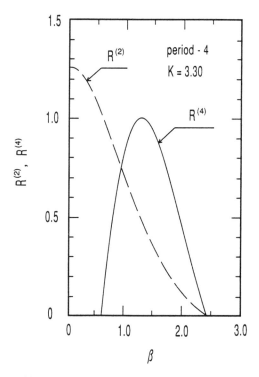

Fig. 4.22. Residues $R^{(2)}$ and $R^{(4)}$ for the Poincaré-Birkhoff period-4 orbit on the X-axis at $K = 3.3$.

4.9 Concluding Remarks on the Standard and Relativistic Maps

We have investigated the nonlinear dynamics of particle acceleration in the relativistic standard map and shown that the relativistic effects suppress the stochastic motion of the particles. Under the action of the coherent spectrum of electrostatic waves whose phase velocities are larger than the speed of light, $\beta = \omega/kc \gg 1$, the particle motion becomes regular with even-numbered periodic orbits being the dominant feature over the phase space. The structure of the periodic orbits is analyzed by constructing families of space inversion and momentum reflection symmetry lines. We show that the odd-numbered periodic orbits are washed out in the stochastic sea, while the even-numbered orbits survive up to quite high periodicity. This difference in behavior is due to the fact that in the high momentum region the phase advancement for the particle motion is asymptotically constant, which is clearly shown by the family of space inversion symmetry lines becoming asymptotically parallel to the P axis in Fig. 4.10.

For the observed phase space structure, we analytically determined the outermost KAM surface with Eq. (4.46), showing that the formula agrees well with the computer experiments for the regime $\beta \gg 1$. For the sub-relativistic regime $\beta < 1$, we show that even though the fixed points of high momentum $P_m \approx m \approx \beta^{-1}$ are allowed to exist, the outermost KAM surface for particles with the initial momentum $P_0 = 0$ is formed as a separatrix between much lower order momentum P_m fixed points, with $m \ll \beta^{-1}$.

Thus, we have established that the relativistic dynamics has no global diffusion, in sharp contrast to the local standard mapping. The relativistic motion, even when it is chaotic, is surrounded by regular motions. The suppression of the global diffusion and dominance of regular motion arises fundamentally from the nonlinear phase advance that allows the particle velocity to have the upper limit given by the speed of light. These properties are favorable for the coherent acceleration of particles with limited heating, or growth of emittance in the language of accelerator physics.

In this connection, Howard and Hohs (1984) have examined the formation of an adiabatic barrier for phase space orbits in a two-parameter Hamiltonian system with a quadratically nonlinear phase advance. They show that the nonlinear phase variation can give rise to a reconnection of the KAM curves and the formation of phase space vortex structures for the period-2 aligned islands. In the Howard-Hohs problem the phase advance function in Eq. (4.2) is $G(P, \alpha) = P - \alpha P^2$ with period-1 fixed points at $P_n^\pm = (1 \pm \sqrt{1 - 8\pi n \alpha})/2\alpha$. In our problem, for small β, the relativistic phase advancement could be approximated by the cubic phase advance $(P - \frac{1}{2}\beta^2 P^3)$, which may partially account for the barrier formation as observed in Figs. 4.8(c) and (d). Yet, it should be noticed that the derivative G' of the phase transformation function $G(P)$ is positive definite in its exact relativistic expression in Eq. (4.10). Howard and Hohs (1984) identify the critical condition for their barrier structure to occur at the point where the phase advance is stationary ($G' = 0$). For predicting the existence of KAM barriers, our observation that the momentum inversion symmetry curves asymptotically approach to the outermost KAM surface provide a theoretical method for determining their structure.

Turning to the super-relativistic regime $\beta > 1$, we have discovered the occurrence of the interchange of the stability of the periodic orbits around the fixed point at the origin with variations of the phase advancement parameter β. By calculating the residue of the period-3 and period-4 orbits, we have shown that the stable-unstable interchange can be explained by the local linear stability properties of the orbits. Study of the period-3 orbits for the logistic twist map examined by Howard and Hohs (1984) indicates that a similar stability interchange can be observed for $K = 3.3$ around $\alpha \approx 0.03$. Thus, we expect that the stability interchange of the periodic orbits upon variation of the phase advancement parameter appears to be a generic features of two parameter Hamiltonian systems containing a nonlinear phase advance.

From the examples and analysis presented in this Chapter it becomes clear that Hamiltonian systems generically have islands of stability with a self-similar structure that produce strong correlations and incomplete chaos (Zaslavsky-Stevens-Weitzner, 1993). These correlations produce deviations, (called anomalous transport) away from the Gaussian diffusion law. The studies of Rechester and White (1980), Cary and Meiss (1981) and Karney (1983) reveal the nature of the correlations. Zaslavsky (1993) has formulated a "fractional Fokker-Planck Kinetic" to account for the anomalous transport.

4.10   *Example 4.1   Fermi Acceleration*

A mechanical analog for the acceleration of charged particles by radio frequency waves is the elastic bouncing of a ball between an oscillation wall at $x = x_w(\omega t)$ and a fixed wall at $x = d$ (Lichtenberg et al., 1980). Due to Fermi's proposal of a similar mechanism to explain the energy gain of protons forming the cosmic rays with energies up to 10^{20} ev, his name is associated with the resulting map. Here we restrict consideration to the nonrelativistic case and ask if there is a limit to the energy the bouncing ball can gain. The answer is that there is a limit and the limiting value can be estimated by use of the value K_c in Eq. (4.13) for the standard map. The Fermi map, as we shall call it, actually has features similar to the relativistic standard map and thus provides an interesting example somewhere between the two maps analyzed in this chapter.

For a particle with an incoming velocity V_n at time t_n reflecting elastically from the moving wall with velocity $U = \dot{x}_w(t_n) = -a\omega \sin(\omega t_n)$ (we take $x_w = a\cos(\omega t)$ for this example), the outgoing velocity is $V' = 2U - V_n$. This is most easily understood by transforming to the frame moving with the wall, ($V_n \rightarrow V_n - U$) reflecting

4.10 Example 4.1 Fermi Acceleration

the velocity, and then transforming back to the lab frame. The travel time from one collision with the moving wall to the next is then $2d/V'$. The actual time t_{n+1} is given by an implicit relation through $t_{n+1} - t_n = [2d + x_w(t_n) + x_w(t_{n+1})]\big/V_{n+1}$. However, for high velocities ($V \gg a\omega$) and large d/a the time lapse is $2d/V_{n+1}$ to a good approximation. Using this simplification we get the explicit, area conserving map

$$V_{n+1} = V_n + 2a\omega \sin(\omega t_n) \quad (4.87)$$

$$\omega t_{n+1} = \omega t_n + \frac{2d\omega}{V_{n+1}}. \quad (4.88)$$

In contrast to the standard map, the phase advance is large for small $u_n = V_n/a\omega$ and small for large u. Thus, the low velocity region is stochastic, and we may expect to find invariant (KAM) curves above a certain velocity. A stability analysis of the fixed points of (4.87) and (4.88) verifies this statement. Numerical evaluation is required to find the invariant curve $u_I = u_I(\psi)$ separating the region of stochastic heating from the region with no net energy gain. However, for large $M = d/a$, a good approximation to the location of the invariant curve may be obtained by linearizing Eq. (4.88) about the unknown u_I as follows:

$$\Delta u_{n+1} = \Delta u_n + 2\sin\psi_n$$

$$\psi_{n+1} = \psi_n + \frac{2d}{au_I} - \frac{2d}{au_I^2}\Delta u_{n+1}.$$

Using $I_n = (2d/au_I^2)\Delta u_n$ gives $\psi_{n+1} = \psi_n + I_{n+1}$ and $I_{n+1} = I_n + K\sin\psi_n$ with $K = (4d/au_I^2)$ so that for $K = K_c$ or

$$u_I = 2\left(\frac{d}{aK_c}\right)^{1/2} \quad (4.89)$$

we expect to find the invariant curve.

In Fig. 4.23 the phase space structure obtained from the map in Eqs. (4.87)–(4.88) is shown for $M = d/a = 100$. Any low velocity initial condition fills the stochastic sea and the periodic island boundaries are readily found. The estimate in Eq. (4.89) is $u_I \simeq 20.0$ while the actual invariant curve ranges from 18.8 to 20.8.

A simple variant of this problem occurs when gravity (or a constant electric field) is used to return the particle to the oscillating wall. The return time is then $\Delta t = V_{n+1}/g$, and the reader can easily show that when the phase equation is written as $\psi_{n+1} = \psi_n + I_{n+1}$ the velocity equation gives the standard map with $K_g = (2\omega^2 a/g)$. For $K_g > K_c$ there is unlimited stochastic heating.

162   Chapter 4   Phase Space Structures in Hamiltonian Systems

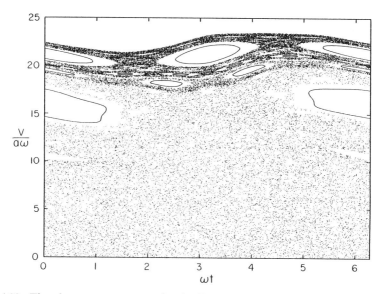

Fig. 4.23. The phase space structure for the Fermi acceleration map in Eqs. (4.87) and (4.88) for $M = d/a = 100$. For large M the limiting velocity is approximated by $U_I = 2M^{1/2}$.

Chapter 5

Solitons in Plasmas

In this Chapter we present the first-principles derivations of several of the well-known integrable nonlinear wave equations of plasma physics. The starting point is the collisionless kinetic equation for the one-particle phase space distribution $F(x, v, t)$ determined in the self-consistent electric field $E(x,t) = -\partial_x \phi(x,t)$. This is the famous Vlasov-Poisson self-consistent field problem. In the first two sections the kinetic form of the Korteweg and deVries (KdV) equation and the cubic nonlinear Schrödinger (NLS) equation are derived from the Vlasov-Poisson field equations. Then the effects of the kinetic particle resonances are discussed, and the theory of the modulational instability (MI) is developed.

With the experience gained from these particular calculations we return to a general analysis of how to balance the nonlinear effects with wave dispersion to obtain the integrable and near integrable wave equations. Conceptually the procedure is related to the secular perturbation theory of Sec. 1.2 where the small amplitude expansion secular terms, here from wave steepening, must be balanced to all orders with the wave dispersion. The procedure is called reductive perturbation and is used to rederive the KdV equation from magnetohydrodynamics equations for the ion acoustic wave and the Alfvén wave.

In Sec. 5.5 and beyond we turn to a brief description of the growth of the large body of knowledge about solitons as solutions of integrable nonlinear wave equations. The term soliton was given to the localized solutions of the nonlinear equations by Zabusky and Kruskal because of their particle-like behavior during collisions. The elastic nature of the interactions of the solitary waves of numerous nonlinear wave equations, including the two-dimensional Kadomtsev-Petviashvili equations, is established analytically by the inverse scattering transform (IST) theory first given for the KdV equation by Gardner et al. (1967). We then outline the IST theory and give its standard general form known as AKNS theory. We show how the soliton theory and IST theory are used to develop new descriptions of the fluctuation dynamics in plasmas. Similar developments have occurred in solid state physics and other areas of science and engineering.

The theory and application of the solitons that occur for a relatively large number of integrable evolution equations is a large and fascinating field of research in itself. There are excellent books devoted to this subject. A very readable account of the development of the field is *Solitons, Nonlinear Evolution Equations and Inverse Scattering*, by Ablowitz and Clarkson (1991). The introductory chapter which gives both historical overview with derivations as well as an account of a soliton alleged to have transported an oil drilling rig one hundred feet in the Andaman Sea off the coast of Burma is strongly recommended. A more mathematical account is found in Novikov, *et al.* (1984), and earlier works include Lonngren and Scott (1978) with experimental examples in plasmas. This area of research has found recent economically important applications in optical pulse propagation, due to the ability of solitons to transport signals over long distances without dispersion and with minimal dissipation (Hasegawa, 1984, 1990; Hasegawa and Tappert, 1973).

5.1 Nonlinear Coherent Modes in Vlasov Plasmas

One-dimensional space-time variation of the collisionless plasma is described by the self-consistent Vlasov-Poisson field equations

$$\frac{\partial F_\alpha}{\partial t} + v \frac{\partial F_\alpha}{\partial x} - \frac{e_\alpha}{m_\alpha} \frac{\partial \phi}{\partial x} \cdot \frac{\partial F_\alpha}{\partial v} = 0 \tag{5.1}$$

$$\frac{\partial^2 \phi}{\partial x^2} = -4\pi \sum_\alpha e_\alpha n_\alpha \int F_\alpha \, dv \tag{5.2}$$

where the index α specifies the species of the plasma particles, $F_\alpha(x,v,t)$ is the particle distribution function normalized to the unity, and $\phi(x,t)$ is the electric potential. Decomposing F_α into the spatially uniform component, time independent component $F_\alpha^{(0)}(v)$ and the fluctuating component $f_\alpha(x,v,t)$, we can expand Eq. (5.1) with the Fourier transformation and solve for the fluctuation distribution function $f_\alpha(v) e^{ikx - i\omega t}$ to obtain

$$f_\alpha(v; k, \omega) = \frac{-e_\alpha}{m_\alpha} \frac{1}{\omega - kv} k\phi(k,\omega) \frac{\partial}{\partial v} F_\alpha^{(0)}(v)$$

$$+ \frac{-e_\alpha}{m_\alpha} \frac{1}{\omega - kv} \sum_{k',\omega'} k' \phi(k',\omega') \frac{\partial}{\partial v} f_\alpha(v; k - k', \omega - \omega'), \tag{5.3}$$

where it is understood that causality requires the analytic continuation to real ω from $\text{Im}\,\omega > 0$. The second term on the right-hand side of Eq. (5.3) determines the generation of phase coherent harmonics producing the coherent structures. To obtain the long wavelength solitary wave equations, we solve Eq. (5.3) by iteration on the mode coupling term (Kono and Sanuki, 1972).

5.1 Nonlinear Coherent Modes in Vlasov Plasmas

Substituting the representation for $f_\alpha(v, k, \omega)$ taken up to the second iteration of Eq. (5.3) into the Fourier transformation of Eq. (5.2), we obtain the mode coupling equation

$$\varepsilon(k,\omega)\phi(k,\omega) = \sum_{k',\omega'} V(k',\omega'; k-k', \omega-\omega')\phi(k',\omega')\phi(k-k', \omega-\omega') \quad (5.4)$$

where the linear dielectric response function is

$$\varepsilon(k,\omega) = 1 + \frac{4\pi}{k} \sum_\alpha \frac{e_\alpha^2 n_\alpha}{m_\alpha} \int \frac{1}{\omega - kv} \frac{\partial}{\partial v} F_\alpha^{(0)} dv \quad (5.5)$$

and the mode coupling coefficients are

$$V(k',\omega'; k-k', \omega-\omega') = \frac{4\pi}{k^2} \sum_\alpha \frac{n_\alpha e_\alpha^3}{m_\alpha^2} k'(k-k')$$

$$\int \frac{1}{\omega - kv} \frac{\partial}{\partial v} \left\{ \frac{1}{\omega - \omega' - (k-k')v} \frac{\partial}{\partial v} F_\alpha^{(0)} \right\} dv. \quad (5.6)$$

In the limit $\operatorname{Im}\omega \to 0$ the propagator $g_{k\omega}(v) = (\omega - kv)^{-1}$ stands for

$$\frac{1}{\omega - kv} = \mathcal{P} \frac{1}{\omega - kv} - i\pi\delta(\omega - kv). \quad (5.7)$$

where $\mathcal{P}$ is the principal value part of the integral. The rule in Eq. (5.7) accounts for the adiabatic switch-on of the disturbance at time $t = -\infty$. The complex function $\varepsilon(k,\omega)$ is the linear plasma dielectric permeability.

Now, we consider the ion acoustic wave in the small wavenumber region, where the dispersion relation is determined by

$$\varepsilon_r(k,\omega) = 1 + \frac{4\pi}{k} \sum_\alpha \frac{e_\alpha^2 n_\alpha}{m_\alpha} \mathcal{P} \int \frac{1}{\omega - kv} \frac{\partial}{\partial v} F_\alpha^{(0)} dv = 0 \quad (5.8)$$

as

$$\omega(k) = \pm c_s k \left(1 - \tfrac{1}{2} \lambda_D^2 k^2\right), \quad k\lambda_D \ll 1 \quad (5.9)$$

with the definition of the ion acoustic velocity $c_s = (\kappa_B T_e/M)^{1/2}$ with $(m_i = M)$ and λ_D is the Debye distance $(\kappa_B T_e/4\pi e^2 n)^{1/2}$ where κ_B is the Boltzmann constant. In the moving coordinate $(x' = x - ut, t' = t)$ with the ion acoustic velocity $u = c_s$, there is a slow temporal variation of the disturbance, which is characterized by the low-frequency Ω defined by

$$\Omega = \omega - ku = \omega - k c_s. \quad (5.10)$$

We can express the left-hand side of Eq. (5.4) as

$$\left.\frac{\partial \varepsilon_r}{\partial \omega}\right|_{\omega=\omega(k)} (\omega - \omega(k))\phi(k,\omega) + i\varepsilon_i(k,\omega(k))\phi(k,\omega)$$

$$= \left.\frac{\partial \varepsilon_r}{\partial \omega}\right|_{\omega=\omega(k)} \left(\Omega - \frac{c_s}{2}\lambda_D^2 k^3\right)\phi(k,\Omega) + i\varepsilon_i(k,c_s k)\phi(k,\Omega) \quad (5.11)$$

while for the right-hand side the coupling coefficient $V(k',\omega';k-k',\omega-\omega')$ is expressed as

$$V(k',\omega';k-k',\omega-\omega') = \frac{4\pi}{k^2} \sum_\alpha \frac{n_\alpha e_\alpha^3}{m_\alpha^2} k'(k-k')$$

$$\frac{1}{k(k-k')} \mathcal{P} \int \frac{1}{(v-c_s)^3} \frac{\partial}{\partial v} F_\alpha^{(0)}(v) dv. \quad (5.12)$$

Hence, we get

$$-i\Omega\phi(k,\Omega) + \frac{c_s \lambda_D^2}{2}(ik)^3 \phi(k,\Omega) + A(c_s)\frac{k}{|k|}\phi(k,\Omega)$$

$$+ B(c_s) \sum_{k',\Omega} ik'\phi(k',\Omega')\phi(k-k',\Omega-\Omega') = 0 \quad (5.13)$$

where

$$A(c_s) = \sum_\alpha \omega_{p\alpha}^2 \int \delta(v-c_s)\frac{\partial F_\alpha^{(0)}}{\partial v}dv \bigg/ \sum_\alpha \omega_{p,\alpha}^2 \mathcal{P}\int \frac{1}{(v-c_s)^2}\frac{\partial F_\alpha^{(0)}}{\partial v}dv \quad (5.14)$$

$$B(c_s) = \sum_\alpha \frac{e_\alpha}{2m_\alpha}\omega_{p\alpha}^2 \int \frac{1}{(v-c_s)^3}\frac{\partial F_\alpha^{(0)}}{\partial v}dv \bigg/ \sum_\alpha \omega_{p,\alpha}^2 \int \frac{1}{(v-c_s)^2}\frac{\partial F_\alpha^{(0)}}{\partial v}dv \quad (5.15)$$

where the principal value part is understood in the integrals in Eq. (5.15). Then, corresponding to the small frequency scale Ω, we introduce the slow time scale τ of the order of Ω^{-1} through the Fourier transform with respect to Ω.

Operating on Eq. (5.13) with the inverse Fourier transformation operator $\sum_{k,\Omega} \exp[i(kx - \Omega\tau)]$, we obtain

kinetic KdV eq.

$$\frac{\partial}{\partial \tau}\phi(x,\tau) + 2B(c_s)\phi(x,\tau)\frac{\partial}{\partial x}\phi(x,\tau)$$

$$+ \tfrac{1}{2}c_s\lambda_D^2 \frac{\partial^3}{\partial x^3}\phi(x,\tau) - A(c_s)\mathcal{P}\int_{-\infty}^{+\infty}\frac{1}{x-x'}\phi(x',\tau)dx' = 0. \quad (5.16)$$

The last nonlocal, but linear term is the space-time domain representation of Landau damping on the fast ions and slow electrons ($v = \omega/k \cong c_s$). Equation (5.16) is the Korteweg-de Vries equation with the inclusion of linear Landau damping. Equation (5.16) has been derived by taking full advantage of the weakly dispersive property of the ion acoustic wave described in Eq. (5.9). In the absence of the ion Landau damping $(A(c_s) \simeq 0)$ the long wavelength ion acoustic waves given by Eq. (5.16) form the solitons (see (5.5)) due to the coherent phase-locked harmonics generated by the mode coupling in Eq. (5.4). In the hydrodynamic description this mode coupling arises from the wave steepening produced by the convective derivative in the ion acceleration.

Turning to the strongly dispersive case, we now extend the method to describe the nonlinear amplitude modulation of the carrier wave having the central wavenumber k_0 with the frequency ω_0.

5.2 Amplitude Modulation of Nearly Monochromatic Waves and the Modulational Instability

Nonlinear modulation of the finite amplitude wave in the dispersive media is one of the general phenomena of importance in the fields of plasma physics, solid state physics and nonlinear optics. Here, let us discuss the nonlinear modulation of a quasi-monochromatic electron plasma wave propagating in the collisionless plasmas. An example of a carrier wave with a modulation in either space $k_0 \pm q$ or time $\omega_0 \pm \Omega$ is shown in Fig. 5.1.

We define the Fourier Laplace transformation of the distribution function $F_\alpha(v; x, t)$ and the electrostatic potential $\phi(x, t)$ as

$$G(x,t) = \sum_{K=(k,\omega)} G(K)\exp[-i(\omega t - kx)] \tag{5.17}$$

$$G(K) = \int_0^\infty dt \int_{-\infty}^{+\infty} dx\, G(x,t) \exp[i(\omega t - kx)] \tag{5.18}$$

where $G(x,t)$ stands for $F_\alpha(v; x, t)$ or $\phi(x, t)$. The variable K is a shorthand notation for (k, ω) and the summation defined by

$$\sum_K \equiv (2\pi)^{-2} \int_{-\infty+i\sigma}^{+\infty+i\sigma} d\omega \int dk. \tag{5.19}$$

168   Chapter 5   Solitons in Plasmas

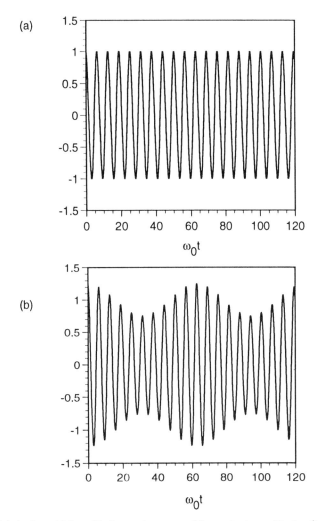

Fig. 5.1. (a) A sinusoidal oscillation at k_0 or ω_0 with constant amplitude. (b) The same sinusoidal oscillation with a slowly varying amplitude $A(x,t)$ at $q \ll k_0$ or $\Omega \ll \omega_0$.

5.2 Amplitude Modulation of Nearly Monochromatic Waves and ...

Carrying out the iteration procedure for $f_\alpha(v, K)$ through third order in powers of $|(e\phi/m)g_K(v)k\,\partial/\partial v|$ in Eq. (5.3), we obtain the mode coupling equation

$$\varepsilon(K)\phi(K) = \sum_{K' \neq K} V(K, K')\phi(K')\phi(K - K')$$

$$- \sum_{K' \neq K} \sum_{K'' \neq K-K'} V(K, K', K'')\phi(K')\phi(K'')\phi(K - K' - K'') \quad (5.20)$$

where $\varepsilon(K)$ is given in Eq. (5.5). The mode coupling coefficients $V(K, K')$ and $V(K, K', K'')$ are expressed explicitly as

$$V(K, K') = \frac{4\pi}{k^2} \sum_\alpha \frac{n_\alpha e_\alpha^3}{m_\alpha^2} \int [k' g_K(v)][(k-k')g_{K-K'}(v)]F_\alpha(v)dv \quad (5.21)$$

and

$$V(K, K', K'') = \frac{4\pi}{k^2} \sum \frac{e_\alpha^4}{m_\alpha^3} \int [k' g_K][k'' g_{K-K'}][(k-k'-k'')g_{K-K'-K''}(v)]F_\alpha(v)dv \quad (5.22)$$

where $g_K(v)$ is the particle propagator

$$g_K(v) = \frac{1}{\omega - kv + i\sigma}\frac{\partial}{\partial v}, \qquad \sigma \to 0^+ \quad (5.23)$$

describing the interaction of the particle v with the fluctuation $K = (k, \omega)$.

In the presence of nonlinear self-modulation of the quasi-monochromatic wave, it is sufficient to consider the three-wave mode-coupling process. The two-mode coupling term in Eq. (5.4), and Eq. (5.20), gives rise to the creation of the modes (k', ω') and $(k - k', \omega - \omega')$, each of them may have large deviations from the quasi-monochromatic mode (k, ω). This creation of the two nonresonant beat waves is obtained by calculating the second order amplitudes from Eq. (5.4) and then reducing Eq. (5.20) to

$$\varepsilon(K)\phi(K) = \sum_{K' \neq K} \sum_{K'' \neq K-K'} \left\{ \frac{[V(K, K') + V(K, K - K')]}{\varepsilon(K - K')} \cdot V(K - K', K'') \right.$$

$$\left. - V(K, K', K'') \right\} \phi(K')\phi(K'')\phi(K - K' - K''). \quad (5.24)$$

Now, for the quasi-monochromatic wave the wavenumber k is concentrated sharply at the specific wavenumber k_0. The nonlinear effect will compete with the dispersion effect associated with the small spread of wavenumbers in the wave packet around the wavenumber k_0. Hence, we may express the electrostatic potential $\phi(x, t)$ as

$$\phi(x, t) = A(x, t)\exp\left[i(k_0 x - \omega_0 t)\right] + (\text{c.c.}) \quad (5.25)$$

where (c.c.) stands for the complex conjugate of the proceeding term. The slow modulation is given by the complex amplitude $A(x,t)$. The carrier frequency ω_0 is defined as the root of the dispersion relation

$$\varepsilon_r(k_0, \omega_0) = 0. \tag{5.26}$$

The Fourier Laplace transformation of Eq. (5.25) has the form

$$\phi(K) = A(K - K_0) + A^*(-K - K_0). \tag{5.27}$$

Equation (5.27) satisfies the condition $\phi(-K) = \phi^*(K)$ required for a representation of the real valued potential $\phi(x,t)$.

The requirement of the slowness of space-time variation of the amplitude $A(x,t)$ is insured by the condition $|K_0 - K| \ll |K_0|$ for appreciable values of ϕ or A. So, we can introduce the slow argument $\delta K = (q, \Omega)$ by the definition of

$$\delta K = (q, \Omega) = (k, \omega) - (k_0, \omega_0). \tag{5.28}$$

Direct calculation of the triple product $(\phi\phi\phi)$ of the potential ϕ gives rise to the terms of $A(\delta K')$, $A(\delta K'')$, $A(\delta K - \delta K' - \delta K'')$ and the terms containing such factors as $A(\delta K - \delta K' - \delta K'' - 2K_0)$ or $A(\delta K - K' - \delta K'' - 4K_0)$. Terms that depend explicitly on the argument K_0 lead to the rapid space-time oscillations, and we can disregard them to describe the slow modulation process. Finally, we obtain

$$\varepsilon(K_0 + \delta K)A(\delta K) = \sum_{\delta K'} \sum_{\delta K''} \Big\{ M_0(\delta K, \delta K', \delta K'')$$

$$+ M_2(\delta K, \delta K', \delta K'') \Big\} A(\delta K') A^*(-\delta K'') A(\delta K - \delta K' - \delta K'') \tag{5.29}$$

where the coupling coefficients M_0 and M_2 are expressed as

$$M_0(\delta K, \delta K', \delta K'') = \frac{1}{\varepsilon(\delta K - \delta K')} \Big\{ V(K_0 + \delta K, K_0 + \delta K')$$

$$+ V(K_0 + \delta K, \delta K - \delta K') \Big\} \Big\{ V(\delta K - \delta K', K_0 + \delta K - \delta K' - \delta K'')$$

$$- V(\delta K - \delta K', -K_0 + \delta K''') \Big\} - V(K_0 + \delta K, K_0 + \delta K', \delta K - \delta K' - \delta K'')$$

$$- V(K_0 + \delta K, K_0 + \delta K', -K_0 + \delta K'') \tag{5.30}$$

and

5.2 Amplitude Modulation of Nearly Monochromatic Waves and ...

$$M_2(\delta K, \delta K', \delta K'') = \frac{1}{\varepsilon(2K_0 + \delta K - \delta K'')} V(2K_0 + \delta K - \delta K'', K_0 + \delta K').$$

$$\{V(K_0 + \delta K, -K_0 + \delta K'') + V(K_0 + \delta K, 2K_0 + \delta K - \delta K'')\}$$
$$- V(K_0 + \delta K, -K_0 + \delta K'', K_0 + \delta K') \quad (5.31)$$

where the suffices 0 and 2 express the contributions of the beat waves with 0− and 2K-wavenumber.

Now, in the left-hand side of Eq. (5.29), we can approximate the real part of the linear dielectric function as

$$\varepsilon_r(K_0 + \delta K, \omega_0 + \Omega) \doteq \varepsilon_r(K_0, \omega_0) + \frac{\partial \varepsilon_r}{\partial \omega_0}\Omega + \frac{\partial \varepsilon_r}{\partial K_0}\delta K + \frac{1}{2}\frac{\partial^2 \varepsilon_r}{\partial \omega_0^2}\Omega^2 + \frac{\partial^2 \varepsilon_r}{\partial \omega_0 \partial K_0}\delta K\Omega$$

$$+ \frac{1}{2}\frac{\partial^2 \varepsilon_r}{\partial K_0^2}\delta K^2 + \cdots = \frac{\partial \varepsilon_r}{\partial \omega_0}\left(\Omega - \lambda\delta K - \frac{1}{2}\frac{\partial \lambda}{\partial K_0}K^2\right) + \frac{1}{2}\frac{\partial^2 \varepsilon_r}{\partial \omega_0^2}(\Omega - \lambda\delta K)^2 + \cdots \quad (5.32)$$

with the definition of the group velocity

$$\lambda = \frac{\partial \omega_0}{\partial k_0}. \quad (5.33)$$

In describing the slow temporal variation, the second term of Eq. (5.32) can be disregarded relative to the first term. Then, we can reduce the left-hand side of Eq. (5.29) as

$$\varepsilon(K_0 + \delta K)A(\delta K) \doteq \frac{\partial \varepsilon_r}{\partial \omega_0}\left(\Omega - \lambda K - \frac{1}{2}\frac{\partial \lambda}{\partial k_0}K^2\right)A(\delta K) + i\varepsilon_i(K_0)A(\delta K) \quad (5.34)$$

where the imaginary part of the dielectric constant $\varepsilon_i(K_0)$ is given as

$$\varepsilon_i(K_0) = \frac{4\pi^2}{k_0}\sum_\alpha \frac{e_\alpha^2}{m_\alpha}n_\alpha \int \delta(\omega_0 - k_0 v)\frac{\partial F_\alpha}{\partial v}dv. \quad (5.35)$$

Applying the inverse Fourier-Laplace transformation Eq. (5.17) and (5.19) to the expression of Eq. (5.34), we obtain

$$\frac{1}{(2\pi)^2}\int dK \int d\Omega \exp\left[-i(\Omega t - Kx)\right]\varepsilon(K_0 + \delta K)A(\delta K)$$

$$= \frac{\partial \varepsilon_r}{\partial \omega_0}\left\{i\left(\frac{\partial}{\partial t} + \lambda\frac{\partial}{\partial x}\right) + \frac{1}{2}\frac{\partial \lambda}{\partial k_0}\frac{\partial^2}{\partial x^2}\right\}A(x,t) + i\varepsilon_i(K_0)A(x,t). \quad (5.36)$$

Now, turn to the right-hand side of Eq. (5.29). As has been discussed above, the slow modulation process of the quasi-monochromatic wave packet requires that

$$k_0 \gg \{\delta K, \delta K', \delta K''\} \quad (5.37)$$

$$\omega_0 \gg \{\Omega, \Omega', \Omega''\}. \quad (5.38)$$

We evaluate the coupling coefficient $M_0(\delta K, \delta K', \delta K'')$ and $M_2(\delta K, \delta K', \delta K'')$ under the condition of Eqs. (5.37) and (5.38).

As for the M_2, we can take simply

$$\varepsilon(2K_0 + \delta K - \delta K') \doteq \varepsilon_r(2K_0) = -3/4 \tag{5.39}$$

$$V(2K_0 + \delta K - \delta K', K_0 + \delta K') \doteq V(2K_0, K_0) \equiv \frac{1}{8} A(K_0) \tag{5.40}$$

$$V(K_0 + \delta K, -K_0 + \delta K'') \doteq V(K_0, -K_0) \equiv -A(K_0) \tag{5.41}$$

$$V(K_0 + \delta K, 2K_0 - \delta K - \delta K'') \doteq V(K_0, 2K_0) \equiv 2A(K_0) \tag{5.42}$$

$$V(K_0 + \delta K, -K_0 + \delta K'', K_0 + \delta K') \doteq V(K_0, -K_0, K_0) \equiv -\frac{K_0}{2} B(K_0) \tag{5.43}$$

where the functions $A(K_0)$ and $B(K_0)$ are defined as

$$A(K_0) = 4\pi \sum_\alpha \frac{e_\alpha^3}{m_\alpha^2} n_\alpha \mathcal{P} \int \frac{1}{\omega_0 - k_0 v} \frac{\partial}{\partial v} \left\{ \frac{1}{\omega_0 - k_0 v} \frac{\partial F_\alpha}{\partial v} \right\} dv \tag{5.44}$$

$$B(K_0) = 4\pi \sum_\alpha \frac{e_\alpha^4}{m_\alpha^3} n_\alpha \mathcal{P} \int \frac{1}{\omega_0 - k_0 v} \frac{\partial}{\partial v}$$

$$\times \left\{ \frac{1}{\omega_0 - k_0 v} \cdot \frac{\partial}{\partial v} \left(\frac{1}{\omega_0 - k_0 v} \frac{\partial F_\alpha}{\partial v} \right) \right\} dv. \tag{5.45}$$

Hence, the coupling coefficient $M_2(\delta K, \delta K', \delta K'')$ is evaluated as

$$M_2(\delta K, \delta K', \delta K'') = K_0 \left\{ \frac{1}{6K_0} [A(K_0)]^2 + \tfrac{1}{2} B(K_0) \right\}. \tag{5.46}$$

We notice that in the above evaluation the contributions of the resonant particle at the phase velocity of the ω_0/k_0 have been neglected throughout. This is equivalent to the requirement that the carrier wave is weakly Landau damped.

Evaluation of the coupling coefficient $M_0(\delta K, \delta K', \delta K'')$ requires a little more detailed consideration. First, in the limit of $\delta K \approx \delta K' \approx \delta K'' \to 0$, we make the reduction

$$\varepsilon(\delta K - \delta K') = 1 + \frac{4\pi}{|K - K'|^2} \sum_\alpha \frac{e_\alpha^2 n_\alpha}{m_\alpha} \int \frac{K - K'}{(\Omega - \Omega') - (K - K')v + i\sigma} \cdot \frac{\partial F_\alpha}{\partial v} dv$$

$$\cong -\frac{1}{|K - K'|^2} \Delta(\lambda) \tag{5.47}$$

with the reduced response function defined by

$$\Delta(\lambda) := 4\pi \sum_\alpha \frac{e_\alpha^2 n_\alpha}{m_\alpha} \int \frac{1}{v - \lambda - i\,\mathrm{sgn}\,(K - K')\sigma} \frac{\partial F_\alpha}{\partial v} dv \tag{5.48}$$

5.2 Amplitude Modulation of Nearly Monochromatic Waves and ...

where we have made use of the following approximation

$$\frac{\Omega - \Omega'}{K - K'} = \frac{\omega - \omega'}{k - k'} = \frac{d\omega}{dk} := \lambda. \tag{5.49}$$

Similarly, we make the following reductions

$$V(K_0 + \delta K, K_0 + \delta K') = 4\pi \sum_\alpha \frac{e_\alpha^2 n_\alpha}{m_\alpha^2} \int \frac{1}{(\omega_0 - k_0 v)^2}$$

$$\times \frac{1}{v - \lambda - i\operatorname{sgn}(K - K')\sigma} \frac{\partial F_\alpha}{\partial v} dv := -W(\lambda) \tag{5.50}$$

$$V(K_0 + \delta K, \delta K - \delta K') = 0 \tag{5.51}$$

$$V(\delta K - \delta K', K_0 + \delta K - \delta K' - \delta K'') + V(\delta K - \delta K', -K_0 + \delta K'')$$

$$\doteq V(\delta K - \delta K', K_0 + \delta K - \delta K') + V(\delta K - \delta K', Y_0)$$

$$\equiv -\frac{k_0^2}{|K - K'|^2} W(\lambda). \tag{5.52}$$

It is worthwhile to remark that the singular factor of $(K - K')^{-2}$ in Eq. (5.52) is cancelled by the same factor in Eq. (5.47).

Similar calculations are carried out for the second and third terms of Eq. (5.30) as

$$V(K_0 + \delta K, K_0 + \delta K', K_0 + \delta K - \delta K' - \delta K'')$$
$$+ V(K_0 + \delta K, K_0 + \delta K, K_0 + \delta K', -K_0 + \delta K'')$$
$$\doteq V(K_0 + \delta K, K_0 + \delta K', K_0 + \delta K - \delta K') + V(K_0 + \delta K, K_0 + \delta K', -K_0)$$

$$= -k_0^2 4\pi \sum_\alpha \frac{e_\alpha^4 n_\alpha}{m_\alpha^3} \int \frac{1}{(\omega_0 - k_0 v)^2} \frac{1}{v - \lambda - i\operatorname{sgn}(K - K')\sigma}$$

$$\cdot \frac{\partial}{\partial v} \left(\frac{v - \lambda}{(\omega_0 - k_0 v)^2} \frac{\partial F_\alpha}{\partial v} \right) dv \equiv k_0^2 C(\lambda). \tag{5.53}$$

Collecting the contributions of Eqs. (5.47), (5.50), (5.51), (5.52), and (5.53), we finally obtain

$$M_0(\delta K, \delta K', \delta K'') = -k_0^2 \left(\frac{W^2}{\Delta} + C \right). \tag{5.54}$$

In order to carry out explicit calculation of the inverse Fourier-Laplace transformation with respect to Ω, the integration path is transformed to the Landau contour, and with the help of the Pelemey's formula

$$\frac{1}{\Omega - kv + i\sigma} = \frac{\mathcal{P}}{\Omega - kv} - i\pi \frac{1}{|k|} \delta\left(v - \frac{\Omega}{k}\right) \tag{5.55}$$

we get

$$\Delta(\lambda) = \Delta_1(\lambda) + i\Delta_2(\lambda)\frac{K - K'}{|K - K'|} \tag{5.56}$$

$$\Delta_1(\lambda) = 4\pi \sum_\alpha \frac{e_\alpha^2 n_\alpha}{m_\alpha} \mathcal{P} \int \frac{1}{v - \lambda} \frac{\partial F_\alpha}{\partial v} dv \tag{5.57}$$

$$\Delta_2(\lambda) = 4\pi^2 \sum_\alpha \frac{e_\alpha^2 n_\alpha}{m_\alpha} \left.\frac{\partial F_\alpha}{\partial v}\right|_{v=\lambda} \tag{5.58}$$

$$W(\lambda) = -W_1(K_0, \lambda) - iW_2(K_0, \lambda)\frac{K - K'}{|K - K'|} \tag{5.59}$$

$$W_1(\lambda) = 4\pi \sum_\alpha \frac{e_\alpha^3 n_\alpha}{m_\alpha^2} \mathcal{P} \int \frac{1}{(\omega_0 - k_0 v)^2} \frac{1}{v - \lambda} \frac{\partial F_\alpha}{\partial v} dv \tag{5.60}$$

$$W_2(\lambda) = 4\pi^2 \sum \frac{e_\alpha^3 n_\alpha}{m_\alpha^2} \frac{1}{(\omega_0 - k_0 \lambda)^2} \left.\frac{\partial F_\alpha}{\partial v}\right|_{v=\lambda} \tag{5.61}$$

and

$$C(K_0, \lambda) = C_1(K_0, \lambda) - iC_2(K_0, \lambda)\frac{K - K'}{|K - K'|} \tag{5.62}$$

$$C_1(K_0, \lambda) = -4\pi \sum \frac{e_\alpha^4 n_\alpha}{m_\alpha^3} \mathcal{P} \int \frac{1}{(\omega_0 - k_0 v)^2} \cdot \frac{1}{v - \lambda}$$

$$\cdot \frac{\partial}{\partial v}\left(\frac{v - \lambda}{(\omega_0 - k_0 v)^2} \frac{\partial F_\alpha}{\partial v}\right) dv \tag{5.63}$$

$$C_2(K_0, \lambda) = 4\pi^2 \sum_\alpha \frac{e_\alpha^4 n_\alpha}{m_\alpha^3} \frac{1}{(\omega_0 - k_0 \lambda)^2} \left.\frac{\partial F_\alpha}{\partial v}\right|_{v=\lambda}. \tag{5.64}$$

The (K, Ω) and (K', Ω') dependence of the coupling coefficient $M_0(\delta K, \delta K', \delta K'')$ come through only the factor $(K - K')/|K - K'|$ multiplied to the resonant term at $v = \lambda$. Therefore, the reduced kinetic theory coupling coefficient $M_0(\delta K, \delta K', \delta K'')$ is expressed as

$$M_0(\delta K, \delta K', \delta K'') = M_0^R + i\,\text{sgn}\,(K - K')M_0^I \tag{5.65}$$

5.2 Amplitude Modulation of Nearly Monochromatic Waves and ...

$$M_0^R(K_0, \lambda) = -k_0^2 \left[\frac{W_1^2}{\Delta_1} + C_1 \right] + k_0^2 \frac{\Delta_2^2}{\Delta_1(\Delta_1^2 + \Delta_2^2)} \cdot \left[W_1 - \frac{\Delta_1}{\Delta_2} W_1 \right]^2 \quad (5.66)$$

$$M_0^I(K_0, \lambda) = k_0^2 \left[\frac{\Delta_2}{\Delta_1^2 + \Delta_2^2} (W_1^2 - W_2^2) - 2 \frac{\Delta_1}{\Delta_1^2 + \Delta_2^2} W_1 W_2 + C_2 \right]. \quad (5.67)$$

For an arbitrary function $H(K, \Omega)$, we have the following important relationships

$$(2\pi)^{-2} \int d\Omega \int dK \exp\left[i(Kx - \Omega t)\right] \mathrm{sgn}(K) H(K, \Omega)$$

$$= i \frac{\mathcal{P}}{\pi} \int \frac{1}{x - x'} H(x', t) dx', \quad (5.68)$$

and also

$$(2\pi)^{-2} \int d\Omega \int dK \exp\left[i(Kx - \Omega t)\right] \sum_{\delta K'} \sum_{\delta K''} \Big\{ M_2(\delta K, \delta K', \delta K'')$$

$$+ M_0(\delta K, \delta K', \delta K'') \Big\} A(\delta K') A^*(-\delta K'') A(\delta K - \delta K' - \delta K'')$$

$$= \Big\{ M_2(K_0) + M_0^R(K_0, \lambda) \Big\} |A(x,t)|^2 A(x,t)$$

$$- M_0^I(K_0, \lambda) \frac{\mathcal{P}}{\pi} \int \frac{|A(x',t)|^2}{x - x'} dx' A(x,t). \quad (5.69)$$

Substituting Eq. (5.69) into the right-hand side of Eq. (5.29) and (5.36) into the left-hand side of Eq. (5.29) gives the final equation for the modulational dynamic equations.

We obtain an equation to describe the nonlinear self-modulation of the complex amplitude as

kinetic modulation instability

$$i \left(\frac{\partial}{\partial t} + \lambda \frac{\partial}{\partial x} \right) A(x,t) + p \frac{\partial^2}{\partial x^2} A(x,t) + q |A(x,t)|^2 A(x,t)$$

$$+ r \frac{\mathcal{P}}{\pi} \int \frac{|A(x',t)|^2}{x - x'} dx' A(x,t) + i s A(x,t) = 0 \quad (5.70)$$

where the coefficients p, q, r, and s are defined by

$$p = \frac{1}{2} \frac{\partial^2 \omega_0}{\partial k_0^2} \quad (5.71)$$

$$q = - \left(\frac{\partial \varepsilon_r}{\partial \omega_0} \right)^{-1} \left(M_2(k_0, \omega_0) + M_0^R(k_0, \omega_0, \lambda) \right) \quad (5.72)$$

$$r = \left(\frac{\partial \varepsilon_r}{\partial \omega_0}\right)^{-1} M_0^I(k_0, \omega_0, \lambda) \quad (5.73)$$

$$s = \left(\frac{\partial \varepsilon_r}{\partial \omega_0}\right)^{-1} \varepsilon_i(k_0, \omega_0) = \gamma_L \quad (5.74)$$

where γ_L is the linear Landau damping rate of the carrier wave (k_0, ω_0). Thus, we see that the modulational dynamics is governed by a generalized nonlinear cubic Schrödinger equation (Ichikawa et al., 1973).

5.3 Modulational Instability and Nonlinear Landau Damping

In the previous section, the contributions of the resonant particles with velocities close to the group velocity of the strongly dispersive plasma wave were incorporated into the cubic nonlinear Schrödinger equation. Now, we discuss the physical consequence of this nonlinear Landau damping term. The analysis begins by deriving the conditions for the modulational instability of the ion plasma wave.

When the linear Landau damping rate (the s-term) is negligibly small we can reduce Eq. (5.70) to the modified nonlinear Schrödinger equation. Transforming to the translating frame $\xi = x - \lambda t$ traveling with the group velocity of the quasi-monochromatic wave, Eq. (5.70) becomes

$$i\frac{\partial}{\partial t} A(\xi, t) + p \frac{\partial^2}{\partial \xi^2} A + q|A|^2 A + r \frac{\mathcal{P}}{\pi} \int \frac{|A(\xi', t)|^2}{\xi - \xi'} d\xi' A(\xi, t) = 0 \quad (5.75)$$

where the coordinate variable ξ stands for the frame moving with the group velocity defined as $\xi = x - \lambda t$.

Physical insight into the form of the modulational instability follows from introducing a representation of Eq. (5.75) with analog fluid variables for the amplitude ρ and phase σ of $A(\xi, t)$. The complex amplitude can be expressed as

$$A(\xi, t) = \rho^{1/2} \exp\left\{\frac{i}{2p} \int^\xi \sigma \, d\xi'\right\} \quad (5.76)$$

with the real functions $\rho(\xi, t)$ and $\sigma(\xi, t)$. Substitution of Eq. (5.76) into Eq. (5.75) leads to the two-coupled equations for ρ and σ,

$$\frac{\partial}{\partial t}\rho + \frac{\partial}{\partial \xi}(\rho\sigma) = 0 \quad (5.77)$$

$$\frac{\partial}{\partial t}\sigma + \sigma\frac{\partial}{\partial \xi}\sigma = 2pq\frac{\partial}{\partial \xi}\rho + p^2 \frac{\partial}{\partial \xi}\left\{\frac{1}{\sqrt{\rho}}\frac{\partial}{\partial \xi}\left(\frac{1}{\sqrt{\rho}}\frac{\partial \rho}{\partial \xi}\right)\right\} + 2pr\frac{\mathcal{P}}{\pi}\int \frac{\rho(\xi', t)}{\xi - \xi'} d\xi'. \quad (5.78)$$

5.3 Modulational Instability and Nonlinear Landau Damping

Observing the close analogy with the hydrodynamic equation, we notice that the first term on the right-hand side of Eq. (5.78) plays the role of the effective hydrodynamic pressure term when $pq < 0$, and thus when $pq > 0$ the system becomes unstable under the negative pressure. In hydrodynamic analogy the mass density ρ is $|A|^2$ and the velocity σ is the space gradient of the phase, i.e. the local wavenumber.

Stability of the system is analyzed by expanding the intensity function $\rho(\xi, t)$ and the phase function $\sigma(\xi, t)$ as

$$\begin{pmatrix} \rho \\ \sigma \end{pmatrix} = \begin{pmatrix} \rho_0 \\ \sigma_0 \end{pmatrix} + \begin{pmatrix} \rho_1 \\ \sigma_1 \end{pmatrix} \exp\{i(k\xi - \Omega t)\} + \text{c.c.} \quad (5.79)$$

where ρ_1, σ_1 are the complex amplitudes of oscillations about the uniform real values ρ_0, σ_0. Linearization of Eqs. (5.77) and (5.78) gives rise to the dispersion relation

$$(\Omega - \sigma_0 k)^2 = -p\left\{2\rho_0(q - i\,\text{sgn}(k)r) - pk^2\right\}k^2 \quad (5.80)$$

determining $\Omega_\pm(k)$ for nontrivial solutions.

To be definite, let us consider the case $p > 0$ — positive dispersion. Setting

$$\Omega = \Delta\omega + i\Gamma \quad (5.81)$$

with $\Delta\omega$ and Γ real for $k > 0$, we get

$$\Delta\omega = k\,\sigma_0 \mp \left(\frac{p}{2}\right)^{1/2} \left\{\left[(2\rho_0 q - pk^2)^2 + (2r\rho_0)^2\right]^{1/2} - (2q\rho_0 - pk^2)\right\}^{1/2} k \quad (5.82)$$

$$\Gamma = \pm\left(\frac{p}{2}\right)^{1/2} \left\{\left[(2q\rho_0 - pk^2)^2 + (2r\rho_0)^2\right]^{1/2} + (2q\rho_0 - pk^2)\right\}^{1/2} k. \quad (5.83)$$

In the absence of the nonlinear Landau damping term, setting $r = 0$, we find that the plane wave is modulationally unstable if $q > 0$. In this case ($r = 0$) the growth rate reduces to

$$\Gamma = p^{1/2} k (2q\rho_0 - pk^2)^{1/2} \quad (5.84)$$

giving the fastest growing mode at $k_m^2 = q\rho_0/p$ with the growth rate $\Gamma_m = \rho_0 q$. Formula (5.84) confirms that instability occurs corresponding to the negative pressure in the analog hydrodynamical system with $pq > 0$.

If the intensity ρ_0 of the unperturbed plane wave is small enough to be

$$pk^2 \gg (2|q|\rho_0, \quad 2r\rho_0), \quad (5.85)$$

then Eqs. (5.82) and (5.83) reduce to

$$\Delta\omega = \sigma_0 k \mp pk^2 \quad (5.86)$$

$$\Gamma = \pm|r|\rho_0. \quad (5.87)$$

The double signs of Eqs. (5.86) and (5.87) describe the process in which the amplitude in the high frequency region is decaying ($\Gamma < 0$), while the amplitude in the lower frequency region is growing ($\Gamma > 0$). The rate of energy transfer is proportional to the wave intensity ρ_0, which is the description of the well-known process of nonlinear Landau damping in terms of the modulational instability.

On the other hand, if the wave intensity is sufficiently large that the relation

$$2|q|\rho_0 \gg pk^2 \tag{5.88}$$

holds, then the plane wave is unstable with the growth rate of the modulation reducing to

$$\Gamma = \pm \left\{ \left(\sqrt{q^2 + r^2} + q \right) p \right\}^{1/2} \rho_0^{1/2} \tag{5.89}$$

approximately independent of k for $k < (2|q|A_0^2/p)^{1/2}$ satisfying Eq. (5.88).

It is particularly interesting to notice that even if $q < 0$, the nonlinear Landau damping process ($r \neq 0$) drives the wave into the modulational instability. The growth rate for $r \neq 0$ and $q \leq 0$ is given by Eq. (5.83).

The same analysis for the case of $p < 0$ (negative dispersion) confirms the same phenomena of the wave modulation takes place under the action on the nonlinear Landau damping process.

5.3.1 NLS solitons with nonlinear Landau damping

Since the cubic nonlinear Schrödinger equation, given by setting $r = 0$ in Eq. (5.75), is known to be the completely integrable soliton equation, it is interesting to investigate how the nonlinear Landau damping term affects the soliton structure determined by the cubic nonlinear Schrödinger equation. This question has been analyzed by Yajima et al. (1978).

In the absence of the nonlinear Landau damping term in Eq. (5.75), the cubic nonlinear Schrödinger equation with $pq > 0$ admits a two-parameter family of envelope-solitary wave solutions, which vanishes at $\xi = \pm\infty$. With the amplitude A_0 and the speed v as the free parameters, the envelope-solitary waves are

$$A(\xi, t) = A_0 \exp\left[i\left\{\left(\frac{v}{p}\right)\xi - \left(\frac{v^2}{2p} - \frac{qA_0^2}{2}\right)t\right\}\right] \operatorname{sech}\left[\sqrt{\frac{q}{p}} A_0(\xi - vt)\right] \tag{5.90}$$

Setting $v = 0$, we have an envelope solitary wave at rest, which satisfies Eq. (5.75) with $r = 0$. When $r \neq 0$ Yajima et al. (1978) carried out numerical integration of Eq. (5.75) for the initial condition,

$$A(x, t = 0) = A_0 \operatorname{sech}\left[\sqrt{\frac{q}{p}} A_0 x\right]. \tag{5.91}$$

5.3 Modulational Instability and Nonlinear Landau Damping

If $r = 0$, the numerical integration of the cubic nonlinear Schrödinger equation confirms that the solitary wave (5.91) stays at rest. For the initial condition of Eq. (5.91), the nonlocal integral term in Eq. (5.75) (Hilbert transform of $|A|^2$) is positive for $\xi > 0$, and negative for $\xi < 0$. This asymmetric contribution to (5.75) induces the deformation of the initially symmetric envelope form. At the same time, as the amplitude increases with x, the phase factor $(qA^2t/2)$ advances more rapidly at larger x, which in turn drives the solitary wave to start to run to $x > 0$. The process is illustrated in Fig. 5.2a for $r = 0.2$ and in Fig. 5.2b for $r = 0.5$. Now, application of the inverse scattering transformation method to the cubic nonlinear Schrödinger equation, Eq. (5.75) with $r = 0$, following Yajima et al., shows that with the initial condition

$$A(\xi, t = 0) = 2A_0 \operatorname{sech}(B\xi) \quad \text{with} \quad B = \left(\frac{q}{p}\right)^{1/2} A_0 \tag{5.92}$$

the cubic nonlinear Schrödinger equation has the pulsating envelope solution

$$A(\xi, t) = 4A_0 \, \exp(iqA_0^2 t/2) \, \frac{\cosh(3B\xi) + 3\cosh(B\xi)\exp(4iqA_0^2 t)}{\cosh(4B\xi) + 4\cosh(2B\xi) + 3\cos(4qA_0^2 t)}. \tag{5.93}$$

Taking the initial condition as Eq. (5.92), Yajima et al. (1978) solved numerically Eq. (5.75) and observed that the nonlinear Landau damping process acts not only to decompose the envelope into the component solitons, but also it drives them to run in the direction determined by the sign of r (to the positive direction for $r > 0$). Figure 5.3 illustrates the observed evolution of the initial envelope pulse under the action on the nonlinear Landau damping process.

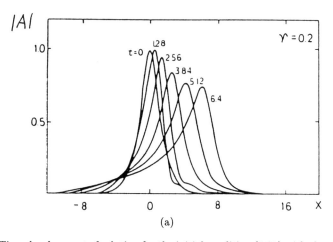

Fig. 5.2. Time development of solution for the initial condition (5.91) with $A = 1$. (a) $r = 0.2$ and (b) $r = 0.5$. In both cases, $p = q = 1$.

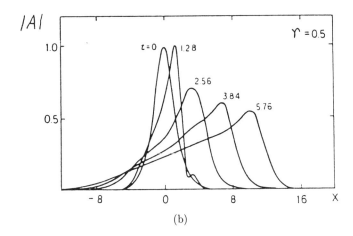

(b)

Fig. 5.2. (*Continued*)

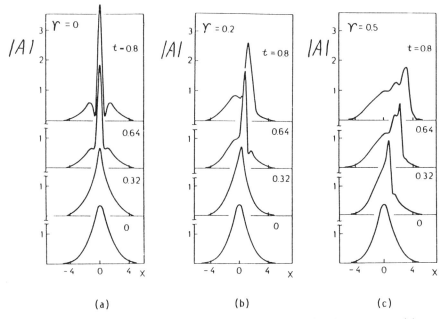

Fig. 5.3. Time development of solution for the initial condition (5.92) with $A = 1$. (a) $r = 0$, (b) $r = 0.2$ and (c) $r = 0.5$. In all cases, $p = q = 1$.

The cubic nonlinear Schrödinger equation is also unstable to transverse perturbations which are studied in the $d = 2$ and $d = 3$ dimensional form of Eq. (5.75) where $\partial_\xi^2 \to \nabla_\xi^2$. The study of the nonlinear Schrödinger equation in higher dimensions $d \geq 2$ is a rich, well-developed field generally involved with the phenomenon of wave collapse and the formation of singularities of the dissipationless equations in a finite time. A special issue of Physica D **52** (1991) is devoted to the study of collapse in the NLS equation and related equations. The article of Kosmatov-Shvets-Zakharov (1991) gives a review and computer simulation of collapse in the NLS.

5.4 Reductive Perturbation Analysis of Nonlinear Wave Propagation

The analysis developed in Secs. 5.1 and 5.2 illustrates the Fourier transformation method to separate the nonlinear slow processes associated with coherent wave distortions. Comparing the analysis in Sec. 5.1 and 5.2 to the well-known mode-mode coupling analysis of the weak turbulence theory where the initial state is a spectrum of random waves, one can identify the critical point of the method which lies in the treatment of *nonlinear self-interaction* of the coherent modes. The coherent self-interactions are lost in the weak turbulence theory but are essential for the formation of the solitons. Although the method developed in Sec. 5.1–5.2 is able to account for the wave-particle resonant interactions, it can not provide further information on the higher order effects. Therefore, it is important to formulate a systematic perturbational approach, in which the lowest order terms account for the nonlinear self-interactions, while the successive higher order terms determine the nonlinear mode-mode coupling effects.

In the early 1960s, Gardner and Morikawa (1960) rediscovered the Korteweg-de Vries equation in studying asymptotic nonlinear propagation of hydromagnetic waves in a cold magnetized plasma. The essential mechanism inherent in their method is to account for the balance for the dispersive wave spreading with the nonlinear steepening of the wave. Taniuti and his collaborators have developed a general formulation called the *reductive perturbation theory* for taking into account this balance. Here, taking the ion acoustic wave as an example, we present the reductive perturbation method in some detail.

5.4.1 Nonlinear ion acoustic wave

One-dimensional wave propagation in a two-component plasma composed of cold ions and isothermal warm electrons is described in terms of the normalized fluid dynamic equation given by

$$\frac{\partial}{\partial t} n(x,t) + \frac{\partial}{\partial x}(nu) = 0 \tag{5.94}$$

$$\frac{\partial}{\partial t} u(x,t) + u \frac{\partial}{\partial x} u = -\frac{\partial}{\partial x} \varphi \qquad (5.95)$$

$$\frac{\partial^2}{\partial x^2} \varphi(x,t) = n_e(\varphi) - n(x,t) \qquad (5.96)$$

where
$$n_e(x,t) = \exp[\varphi(x,t)]. \qquad (5.97)$$

In Eqs. (5.94)–(5.97) the space-time variables x and t are normalized by the electron Debye length $\lambda_D = (\kappa_B T_e/4\pi e^2 n_0)^{1/2}$, and the inverse of the ion plasma frequency $\Omega_p = (4\pi e^2 n_0/M)^{1/2}$. The ion density n and the electron density n_e are normalized by the unperturbed average number density n_0, while the velocity u is measured in units of the ion sound velocity $c_s = (\kappa_B T_e/M)^{1/2}$. The electrostatic potential φ is normalized by $\kappa_B T_e/e$. Here κ_B is the Boltzmann constant.

For the given set of nonlinear partial differential equations of Eqs. (5.94)–(5.97), let us begin with the analysis of their linearized structure. The linearized equation takes the form of

$$\frac{\partial^2}{\partial t^2} \varphi - \frac{\partial^2}{\partial x^2} \varphi - \frac{\partial^4}{\partial t^2 \partial x^2} \varphi = 0 \qquad (5.98)$$

which gives rise to the linear dispersion relation

$$\omega(k) = \pm \frac{k}{(1+k^2)^{1/2}} \qquad (5.99)$$

for ion acoustic waves. In the long wavelength limit $k^2 \ll 1$, the dispersion becomes weak with $\omega(k)$ becoming

$$\omega(k) = \pm k \left(1 - \frac{1}{2}k^2 + \frac{3}{8}k^4 - \cdots \right). \qquad (5.100)$$

The space-time variation of nonlinear wave is characterized by the complex phase factor as

$$\exp\{i(kx - \omega t)\} \approx \exp\left\{i\left[k(x-t) + \frac{1}{2}k^3 t + \cdots + (\text{nonlinear effects})\right]\right\}. \qquad (5.101)$$

Thus, identifying the wavenumber k as the small parameter of the order of $\varepsilon^{1/2}$, we introduce the new space-time coordinates

$$\xi = \varepsilon^{1/2}(x - t) \qquad (5.102)$$

$$\tau = \varepsilon^{3/2} t. \qquad (5.103)$$

Equations (5.102)–(5.103) indicate that in the moving coordinate with the ion acoustic velocity, the nonlinear effect manifests itself as the slow temporal variation.

5.4 Reductive Perturbation Analysis of Nonlinear Wave Propagation

Transforming Equations (5.94)–(5.97) into the new (ξ, τ) variables, we get

$$\varepsilon \frac{\partial}{\partial \tau} n - \frac{\partial}{\partial \xi} n + \frac{\partial}{\partial \xi}(nu) = 0 \tag{5.104}$$

$$\varepsilon \frac{\partial}{\partial \tau} u - \frac{\partial}{\partial \xi} u + u \frac{\partial}{\partial \xi} u = -\frac{\partial}{\partial \xi} \varphi \tag{5.105}$$

$$\varepsilon \frac{\partial^2}{\partial \xi^2} \varphi = \exp(\varphi) - n. \tag{5.106}$$

As for the dynamical variables n, u, and φ, we expand them around the unperturbed constant state as

$$\begin{pmatrix} n \\ u \\ \varphi \end{pmatrix} = \begin{pmatrix} 1 \\ 0 \\ 0 \end{pmatrix} + \sum_{\alpha=1}^{\infty} \varepsilon^{\alpha} \begin{pmatrix} n^{(\alpha)} \\ u^{(\alpha)} \\ \varphi^{(\alpha)} \end{pmatrix}. \tag{5.107}$$

Substitution of Eq. (5.107) into Eqs. (5.104)–(5.106) gives rise to a relationship among $n^{(1)}, u^{(1)}$, and $\varphi^{(1)}$ given as

$$n^{(1)}(\xi, \tau) = u^{(1)}(\xi, \tau) = \varphi^{(1)}(\xi, \tau) + u(\infty) \tag{5.108}$$

where $u(\infty)$ stands for the asymptotic velocity at $\xi \to \infty$. Unlike the ordinary perturbation theory, the first order quantities are not determined at the first order but in second order.

Proceeding up to the second order in the expansion in Eq. (5.107), Eqs. (5.104) to (5.106) become

$$\frac{\partial}{\partial \tau} n^{(1)} - \frac{\partial}{\partial \xi} n^{(2)} + \frac{\partial}{\partial \xi} u^{(2)} + \frac{\partial}{\partial \xi} \left(n^{(1)} u^{(1)}\right) = 0 \tag{5.109}$$

$$\frac{\partial}{\partial \tau} u^{(1)} - \frac{\partial}{\partial \xi} u^{(2)} + u^{(1)} \frac{\partial}{\partial \xi} u^{(1)} = -\frac{\partial}{\partial \xi} \varphi^{(2)} \tag{5.110}$$

$$\frac{\partial^2}{\partial \xi^2} \varphi^{(1)} = \varphi^{(2)} + \frac{1}{2} \varphi^{(1)} \varphi^{(1)} - n^{(2)}. \tag{5.111}$$

Adding both sides of Eqs. (5.109)–(5.110), we can write down an expression for the quantity $\partial \left(\varphi^{(2)} - n^{(2)}\right)/\partial \xi$, which in turn is calculated from Eq. (5.111) as $\partial^3 \varphi^{(1)}/\partial \xi^3 - \varphi^{(1)} \partial \varphi^{(1)}/\partial \xi$. Thus, with the help of Eq. (5.108) with $u^{(1)}(\infty) = 0$, we obtain the Korteweg-de Vries equation. This procedure has eliminated the second order quantities $n^{(2)}, u^{(2)}$, and $\varphi^{(2)}$ and gives rise to the Korteweg-de Vries equation

$$\frac{\partial}{\partial \tau} n^{(1)} + n^{(1)} \frac{\partial}{\partial \xi} n^{(1)} + \frac{1}{2} \frac{\partial^3}{\partial \xi^3} n^{(1)} = 0. \tag{5.112}$$

At the same time, from the elimination procedure we get the following relationships among the second order quantities $n^{(2)}$, $u^{(2)}$, and $\varphi^{(2)}$,

$$u^{(2)} = n^{(2)} + \frac{\partial^2}{\partial \xi^2} n^{(1)} - \frac{3}{4}(n^{(1)})^2 \tag{5.113}$$

$$\varphi^{(2)} = n^{(2)} + \frac{\partial^2}{\partial \xi^2} n^{(1)} - \frac{1}{2}(n^{(1)})^2. \tag{5.114}$$

Going up to the third order in ε, we then obtain the equation for the second order density perturbation as

$$\frac{\partial}{\partial \tau} n^{(2)} + \frac{1}{2} \frac{\partial^3}{\partial \xi^3} n^{(2)} + \frac{\partial}{\partial \xi}\left(n^{(1)} n^{(2)}\right) = S(n^{(1)}) \tag{5.115}$$

where

$$S(n^{(1)}) = -\frac{3}{8} \frac{\partial^5}{\partial \xi^5} n^{(1)} + \frac{1}{2} n^{(1)} \frac{\partial^3}{\partial \xi^3} n^{(1)} - \frac{5}{8} \frac{\partial}{\partial \xi}\left(\frac{\partial}{\partial \xi} n^{(1)}\right)^2. \tag{5.116}$$

We emphasize that Eq. (5.115) is linear in the unknown second order density, and contribution of various nonlinear terms in the original set of Eqs. (5.94)–(5.97) are isolated into the nonlinear term of Eq. (5.116). Also, it is worthwhile to observe that the first term of the right-hand side of Eq. (5.116) stands for the next order dispersive term of the linear dispersion relation given in Eq. (5.100). The second and third terms on the right-hand side give nonlinear modifications to the K-dV solitons.

In the higher order contributions the nonlinear effects describe interaction among the first order, yet essentially nonlinear soliton solution of the Korteweg-de Vries equation (5.112), with account of the higher order dispersive effects. The fundamental properties of the soliton solution of Eq. (5.112) will be discussed in the following sections. Thus, in the framework of the reductive perturbation theory, the higher order nonlinear effects describe the interaction among the lowest order nonlinear mode $\varphi^{(1)}$ as expressed in Eq. (5.116), and in turn it leads to the higher order modification $\varphi^{(2)}$ through Eq. (5.115) and Eqs. (5.114). At the same time, the higher order dispersive term of Eq. (5.100) competes with the nonlinear interaction effect among the lowest order nonlinear mode $\varphi^{(1)}$. Since the solution $\varphi^{(1)}$ of Eq. (5.112) is characterized by the novel physical structure called the *soliton*, we conclude that the lowest order analysis in the reductive perturbation theory accounts for the critical aspect of nonlinearity, giving rise to the soliton mode. The higher order contributions account for the distortion of the soliton structure as the result of the soliton-soliton interactions.

5.4.2 Parallel propagating right and left polarized Alfvén waves

As another illustration of the reductive perturbation theory, we discuss the nonlinear propagation of magnetohydrodynamic waves along the uniform external magnetic field, $\mathbf{B}_0$ which we choose to be the x-axis. Under the quasi-neutrality condition, the magnetodydrodynamic equations are

$$\frac{\partial}{\partial t} n + \frac{\partial}{\partial x}(nu) = 0 \tag{5.117}$$

$$\left(\frac{\partial}{\partial t} + u\frac{\partial}{\partial x}\right) u + \frac{1}{n}\frac{\partial}{\partial x}\left(\frac{1}{2}\left(B_y^2 + B_z^2\right)\right) = 0 \tag{5.118}$$

$$\left(\frac{\partial}{\partial t} + u\frac{\partial}{\partial x}\right) v - \frac{1}{n} B_x \frac{\partial}{\partial x} B_y = -\frac{1}{\omega_{ce}} \left(\frac{\partial}{\partial t} + u\frac{\partial}{\partial x}\right)\left(\frac{1}{n}\frac{\partial}{\partial x} B_z\right) \tag{5.119}$$

$$\left(\frac{\partial}{\partial t} + u\frac{\partial}{\partial x}\right) w - \frac{1}{n} B_x \frac{\partial}{\partial x} B_z = +\frac{1}{\omega_{ce}} \left(\frac{\partial}{\partial t} + u\frac{\partial}{\partial x}\right)\left(\frac{1}{n}\frac{\partial}{\partial x} B_y\right) \tag{5.120}$$

$$\left(\frac{\partial}{\partial t} + u\frac{\partial}{\partial x}\right) B_y - B_x \frac{\partial}{\partial x} v + B_y \frac{\partial}{\partial x} u = +\frac{1}{\omega_{ci}} \frac{\partial}{\partial x}\left(\left(\frac{\partial}{\partial t} + u\frac{\partial}{\partial x}\right) w\right) \tag{5.121}$$

$$\left(\frac{\partial}{\partial t} + u\frac{\partial}{\partial x}\right) B_z - B_x \frac{\partial}{\partial x} w + B_z \frac{\partial}{\partial x} u = -\frac{1}{\omega_{ci}} \frac{\partial}{\partial x}\left(\left(\frac{\partial}{\partial t} + u\frac{\partial}{\partial x}\right) v\right) \tag{5.122}$$

where the plasma density n, the magnetic flux (B_x, B_y, B_z), the ion velocity (u, v, w), time t, and space coordinate x are normalized in terms of the uniform plasma number density n_0, the constant uniform magnetic field B_0, the Alfvén velocity v_A, the inverse of the characteristic frequency ω_0 and the length $L = v_A\,\omega_0^{-1}$. The normalized electron and ion cyclotron frequencies are given as $\omega_{ce} = (eB_0/m_e\,c\omega_0)$, and $\omega_{ci} = (eB_0/m_i\,c\omega_0)$, respectively.

When the wave is propagating along the external magnetic field, $B_x = B_0$, the space-time units are equivalent to setting B_0 to unity. The linear dispersion relation is given as

$$\omega(k) = k \pm \mu k^2, \qquad \mu = \frac{1}{2}\left(\omega_{ci}^{-1} - \omega_{ce}^{-1}\right) \tag{5.123}$$

where the upper sign stands for the right and the lower sign for the left circularly polarized Alfvén waves. Transforming the independent variables (x, t) into

$$\xi = \varepsilon(x - t), \qquad \tau = \varepsilon^2 t \tag{5.124}$$

we expand the dynamical variables as

$$\begin{cases} n = 1+ \varepsilon n^{(1)} + \varepsilon^2 n^{(2)} + \cdots \\ u = \quad \varepsilon u^{(1)} + \varepsilon^2 u^{(2)} + \cdots \\ v = \quad \varepsilon^{1/2} \left(v^{(1)} + \varepsilon v^{(2)} + \cdots \right) \\ w = \quad \varepsilon^{1/2} \left(w^{(1)} + \varepsilon w^{(2)} + \cdots \right) \\ B_y = \quad \varepsilon^{1/2} \left(B_y^{(1)} + \varepsilon B_y^{(2)} + \cdots \right) \\ B_z = \quad \varepsilon^{1/2} \left(B_z^{(1)} + \varepsilon B_z^{(2)} + \cdots \right) \end{cases} \quad (5.125)$$

where we notice that the ordering of the transverse quantities v, w, B_y, and B_z must start with $\varepsilon^{1/2}$. The straightforward analysis leads to the following coupled equations for $B_y^{(1)}$ and $B_z^{(1)}$,

$$\begin{cases} \dfrac{\partial}{\partial \tau} B_y^{(1)} + \dfrac{1}{4} \dfrac{\partial}{\partial \xi} \left[\left(\left(B_y^{(1)} \right)^2 + \left(B_z^{(1)} \right)^2 \right) B_y^{(1)} \right] - \mu \dfrac{\partial^2}{\partial \xi^2} B_z^{(1)} = 0 \\ \dfrac{\partial}{\partial \tau} B_z^{(1)} + \dfrac{1}{4} \dfrac{\partial}{\partial \xi} \left[\left(\left(B_y^{(1)} \right)^2 + \left(B_z^{(1)} \right)^2 \right) B_z^{(1)} \right] + \mu \dfrac{\partial^2}{\partial \xi^2} B_y^{(1)} = 0. \end{cases} \quad (5.126)$$

The coupled equations (5.126) can be expressed by the single complex equation
derivative NLS

$$i \frac{\partial}{\partial t} q \pm \mu \frac{\partial^2}{\partial \xi^2} q + i \frac{1}{4} \frac{\partial}{\partial \xi} \left\{ |q|^2 q \right\} = 0, \quad (5.127)$$

upon introducing the complex magnetic field

$$B_R = B_y^{(1)} - i B_z^{(1)}, \qquad B_L = B_y^{(1)} + i B_z^{(1)} \quad (5.128)$$

and $q = B_R$ and $q^* = B_L$ where the $\pm$ in Eq. (5.127) apply to the right- and left-hand polarizations, respectively. Comparing the structure of Eq. (5.127) with Eq. (5.75) with $r = 0$, we call Eq. (5.127) the derivative (cubic) nonlinear Schrödinger equation.

5.4.3 General nonlinear equation for oblique waves

For the magnetohydrodynamic wave propagating in a direction of finite angle with respect to the external magnetic field, Kakutani et al. (1968) developed another scheme of the reductive perturbation analysis for the set of Eqs. (5.117)–(5.122), deriving the modified Korteweg-de Vries equation
modified KdV

$$\frac{\partial}{\partial t} n^{(1)} + \alpha (n^{(1)})^2 \frac{\partial}{\partial \xi} n^{(1)} + \mu \frac{\partial^3}{\partial \xi^3} n^{(1)} = 0. \quad (5.129)$$

5.4 Reductive Perturbation Analysis of Nonlinear Wave Propagation

Turning to the strongly dispersive wave, Asano et al. (1969) have developed the reductive perturbation theory for a general nonlinear partial differential equation given as

$$\frac{\partial}{\partial t} U + A(U) \frac{\partial}{\partial x} U + B(U) = 0, \qquad (5.130)$$

where U is an n-dimension vector with components $(u_1, u_2, \ldots, u_n)$ and $A(U), B(U)$ are the $n \times n$ matrices. The unperturbed constant state is determined by

$$B(U_0) = 0. \qquad (5.131)$$

When the linearization of Eq. (5.130) around the constant state U_0 leads to the dispersion relation

$$\omega(k) = \omega_0 + \lambda k^2, \qquad (5.132)$$

the physical quantity U is expanded into harmonics of the fundamental carrier wave $(k, \omega(k))$ given by

$$U = U_0 + \sum_{\ell=1}^{\infty} U_\ell(\xi, \tau) \exp\{i\ell(kx - \omega(k)t)\}, \qquad (5.133)$$

where the slow variables (ξ, τ) are defined to be

$$\xi = \varepsilon(x - \lambda t), \qquad \tau = \varepsilon^2 t. \qquad (5.134)$$

The complex amplitude modulation $U_\ell(\xi, \tau)$ is investigated on the basis of perturbational expansion

$$U^{(1)} = \begin{pmatrix} U_1 \\ U_2 \\ \vdots \end{pmatrix} q(\xi, \tau)$$

$$U_\ell(\xi, \tau) = \sum_\alpha \varepsilon^\alpha U_\ell^{(\alpha)}(\xi, \tau). \qquad (5.135)$$

It is instructive to remark that Eq. (5.133) with Eq. (5.134) suggests the original differential equation Eq. (5.130) is transformed to

$$\left(\frac{\partial}{\partial t} - \varepsilon \frac{\partial}{\partial \xi} + \varepsilon^2 \frac{\partial}{\partial \tau} \right) U + A(U) \left(\frac{\partial}{\partial x} + \varepsilon \frac{\partial}{\partial \xi} \right) U + B(U) = 0. \qquad (5.136)$$

Decomposing Eq. (5.136) up to the third order in ε, we obtain the cubic nonlinear Schrödinger equation

NLS eq.

$$i \frac{\partial}{\partial \tau} q + \frac{1}{2} \frac{\partial \lambda}{\partial k} \frac{\partial^2}{\partial \xi^2} q + \alpha |q|^2 q = 0, \qquad (5.137)$$

where the complex amplitude $q(\xi, \tau)$ describes the nonlinear modulation of the carrier wave of the wavenumber k.

5.5 Birth of the Soliton

In the early 1950s the first high speed electronic computer became available. Fermi, Pasta, and Ulam have carried out the numerical experiment to search for the conditions for the *ergodic behavior* of the nonlinear lattice. They examined the temporal evolution of a one-dimensional nonlinear lattice composed of 16, 32, and 64 particles. Taking the initial mode as the shape of the fundamental linear sinusoidal mode, but with large displacement, they followed the process of harmonic generation through the nonlinear coupling. Contrary to their expectation of seeing equal excitation of all the modes after a long time, after the elapse of a finite time the system returned to its original state. This dynamical process is named the Fermi-Pasta-Ulam recurrence phenomena.

Several years of struggle to understand this unexpected behavior of the system led Zabusky and Kruskal (1965) to explore the secret of the recurrence phenomena by reducing the original set of discrete coupled equations of motion for the lattice to the Korteweg-de Vries equation in the continuum limit. Let us introduce the Korteweg-de Vries (KdV) equation by the operator $K[q]$ where

$$K[q] \equiv \frac{\partial}{\partial t} q + \frac{\partial^3}{\partial x^3} q + \varepsilon q \frac{\partial}{\partial x} q = 0, \tag{5.138}$$

which is reduced to Eq. (5.112) by appropriate scaling of the coordinate x and the dynamical variable q. Now, since 1895, it was known that Eq. (5.138) has a stationary solitary wave solution

$$q(x,t) = \frac{12}{\varepsilon} \kappa^2 \operatorname{sech}^2 \left(\kappa(x - 4\kappa^2 t) \right), \tag{5.139}$$

where the constant κ characterizes the size $\Delta x = 1/\kappa$ of the solitary wave. It should be emphasized here that since Eq. (5.139) becomes singular for the limit of $\varepsilon \to 0$, the solitary wave solution Eq. (5.139) is the genuine nonlinear entity, and it can not be constructed by the perturbational approach of the nonlinear term in Eq. (5.138). Equation (5.139) tells us that the large solitary wave has a sharper localization and moves faster $(v = 4\kappa^2)$ than a small solitary wave.

Zabusky and Kruskal carried out numerical integration of Eq. (5.138) by letting a small solitary wave go first; after a while a larger solitary wave follows it. Since Eq. (5.138) is a nonlinear equation, when the larger solitary wave catches up to the slow-going smaller solitary wave, the nonlinear interaction between them will induce strong distortion of wave forms. To their surprise, however, these two solitary waves pass through each other emerging with their same shapes. In other words, the nonlinear term of Eq. (5.138) allows the superposition of nonlinear solitary waves with different amplitudes. Zabusky and Kruskal proposed calling these solitary waves "solitons," and gave the coherent superposition of the solitons as the explanation for the Fermi-Pasta-Ulam recurrence phenomena by pointing out that the initial nonlinear distortion contains a finite number of solitons, and as time elapses, the larger solitons emerge and run faster. Since the system is periodic the large solitons catch up to the slow-going smaller solitons, and thus the system returns to its initial state.

Why does the solitary wave of the Korteweg-de Vries equation, Eq. (5.138), exhibit such peculiar behavior? The answer for this question is given by Gardner et al. (1967) by the discovery of the inverse scattering transformation.

5.6 The Inverse Scattering Transformation Method

In the field of fluid dynamics the Burger's equation

$$\frac{\partial}{\partial t} q + q \frac{\partial}{\partial x} q = \delta^2 \frac{\partial^2}{\partial x^2} q \qquad (5.140)$$

describes the one-dimensional shock wave and turbulence phenomena, where the nonlinear effect competes with the dissipative effect in the right-hand side of Eq. (5.140). It has been known that the nonlinear Burgers equation can be put into a linear form by the Hopf-Cole transformation

$$q = -2\delta^2 \frac{\psi_x}{\psi} \qquad (5.141)$$

to the

$$\frac{\partial}{\partial t} \psi = \delta^2 \frac{\partial^2}{\partial x^2} \psi. \qquad (5.142)$$

Observing the similarity of the Burgers equation, Eq. (5.140), and the Korteweg-de Vries equation, Eq. (5.138), Miura was challenged to find a transformation to linearize the Korteweg-de Vries equation, Eq. (5.138). He found, however, that the nonlinear transformation

$$q = r^2 + r_x \qquad (5.143)$$

reduces the Korteweg-de Vries equation as

$$K[q] = \left(2r + \frac{\partial}{\partial x}\right) M[r] \qquad (5.144)$$

where
<u>modified KdV</u>

$$M[r] \equiv \frac{\partial}{\partial t} r + \frac{\partial^3}{\partial x^3} r - 6r^2 \frac{\partial}{\partial x} r = 0. \qquad (5.145)$$

Equation (5.145) is the modified Korteweg-de Vries equation. Equation (5.144) and (5.145) tell us that if we could obtain a solution r of the modified Korteweg-de Vries equation, Eq. (5.145), then we could construct a solution q of the Korteweg-de Vries equation, Eq. (5.138) through the transformation of Eq. (5.143). Now, this is just opposite to what we had aimed to achieve.

However, at this stage, Kruskal noticed that the Riccati equation for r, Eq. (5.143) is linearized by a transformation

$$r = \frac{\psi_x}{\psi} \tag{5.146}$$

that brings Eq. (5.143) to the form of

$$\psi_{xx} - q\psi = 0. \tag{5.147}$$

Furthermore, noticing that the Korteweg-de Vries equation, Eq. (5.138), is invariant for the Galilei transformation $q \to q - \lambda$, Kruskal and his collaborator noticed that Eq. (5.147) could be extended to

$$\psi_{xx} - (q - \lambda)\psi = 0. \tag{5.148}$$

Now, when q is evolving in time in accord with the Korteweg-de Vries equation, Eq. (5.138), how does the eigenvalue λ and the eigenfunction ψ evolve in time? Rewriting Eq. (5.148) as

$$q = \lambda + \frac{\psi_{xx}}{\psi}, \tag{5.149}$$

we may substitute Eq. (5.149) into the Korteweg-de Vries equation, Eq. (5.138), and obtain

$$\lambda_t \psi^2 + [\psi Q_x - \psi_x Q]_x = 0 \tag{5.150}$$

where Q is defined as

$$Q \equiv \frac{\partial}{\partial t}\psi + 4\frac{\partial^3}{\partial x^3}\psi - 3q\frac{\partial}{\partial x}\psi - 3\frac{\partial}{\partial x}(q\psi). \tag{5.151}$$

Therefore, for the boundary condition of $\psi \to 0$ at $x \to \pm\infty$, or for the periodic boundary condition, since the integral of ψ^2 does not vanish, we obtain the wonderful result of

$$\frac{\partial \lambda}{\partial t} = \lambda_t \equiv 0 \tag{5.152}$$

from Eq. (5.149).

Then, Eq. (5.150) in turn tells us that

$$\psi Q_x - \psi_x Q = C(t). \tag{5.153}$$

Dividing both sides by ψ^2, and integrating with respect to x, we obtain an expression for Q to within an integration constant, which should be identically equal to zero.

5.6 The Inverse Scattering Transformation Method

In other words, we get $Q = 0$, namely we get from Eq. (5.151) that

$$\frac{\partial}{\partial t}\psi = \left(-\frac{\partial^3}{\partial x^3}\psi + 3(q+\lambda)\frac{\partial}{\partial x}\psi\right) \tag{5.154}$$

from the condition $\lambda_t = 0$ and the boundary condition described above.

What we have accomplished here is to show that we have the eigenvalue problem Eq. (5.148) whose potential $q(x,t)$ is evolving in time in accord with the Korteweg-de Vries equation, Eq. (5.138), and when the potential $q(x,t)$ is changing in time, the eigenfunction $\psi(x,t)$ evolves in time in accord with Eq. (5.154). Yet the eigenvalue λ remains constant as confirmed in Eq. (5.152).

Now, in order to obtain a solution of $q(x,t)$ of the Korteweg-de Vries equation, Eq. (5.138) for a given initial condition $q(x,t=0)$, we can follow these steps to determine the eigenfunction $\psi(x,t=0)$ and the eigenvalue λ for this given potential $q(x,t=0)$. Then, even though we are ignorant about $q(x,t)$, we can determine the eigenfunction $\psi(x,t)$ in the region where $q(x,t)$ vanishes, namely in the region far from the center of the localized potential $q(x,t)$. The last step is then to construct the potential $q(x,t)$ from the knowledge of the asymptotic eigenfunction $\psi(x,t)$ in the scattering region of $|x| \to \infty$ where $|q| \to 0$.

Contrary to the ordinary scattering problem, where we are obtaining the eigenstates of the given potential $q(x,t)$, here we have to determine the potential $q(x,t)$ from the scattering data. Thus, we are dealing with the *inverse scattering problem*. Since the procedure is central to solving nonlinear integrable systems let us repeat the steps more specifically.

The inverse scattering method of solving the initial value problem of the KdV equation, (5.138) proceeds as follows:

(i) for a given initial condition $q(x,t=0)$, solve the eigenvalue problem, (5.148). As the discrete eigenstates, we can determine the eigenvalues,

$$\lambda_n = -\kappa_n^2 \quad (\kappa_n \text{ is real}, n = 1, 2, \ldots, N) \tag{5.155}$$

with the asymptotic eigenfunctions,

$$\psi_n(x) = c_n(0)\exp(\mp\kappa_n x), \quad x \to \pm\infty \tag{5.156}$$

where $c_n(0)$ is determined from the normalization condition of the eigenfunction $\psi(x,t=0)$. For the continuum eigenstates, we have the eigenvalues

$$\lambda = k^2 \quad (\text{for all real } k) \tag{5.157}$$

and the asymptotic wave functions corresponding to the incident wave coming from the left, given as

$$\psi(x) = \begin{cases} a(k)\exp(ikx), & x \to \infty \\ \exp(ikx) + b(k)\exp(-ikx), & x \to -\infty \end{cases} \tag{5.158}$$

where $a(k, t = 0)$ and $b(k, t = 0)$ are the transmission and reflection coefficient, respectively.

(ii) in the course of temporal evolution $t > 0$, the initial potential $q(x, t = 0)$ changes its shape according to the KdV equation, Eq. (5.138), and evolves into $q(x, t)$. During this temporal evolution, however, the eigenvalues $\lambda_n = -\kappa_n^2$ and $\lambda = k^2$ remain constant, as insured by Eq. (5.152). Thus, temporal variation of the coefficients of the wave function are determined by substitution of Eqs. (5.156) and (5.158) into Eq. (5.154) as

$$c_n(t) = c_n(0) \exp(-4\kappa_n^3 t) \tag{5.159}$$

$$a(k, t) = a(k, 0) \tag{5.160}$$

$$b(k, t) = b(k, 0) \exp(-9i\, k^3 t) \tag{5.161}$$

(iii) lastly, the potential $q(x, t)$ at time t can be constructed from the time evolved scattering data as

$$q(x, t) = 2 \frac{d}{dx} K(x, x) \tag{5.162}$$

where the function $K(x, x)$ is a solution of the Gelfand-Levitan equation
G-L eq

$$F(x + y) + K(x, y) + \int_{-\infty}^{x} F(y + z) K(x, z) dz = 0 \tag{5.163}$$

with the definition of

$$F(x, y) = \sum_n c_n^2(t) \exp(\kappa_n(x + y)) + \frac{1}{2\pi} \int b(k, t) \exp(-ik(x + y)) dk. \tag{5.164}$$

5.7 Gelfand-Levitan Equation for the Scattering Potential

The Gelfand-Levitan equation is a linear integral equation, which can be solved to yield the N-soliton solution in the case of reflectionless $b(k) = 0$ potentials $q(x, t = 0)$.

As an illustrative example of how the Gelfand-Levitand equation is solved, let us consider temporal evolution of the initial pulse,

$$q(x, t = 0) = -2 \operatorname{sech}^2(x). \tag{5.165}$$

The eigenvalue problem Eq. (5.148) for this initial potential gives rise to one bound state having an eigenvalue of

$$\lambda = -\kappa_1^2 = -1 \tag{5.166}$$

5.7 Gelfand-Levitan Equation for the Scattering Potential

with the eigenfunction
$$\psi_1(x) = \frac{1}{\sqrt{2}} \operatorname{sech}(x). \tag{5.167}$$

Hence, we have $c_1(0) = \sqrt{2}$. For the continuum state, $b(k) = 0$, that is there is only transmission, no reflection. Equation (5.159) gives the scattering data at time t as $c_1(t) = \sqrt{2} \exp(-4t)$. To find the potential $q(x,t)$, we calculate the kernel of Eq. (5.164) as
$$F(x,y) = 2 \exp(-8t + (x+y)) \tag{5.168}$$
giving rise to the Gelfand-Levitan equation, Eq. (5.163), as
$$K(x+y) + 2e^{-8t+x+y} - 2e^{-8t+y} \int_{-\infty}^{x} e^{z} K(x,z) dz = 0. \tag{5.169}$$

Setting
$$K(x,y) = L(x) \exp(y) \tag{5.170}$$
we obtain an algebraic equation for $L(x)$,
$$\left\{1 + e^{-(8t-2x)}\right\} L(x) + 2e^{(-8t-2x)} = 0. \tag{5.171}$$

Then, we finally obtain
$$K_t(x, y=x) = -2 \left(e^{8t-2x} + 1\right)^{-1}, \tag{5.172}$$

which gives rise to the potential $q(x,t)$ as
$$q(x,t) = -2 \operatorname{sech}^2(x - 4t). \tag{5.173}$$

For a deeper potential at $t = 0$, given as
$$q(x,t) = -6 \operatorname{sech}^2(x) \tag{5.174}$$

the eigenvalue problem, Eq. (5.148), determines two bound states with the eigenvalues and the corresponding eigenfunctions as
$$\lambda_1 = -\kappa_1^2 = -4, \qquad \psi_1 = \frac{\sqrt{3}}{2} \operatorname{sech}^2(x) \tag{5.175}$$

and
$$\lambda_2 = -\kappa_2^2 = -1, \qquad \psi_2 = \sqrt{\frac{3}{2}} \operatorname{sech}(x) \tanh(x). \tag{5.176}$$

Solving the Gelfand-Levitan equation, Eq. (5.163), we obtain the potential $q(x,t)$ as
$$q(x,t) = -12 \left(\frac{3 + 4 \cosh(2x - 8t) - \cosh(4x - 64t)}{[\cosh(3x - 36t) + 3\cosh(x - 28t)]^2} \right) \tag{5.177}$$

which describes the collision process of two solitons

$$q_1(x,t) = -8\,\text{sech}^2(2x - 32t), \quad x \to -\infty \tag{5.178}$$

$$q_2(x,t) = -2\,\text{sech}^2(x - 4t + \delta), \quad x \to -\infty. \tag{5.179}$$

The student may wish to confirm that there occurs a finite phase jump during the collision process $\left(\delta = \frac{1}{2}\log 3\right)$.

The above analysis reveals that the preservation of the individual identity of the solitary wave solution of the KdV equation is insured by the time invariance of the eigenvalues of the associated eigenvalue equation, and that the soliton is characterized as the bound state of the eigenvalue problem, in which the localized solitary wave acts as the potential.

5.8 Inverse Scattering Transform (IST) for the NLS Equation

For nearly five years after the discovery of Gardner et al. (1967), it was a shared belief that the inverse scattering method would only be applicable to the elite KdV equation. In 1972, however, Zakharov and Shabat (1972) showed that the cubic nonlinear Schrödinger equation

$$i\frac{\partial}{\partial t}q + \frac{1}{2}\frac{\partial^2}{\partial x^2}q + \varepsilon^2|q|^2 q = 0 \tag{5.180}$$

can also be decomposed into an eigenvalue problem. The method used to find the inverse scattering transform for the NLS equation is to introduce a pair of potentials $u_1(x,t)$ and $u_2(x,t)$ in a manner given by Lax (1968). These potentials are called Lax pairs and will be important for finding a general procedure in the next section.

The decomposition of the NLS equation is performed in terms of u_1 and u_2 where

$$\frac{\partial}{\partial x}u_1 = -i\lambda\,u_1 + \varepsilon q\,u_2 \tag{5.181}$$

$$\frac{\partial}{\partial x}u_2 = -\varepsilon q^*\,u_1 + i\lambda\,u_2, \tag{5.182}$$

and the temporal evolution equations are

$$\frac{\partial}{\partial t}u_1 = -i\left(\lambda^2 - \frac{\varepsilon^2}{2}|q|^2\right)u_1 - \varepsilon\left(\lambda q + \frac{i}{2}q_x\right)u_2 \tag{5.183}$$

$$\frac{\partial}{\partial t}u_2 = \varepsilon\left(-\lambda q^* - \frac{i}{2}q_x^*\right)u_1 + i\left(\lambda^2 - \frac{\varepsilon^2}{2}|q|^2\right)u_2. \tag{5.184}$$

5.8 Inverse Scattering Transform (IST) for the NLS Equation

It is crucial to observe that the compatibility condition of Eqs. (5.181) through (5.184)

$$u_i)_{xt} = u_i)_{tx} \quad (i = 1, 2) \tag{5.185}$$

with the requirement of the isospectral condition

$$\frac{\partial}{\partial t}\lambda = 0 \tag{5.186}$$

leads to the cubic nonlinear Schrödinger equation (5.180).

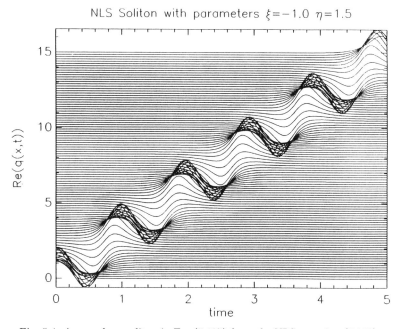

Fig. 5.4. An envelope soliton in Eq. (5.187) from the NLS equation (5.180).

Constructing the Gelfand-Levitan equation for the 2×2 Dirac's eigenvalue problem, Eqs. (5.181) through (5.185), Zakharov and Shabat obtained a soliton solution

$$q(x,t) = \frac{2\eta}{\varepsilon} \frac{\exp\left\{-2i\xi(x - x_0) - 2i(\xi^2 - \eta^2)t\right\}}{\cosh\left\{2\eta(x - x_0 + 2\xi t)\right\}} \tag{5.187}$$

where ξ and η are the real and imaginary part of the eigenvalue λ defined as

$$\lambda = \xi + i\eta, \tag{5.188}$$

The imaginary part is positive for the bound state, $\eta > 0$. The solution Eq. (5.187) is illustrated in Fig. 5.4, which suggests the name of an envelope soliton. It should be noticed that Eq. (5.187) is inversely proportional to the coupling constant ε. Thus, the envelope soliton is a genuine nonlinear entity, which can not be constructed in terms of perturbational approach. Soon after Zakharov and Shabat, Wadati (1972) has shown that the modified KdV equation, Eq. (5.145) is also solvable by the inverse scattering transformation.

5.9 Generalization of the Integrability Conditions

Conversely, to consider looking for other nonlinear evolution equations, Ablowitz et al. (1974) have posed the following question: what types of nonlinear evolution equations are associated with a set of 2×2 matrix forms of the linear eigenvalue problem and the linear temporal evolution equation, under the requirement of time independence of the eigenvalue λ?

Generalizing the scheme of Ablowitz-Kaup-Newell-Segur, we (Wadati et al., 1979a, 1979b) consider the eigenvalue problem,

$$\frac{\partial}{\partial x} u_1 + F(\lambda) u_1 = G(\lambda) q(x,t) u_2 \qquad (5.189)$$

$$\frac{\partial}{\partial x} u_2 - F(\lambda) u_2 = G(\lambda) r(x,t) u_1 \qquad (5.190)$$

where $F(\lambda)$ and $G(\lambda)$ are polynomials of the eigenvalue parameter λ. Together with Eqs. (5.189) and (5.190), we assume that the eigenfunctions u_1 and u_2 evolve in time according to the temporal evolution,

$$\frac{\partial}{\partial t} u_1 = A(\lambda; q, r) u_1 + B(\lambda; q, r) u_2 \qquad (5.191)$$

$$\frac{\partial}{\partial t} u_2 = C(\lambda; q, r) u_1 - A(\lambda; q, r) u_2 \qquad (5.192)$$

where A, B and C depend on the parameter λ and functionals of the potential q and r and of their spatial derivatives in the arbitrary order. It is natural to postulate that

$$\frac{\partial}{\partial t}\left(\frac{\partial}{\partial x} u_i\right) = \frac{\partial}{\partial x}\left(\frac{\partial}{\partial t} u_i\right) \qquad i = 1, 2. \qquad (5.193)$$

The key point to squeeze out the soliton equations is to postulate that the eigenvalue λ remains invariant in time while the potentials q and r change their shapes in time. Namely, we set the condition

$$\frac{\partial}{\partial t} \lambda = 0. \qquad (5.194)$$

5.9 Generalization of the Integrability Conditions

Then Eq. (5.193) gives rise to the following set of equations for the functionals A, B and C

$$\frac{\partial}{\partial x} A - G(\lambda)(rB - qC) = 0 \qquad (5.195)$$

$$G(\lambda) \frac{\partial}{\partial t} q - \frac{\partial}{\partial x} B - 2F(\lambda)B - 2G(\lambda)qA = 0 \qquad (5.196)$$

$$G(\lambda) \frac{\partial}{\partial t} r - \frac{\partial}{\partial x} C - 2F(\lambda)C - 2G(\lambda)rA = 0. \qquad (5.197)$$

For arbitrarily chosen expressions of $F(\lambda)$ and $G(\lambda)$, we construct the functionals A, B and C from Eqs. (5.195)–(5.197), and we can obtain the nonlinear evolution equation for q and r. These equations for the potentials q and r are the soliton equations.

Indeed, we notice that for the choice of the set of expressions

$$F(\lambda) = \lambda \qquad G(\lambda) = 1, \qquad (5.198)$$

Eqs. (5.195)–(5.197) reduce to the A-K-N-S scheme. Setting $r =$ const, and determining expressions of A, B and C, we get the Korteweg-de Vries equations (5.138), and for the choice of $r = -q$ (real), Eqs. (5.195)–(5.197) give rise to the modified Korteweg-de Vries equation (5.145). For the choice of $r = -q^*$ (complex), we get the cubic nonlinear Schrödinger equation (5.180).

Another important integrable equation is obtained by setting

$$A = \frac{1}{4\lambda} \cos\theta \qquad B = C = \frac{1}{4\lambda} \sin\theta \qquad (5.199)$$

and

$$q = -r = -\tfrac{1}{2} \theta_x \qquad (5.200)$$

we obtain

$$\frac{\partial^2}{\partial t \, \partial x} \theta = \sin\theta \qquad (5.201)$$

which is the sine-Gordon equation in the light cone coordinates $x = (\xi - \tau)/2$, $t = (\xi + \tau)/2$. The sine-Gordon equation in the laboratory coordinate

sine-Gordon eq.

$$\frac{\partial^2}{\partial \xi^2} \theta - \frac{\partial^2}{\partial \tau^2} \theta = \sin\theta \qquad (5.202)$$

is one of the most important soliton equations, which describes the self-induced transparency of intense optical pulse, the propagation of dislocation in solids, and the propagation of magnetic flux in the Josephson junction.

For the choice of

$$F(\lambda) = -i\alpha\lambda^2 - \sqrt{2\beta}\,\lambda \quad (5.203)$$

$$G(\lambda) = \alpha\lambda - i\sqrt{\beta/2} \quad (5.204)$$

with $r = \pm q^*$, we can determine the functionals A, B and C and the integrable nonlinear evolution equation given by

$$i\frac{\partial}{\partial t}q + \frac{\partial^2}{\partial x^2}q - i\alpha\frac{\partial}{\partial x}(|q|^2 q) \pm \beta |q|^2 q = 0 \quad (5.205)$$

which is called the modified nonlinear Schrödinger equation. Equation (5.205) describes nonlinear propagation of the Alfvén wave as given in Sec. 5.4.2.

The modified NLS equation (5.204) is also relevant in the discussion of a deformed continuous Heisenberg ferromagnet, in the study of two-photon self-induced transparency and ultrashort light pulse propagation in an optical fiber.

Furthermore, we have still other integrable nonlinear evolution equations for the choice of

$$F(\lambda) = i\lambda \quad (5.206)$$

$$G(\lambda) = \lambda \quad (5.207)$$

we obtain a highly nonlinear evolution equation

$$\frac{\partial}{\partial t}q_x + \mathrm{sgn}\left(\frac{dx}{ds}\right)\frac{\partial^2}{\partial x^2}\left(-i\frac{q_x}{\Phi} + w\frac{q_{xx}}{\Phi^2}\right) = 0 \quad (5.208)$$

with the abbreviation of

$$\Phi = \left\{1 + |q_x|^2\right\}^{1/2} \quad (5.209)$$

which describes nonlinear dynamics of the vortex filament with axial flow (Konno and Ichikawa, 1992).

The above choices do not exhaust the possible existence of still other integrable nonlinear evolution equations. For the choice of

$$F(\lambda, q) = \frac{i}{2}\lambda^2 + |q|^2 \quad (5.210)$$

$$G(\lambda) = \lambda \quad (5.211)$$

with $r = \pm q^*$, we can construct another kind of derivative nonlinear Schrödinger equation

derivative NLS

$$i\frac{\partial}{\partial t}q + \frac{\partial^2}{\partial x^2}q \pm 2i|q|^2\frac{\partial}{\partial x}q = 0 \quad (5.212)$$

which was shown to be integrable by Chen, Lee, and Liu (1979) some years ago.

We conclude the present section by referring to a generalization of Morris and Dodd (1980) to extend the two-component inverse scattering problem to the 3×3 matrix formalism. In connection with propagation of the nonlinear pulse in optical fiber, Sasa and Satsuma (1991) developed the three components inverse scattering problem.

5.10 Alfvén Soliton

Laboratory investigations of the properties of finite amplitude Alfvén waves in plasmas are of particular interest in connection with radio frequency plasma heating. In the problems of space physics, large amplitude incompressible magnetic perturbation observed in the solar wind are attributed to the propagation of the Alfvén waves (Kennel, 1988). The finite amplitude Alfvén wave propagating along the magnetic field is described by the derivative nonlinear Schrödinger equation, Eq. (5.127),

$$i\frac{\partial}{\partial \tau} q - \frac{\partial^2}{\partial \xi^2} q + i \frac{\partial}{\partial \xi} (|q|^2 q) = 0 \tag{5.213}$$

where q is defined as

$$q = \frac{1}{2B_0} (B_y + i B_z), \tag{5.214}$$

and the variables ξ and τ are related to the space and time by

$$\xi = 2(x - v_A t)/d \tag{5.215}$$

$$\tau = 2\Omega_i t \tag{5.216}$$

where v_A is the Alfvén velocity, Ω_i is the ion gyro-frequency and $d = v_A/\Omega_i = c/\omega_{pi}$.

First, let us discuss the modulational instability of a finite amplitude Alfvén wave. Equation (5.212) admits a special solution

$$q(\xi, \tau) = q_0 \exp\{i(K_0 \xi - \Omega_0 \tau)\} \tag{5.217}$$

with

$$\Omega_0^2 = -K_0^2 + q_0^2 K_0. \tag{5.218}$$

Now, in order to examine the stability of such a finite amplitude wave, we modulate its amplitude and phase as

$$q(\xi, \tau) = (q_0 + a(\xi, \tau)) \exp\{i(K_0 \xi - \Omega_0 \tau + \phi(\xi, \tau))\}. \tag{5.219}$$

Linearizing Eq. (5.212) under the assumption of $q_0 \gg a$, we obtain

$$\frac{\partial}{\partial \tau} a - (-2K_0 + 3q_0^2) \frac{\partial}{\partial \xi} a - q_0 \frac{\partial^2}{\partial \xi^2} \phi = 0 \tag{5.220}$$

$$\frac{\partial}{\partial \tau}\phi + (-2K_0 + q_0^2)\frac{\partial \phi}{\partial \xi} + 2q_0 K_0 a + \frac{1}{q_0}\frac{\partial^2}{\partial \xi^2}a = 0. \quad (5.221)$$

Seeking solutions of the linear equations (5.220) and (5.221) of the form

$$(a, \phi) \propto \exp\{i(k\xi - \omega\tau)\} \quad (5.222)$$

we obtain the dispersion relation

$$\omega = k(q_0^2 - 2K_0) \pm k\left[q_0^4 - 2q_0^2 K_0 + k^2\right]^{1/2}. \quad (5.223)$$

Thus, if $K_0 > q_0^2/2$, Eq. (5.223) predicts a modulational instability with the growth rate of

$$\Gamma(k) = k\left[2q_0^2 K_0 - q_0^4 - k^2\right]^{1/2}. \quad (5.224)$$

From the definition of Eq. (5.212), with Eqs. (5.213) and (5.216), the condition of $K_0 > 0$ stands for the left circular polarized Alfvén wave. Here, we have shown that the left circular polarized Alfvén plane waves are modulationally unstable.

As in the case of the Korteweg-de Vries equation, we turn to study whether the derivative nonlinear Schrödinger equation, Eq. (5.212), bears a stationary solitary wave solution. Let us express the complex amplitude $q(\xi, \tau)$ in terms of a real amplitude $\rho(\xi, \tau)$ and a phase $\phi(\xi, \tau)$ as

$$q(\xi, \tau) = \rho(\xi, \tau)\exp[i\phi(\xi, \tau)]. \quad (5.225)$$

Equation (5.212) is separated into the coupled equations:

$$\rho_\tau = 2\rho_\xi \phi_\xi - \rho\phi_{\xi\xi} - 3\rho^2 \rho_\xi \quad (5.226)$$

$$\rho\phi_\tau = -\rho_{\xi\xi} + \rho\phi_\xi^2 - \rho^3 \phi_\xi. \quad (5.227)$$

Introducing a moving coordinate

$$z = \xi - v\tau, \quad (5.228)$$

we seek a solution of the form of

$$\rho(\xi, \tau) = \rho(z) \quad (5.229)$$

$$\phi(\xi, \tau) = K\xi - \Omega\tau + \theta(z) \quad (5.230)$$

with boundary conditions at $z \to -\infty$,

$$\frac{d\rho}{dz} = \frac{d^2\rho}{dz^2} = 0, \quad (5.231)$$

$$\frac{d\theta}{dz} = 0. \quad (5.232)$$

5.10 Alfvén Soliton

Then, Eq. (5.227) for the phase function integrates to give

$$\frac{d\theta}{dz} = -\frac{1}{2}(v + 2K) + \frac{3}{4}\rho^2, \quad (5.233)$$

which gives rise to the nonlinear phase modulation. This phase modulation is crucial to have the stationary solitary wave. Imposing a boundary condition,

$$\rho = 0 \quad \text{at} \quad z \to \infty \quad (5.234)$$

together with Eq. (5.233), we can determine the arbitrary velocity to be

$$v = -2K. \quad (5.235)$$

Substitution of Eq. (5.233) with Eq. (5.235) back into the equation for the real amplitude ρ, we can carry out integration to get

$$\rho^2 = \frac{4(\Omega + K^2)}{(\Omega + 2K^2)^{1/2}\cosh(z/\delta) + K} \quad (5.236)$$

where the soliton width δ is given by

$$\delta^{-1} = 2(\Omega + K^2)^{1/2}. \quad (5.237)$$

Equation (5.225) with Eq. (5.236) represents a stationary solitary wave of the derivative nonlinear Schrödinger equation. Here we notice that the wavenumber K may either be positive or negative. Equation (5.235) tells us that the left polarized Alfvén solitary wave propagates with sub-Alfvén velocity, while the right polarized Alfvén solitary wave propagates with super-Alfvén solitary velocity. We remark here that a similar analysis has been carried out by Spangler et al. (1985), presenting one parameter family of Alfvén wave solitary waves, while Mjølhus and Wyller (1986) have emphasized important aspects of the two parameters family of Alfvén solitary waves. At this stage of analysis, however, we can not identify the solitary wave as the Alfvén soliton.

Kaup and Newell (1978) have discovered that the derivative nonlinear Schrödinger equation, Eq. (5.213) is transferred to a pair of linear eigenvalue equations

$$\frac{\partial}{\partial x} u_1 = i\lambda u_1 + \lambda^{1/2} q\, u_2$$

$$\frac{\partial}{\partial x} u_2 = \lambda^{1/2} q^* u_1 + i\lambda u_2 \quad (5.238)$$

and a linear temporal evolution equation

$$\frac{\partial}{\partial t} u_1 = A\, u_1 + B\, u_2$$

$$\frac{\partial}{\partial t} u_2 = C\, u_1 - A\, u_2 \quad (5.239)$$

with the abbreviation of

$$A = 2i\lambda^2 + i|q|^2\lambda \tag{5.240}$$

$$B = -2\lambda^{3/2}q - \lambda^{1/2}(iq_x + |q|^2 q) \tag{5.241}$$

$$C = -2\lambda^{3/2}q^* + \lambda^{1/2}(iq_x^* - |q|^2 q). \tag{5.242}$$

Kaup and Newell have shown that the bond state of the eigenvalue problem of Eq. (5.238) is given as

$$\lambda = u + i\gamma, \quad \text{Im}(\lambda) = \gamma > 0, \tag{5.243}$$

and the soliton solution is indeed constructed as Eq. (5.243), where the parameters K and Ω are related to the eigenvalue λ, as

$$\mu = -\tfrac{1}{2} K \tag{5.244}$$

$$\gamma = \tfrac{1}{2} (K^2 + \Omega)^{1/2}. \tag{5.245}$$

The velocity, width and height of the soliton are given as

$$v = 4\mu \tag{5.246}$$

$$\delta = (4\gamma)^{-1} \tag{5.247}$$

$$\rho_m = \sqrt{8} \left\{ \sqrt{\gamma^2 + \mu^2} + \mu \right\}^{1/2}. \tag{5.248}$$

Applying the inverse scattering method, Mjølhus (1978) examined the process of soliton formation as the result of modulational instability of the large amplitude plane wave, Eq. (5.217) and (5.224). Ichikawa and Abe (1988) have solved explicitly the eigenvalue problem for the initial condition

$$q(\xi, t = 0) = q_0 \operatorname{sech}(\xi/D) \exp(iK_0 \xi), \tag{5.249}$$

obtaining the real and imaginary parts of the eigenvalue as

$$\mu = \frac{q_0^2}{2} - \frac{K_0}{2} - q_0 \sqrt{\frac{1}{2} \left[S_n(q_0, K_0) + \frac{q_0^2}{4} - \frac{K_0}{2} \right]} \tag{5.250}$$

$$\gamma = \sqrt{\frac{1}{2} \left[S_n(q_0, K_0) - \frac{q_0^2}{4} + \frac{K_0}{2} \right]} - \frac{4n - 1}{2D} \tag{5.251}$$

with the abbreviation of

$$S_n(q_0, K_0) = \sqrt{\left(\frac{q_0^2}{4} - \frac{K_0}{2}\right)^2 - \left(\frac{4n-1}{2D}\right)^2}. \tag{5.252}$$

5.10 Alfvén Soliton

The integer n specifies the number of bound states. The condition $\gamma > 0$ determines the number of bound states, namely, the number of solitons N is given by

$$N = \frac{1}{2\sqrt{2}} D\, q_0\, K_0^{1/2} + \frac{1}{4}. \tag{5.253}$$

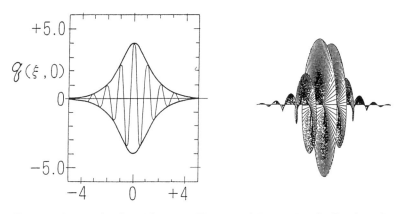

Fig. 5.5. An initial pulse with $q_0 = 4$, $K_0 = 6$, and $D = 1$ given by Eq. (5.249).

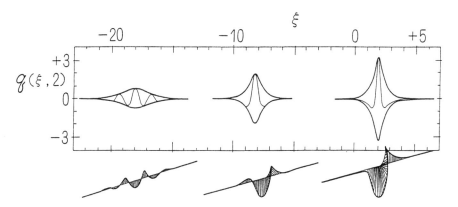

Fig. 5.6. The three solitons at $\tau = 2$ that evolve out of the initial pulse shown in Fig. 5.5.

In Fig. 5.5 we illustrate the initial pulse of Eq. (5.249) with the values of $q_0 = 4$, $K_0 = 6$, and $D = 1$. Equation (5.253) tells us this initial pulse splits into three solitons. Figure 5.6 illustrates these three solitons decomposed from the initial pulse after the time of $\tau = 2$.

There is active interest in the problem of nonlinear Alfvén wave propagation in astrophysical plasma phenomena, and there have been some numerical studies of

Alfvén soliton formation. Dawson and Fontane (1983) have carried out numerical studies with the initial pulse of the Gaussian shape. Since the Gaussian shape is close to the sech-function, we estimate the number of solitons observed by Dawson and Fontane by the condition

$$\frac{A_0 \ell}{\sqrt{\lambda}} \geqq n \quad \text{(integer n)} \tag{5.254}$$

for the values of parameters given in their numerical observation. We confirm that Eq. (5.254) gives a correct estimate of the observed number of solitons.

To conclude the present section, we call attention to the fact that the modified nonlinear Schrödinger equation (5.205) is relevant to describe the rapid modulation of the nonlinear pulse propagating through the optical fiber (Tzoar and Jain, 1981). Ohkuma et al. (1987) have carried out rigorous analysis based on the fact that Eq. (5.205) is an integrable soliton equation. We emphasize that the soliton aspects of nonlinear wave phenomena have indeed the interdisciplinary character, in which plasma physics is playing the key role in exploring the fundamental properties of the forms of energy propagation.

5.11 One-Dimensional Soliton Gas Models

The existence of the inverse of scattering transform for solving the integrable nonlinear wave equations as described in the preceding sections gives rise to a new description of collective modes of systems. In the new description, which may be called the soliton gas, the basic collective modes are the single particle-like states of the solitons rather than the small amplitude waves (plasmons and phonons). Due to the integrability of the KdV and cubic Schrödinger equation we know that the elementary nonlinear structures emerge from scattering events with the same form as before but carrying phase shifts from the effect of change of their relative speed during time of interaction. The development of the N-soliton solutions of the KdV equation are now well known (Lamb, 1980) (pp. 125–126), Whitham (1974), and Dodd et al. (1982).

5.12 Example 5.1 — Elastic Collisions of Two Solitons

Here we give the well-known example of the two-soliton solution and suggest the student develop a computer code to study this and more exotic soliton/wave interactions.

For two KdV solitons governed by Eq. (5.138) with parameters K_1 and K_2 it is convenient to write the two soliton solutions in terms of the phase variables

$$\gamma_1 = K_1 x - 4K_1^3 t + \delta_1 \tag{5.255}$$

5.12 Example 5.1 — Elastic Collisions of Two Solitons

$$\gamma_2 = K_2 \, x - 4K_2^3 \, t + \delta_2. \tag{5.256}$$

Then well before the collision the two-soliton solution is

$$q(x, t \to -\infty) = 12K_1^2 \operatorname{sech}^2 \gamma_1 + 12K_2^2 \operatorname{sech}^2 \gamma_2 \tag{5.257}$$

and well after the soliton collision the solution is

$$q(x, t \to +\infty) = 12K_1^2 \operatorname{sech}^2(\gamma_1 + 2\Delta) + 12K_2^2 \operatorname{sech}^2(\gamma_2 - 2\Delta) \tag{5.258}$$

where the phase shift Δ is given by

$$\Delta = \tan^{-1}\left(\frac{K_1}{K_2}\right) = 2\ln\left(\frac{K_2 + K_1}{K_2 - K_1}\right). \tag{5.259}$$

To simplify the exact 2-soliton solution it is convenient to symmetrize (5.255) and (5.256) with respect to the phase shift by letting $\delta_1 \to \delta_1 - \Delta$ and $\delta_2 \to \delta_2 + \Delta$. Then, the exact 2-soliton solution found by the inverse transform method (Sec. 5.7) is

$$q(x,t) = 12(K_2^2 - K_1^2)\left[\frac{K_2^2 \operatorname{csch}^2 \gamma_2 + K_1^2 \operatorname{csch}^2 \gamma_1}{(K_2 \coth \gamma_2 - K_1 \tanh \gamma_1)^2}\right] \tag{5.260}$$

which one readily finds has the asymptotic limits before and after the collision of two isolated solitons (Lamb, 1980).

The existence of the multisoliton solutions provides an alternative theoretical framework for interpretation of the scattering experiments, based on the concept of an (nearly) ideal gas of plasma solitons as the form of the fluctuations spectrum rather than a spectrum of small amplitude waves. These ideas have also been studied in the context of condensed matter physics (Currie, 1980). The condensed matter work shows that solitons contribute to the free energy of nonlinear lattices in equilibrium, and, furthermore, that their effects are experimentally detectable (Bishop, 1978).

For electron plasma waves in an unmagnetized plasma Kingsep et al. (1973) have studied the wavenumber spectrum generated by a gas of Langmuir solitons. Zakharov (1971) has obtained a kinetic equation for a given set of Korteweg-de Vries (KdV) solitons. Zakharov shows that the pairwise interaction of solitons (which causes a phase shift) leads to an effective renormalization of the soliton velocity. A review by Makhankov (1978) emphasizes the importance in physical system of the non-integrable soliton equations — in contrast to the integrable KdV soliton equation. The drift wave solitary waves are non-integrable but have the robust features emphasized by Makhankov.

Meiss and Horton (1982a, 1982b) have developed the soliton gas description of drift wave turbulence. The electromagnetic scattering experiments of Mazzucato (1976, 1978), Slusher and Surko (1978), TFTR Group (1977, 1980), and Brower *et al.* (1985, 1987) have led to the identification of the microturbulence in tokamaks with drift wave turbulence. The general features of the dynamical form factor $S(\mathbf{k}, \omega)$ for the electron density fluctuations $\langle |\delta n_e(\mathbf{k}, \omega)|^2 \rangle$ are interpreted in terms of the frequency $k_y v_{de}$, where $v_{de} = cT_e/(eBL_n)$ is the electron diamagnetic drift velocity, and the most unstable wavenumbers of drift wave theory $k_\perp \rho_s \lesssim 1$, where ρ_s is the ion inertial scale length, $\rho_s = c(m_i T_e)^{1/2}/eB$.

Efforts to make a detailed comparison of the scattering data with theory, however, have been frustrated by the fact that for a well-defined $\mathbf{k}$ and scattering volume, the distribution of the scattered power has a peak at a frequency which is two to three times larger than the linear drift frequency $\omega^\ell(k)$ (Mazzucato (1976) and Horton (1976)). Weak turbulence theory use an assumption of weak correlations e.g., "maximal randomness" of the direct interaction approximation (Kraichnan, 1959) which predicts, for moderate levels, nonlinear frequency shifts $\Delta \omega_{n\ell}$ proportional to integrals over $I(\mathbf{k}) \propto \langle |\delta n_e(\mathbf{k})|^2 \rangle$.

Meiss and Horton (1982a, 1982b) derived the wavenumber spectrum due to KdV solitons from a particular initial configuration, assuming that soliton overlap can be neglected. They obtain the form factor, $S(\mathbf{k}, \omega)$ given here, due to solitons which arise from an ensemble of initial conditions with a given mean square fluctuation level, $\langle \delta n_e^2 \rangle$, of the electron density.

The Meiss-Horton theory uses a one-dimensional drift wave equation given by Petviashvili (1977) described in Sec. 5.12.1. This Petviashvili drift wave equation is now thought to be inadequate to describe the actual confinement experiment drift waves. However, the concept of a drift wave soliton gas applies to many other systems with long-lived, localized, coherent structures in any number of dimensions. For the 2D vortex system developed in Chapter 6 the concepts developed for soliton gas are often taken to form a point of view for analysis of the numerical and laboratory experiments.

A principal result from the soliton gas description is a formula for the density fluctuation spectrum, $S(\mathbf{k}, \omega)$, which is qualitatively different from formulas based on weakly correlated linear normal modes. The difference follows from the soliton "dispersion" relation, $\omega = ku$, where the soliton velocity u depends linearly upon its amplitude and is independent of k. In a system where a large number of solitons are excited with varying amplitudes, the frequency spread for a given k is $\Delta \omega \sim k \langle (\Delta u)^2 \rangle^{1/2}$, where Δu is the width of the soliton velocity distribution $f_s(u)$. This spectral density contrasts with that obtained by renormalized weak turbulence theory in Chapter 7.

Furthermore, the allowed soliton and vortex velocities u fall in a range complementary to the phase velocity of the linear modes. In particular, localized drift wave solitons and vortices have $u > v_d$ or $uv_d < 0$. These features give rise to the qualitative shape of the spectral contours shown in Fig. 5.7. Here the spectrum for fixed ω/k is that of a single soliton, and if a large number of solitons with different velocities are excited then the integrated spectrum becomes broad. For moderate

5.12 Example 5.1 — Elastic Collisions of Two Solitons

fluctuation levels, the $u > v_d$ solitons are preferentially excited, leading to a spectra peaked at $\omega > kv_d$. For larger fluctuation levels the $uv_d < 0$ solitons become more important, giving spectra peaked at $\omega\omega_{*e} \leq 0$.

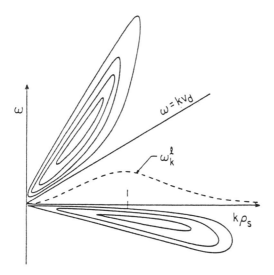

Fig. 5.7. Contours of $S(k,\omega)$ for the 1D solitary drift wave showing the complementary regions of wave and soliton propagation.

5.12.1 Petviashvili drift wave solitons and the soliton gas

For a simplified two-fluid description of the plasma, where the electrons are adiabatic

$$n_e(\mathbf{r},t) = n(x)\exp\left[e\Phi(\mathbf{r},t)/T_e(x)\right],$$

and the ions are pressureless, Tasso et al. (1967) and Petviashvili (1977) derive and study a nonlinear wave equation closely related to the Korteweg-deVries equation for drift waves. The equation may apply only for the limit $(k_x\rho_s)(k\rho_s)^2 \ll \eta_e(\rho_s/L_n)$ where $\eta_e = -L_n(\partial_x T_e/T_e)$ so that the temperature gradient nonlinearity dominates the $\mathbf{E}\times\mathbf{B}$ convective nonlinearity of the vorticity equation derived from the continuity of the plasma currents $\nabla\cdot\mathbf{j} = 0$. Even in this limit recent studies show that the model is inadequate due to the unstable growth of a transverse modulation of the one-dimensional planar solitary wave (Horton and Hasegawa, 1994).

The temperature gradient drift wave is governed by the equation
Petviashvili eq.

$$(1 - \rho_s^2\nabla_\perp^2)\partial_t\varphi + v_d\partial_y\varphi - v_d\varphi\partial_y\varphi = 0, \tag{5.261}$$

where $\varphi = \eta_e(e\Phi/T_e)$ and $v_d = \rho_s c_s/L_n$. The Petviashvili equation (5.261) possesses two-dimensional solutions that are described in Sec. 6.9. The one-dimensional limit of Eq. (5.261) has been studied extensively as a "regularized" version of the KdV equation, Benjamin et al. (1972) called the regularized long-wave (RLW) equation and seems to have been first derived by Peregrine (1966) in the context of tidal waves. Morrison et al. (1984) analyze inelastic head-on collisions and essentially elastic overtaking collisions with simulations and a reduced description derived from the least action principle and the Lagrangian density for Eq. (5.261). The radial dimension of the quasi-one-dimensional solutions is limited by the scale of variation of $v_d(x)$ in Eq. (5.261). Balancing $v_d \rho^2 \partial_x^2$ with the variation of $v_d(x)$ about its maximum, $\Delta x^2 \partial_x^2 v_d$, gives

$$(\Delta x)^2 \sim \rho_s L_n. \quad (5.262)$$

Thus, the one-dimensional drift wave solitons are taken to extend over the radial region Δx, centered at the maximum of v_d, and in addition we assume an axial length $\Delta z = L_c$. Like the KdV equation, Eq. (5.261) has solitary wave solutions

$$\varphi_s(y, t: y_0, u) = -3(u/v_d - 1)\operatorname{sech}^2\left[\frac{1}{2\rho_s}\left(1 - \frac{v_d}{u}\right)^{1/2}(y - y_0 - ut)\right], \quad (5.263)$$

where the velocity u is restricted to the ranges

$$u > v_d \quad \text{or} \quad uv_d < 0. \quad (5.264)$$

Unlike the KdV equation, these solutions are not solitons in the pure sense, since they are not preserved upon collision. However for collisions of moderate amplitude waves traveling in the same direction, the inelasticity of collisions is extremely difficult to detect. In head-on collisions the elasticity is more pronounced (Abdullaev et al., 1974 and Morrison et al., 1984)).

Therefore, solitary waves of Eq. (5.263) persist for long times and through many collisions. As has been emphasized by Currie (1980) solitary waves which are not strictly solitons still can have an important contribution to the statistical properties of the turbulent fields.

To determine statistical properties we will need the drift wave energy

$$E = \tfrac{1}{2}\int \left[\varphi^2 + (\rho_s \partial_y \varphi)^2\right]\frac{dy}{\rho_s}, \quad (5.265)$$

which is the physical energy in units of $n_e T_e \rho_s \Delta x L_c/\eta_e^2$. The energy of the solitary wave, Eq. (5.263), is

$$E_s = \frac{12}{5}\left(\frac{u}{v_d}\right)^2\left(1 - \frac{v_d}{u}\right)^{3/2}\left(6 - \frac{v_d}{u}\right), \quad (5.266)$$

which is to be compared with Eqs. (6.71) and (6.72) for the 2D dipole vortex structures. For $u/v_d \gg 1$, E_s increases quadratically, while as $u \to 0^-$ Eq. (5.226) reduces

5.12 Example 5.1 — Elastic Collisions of Two Solitons

to $E_s \simeq 12/5(v_d/|u|)^{1/2}$. The minimum E_s for $u < 0$ occurs at $u = v_d(2-\sqrt{10})/12 = -0.096 v_d$, where $E_s = 14.01$. The graph of $E_s(u)$ from Eq. (5.266) is shown in Fig. 5.8.

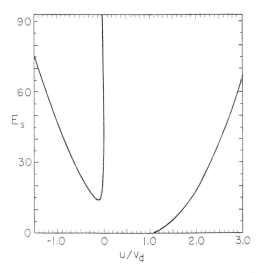

Fig. 5.8. One-dimensional drift wave soliton energy as a function of u/v_{de} from the Petviashvili model in Eq. (5.261).

(ii) Solitary Wave Spectrum

A turbulent state described by Eq. (5.261) will consist of a broad wavenumber spectrum of small-amplitude modes together with an ensemble of solitary waves. For each linear mode, the frequency spectrum will be peaked about $\omega_k^\ell = kv_d/[1+(k\rho_s)^2]$ with some width determined, for example, by resonance broadening theory. Each solitary wave, however, contributes frequencies which depend upon its velocity (and hence its amplitude) through $\omega = ku$. By virtue of Eq. (5.264) these frequencies will range over $\omega > \omega_{*e} = kv_d$ and $\omega\omega_{*e} < 0$, which is complementary to the range of the linear dispersion relation $0 < \omega < \omega_{*e}$.

As a first approximation we ignore the small-amplitude component supposing that its spectrum can be merely added to that for the solitary waves. The interaction between solitary waves and linear modes may act to renormalize the solitary wave parameters giving "dressed solitons" (Currie, 1980).

The potential is written as a superposition of solitary waves

$$\varphi(y,t) = \sum_{n=1}^{N_s} \varphi_s(y,t;\ y_n, u_n). \tag{5.267}$$

The assumption in writing Eq. (5.267) ignores the strong interactions between these essentially nonlinear objects which occur whenever they overlap. To the extent that the solitary waves act as KdV solitons, ($\overline{k}^2 \rho_s^2 \ll 1$), the only effect of this interaction is a phase shift of the soliton positions. Zakharov and Faddeev (1971) show that the phase shifts act to renormalize the soliton velocities. This effect is proportional to the soliton number density which we assume small, and verify *a posteriori*.

The spectral density is the Fourier transform of the two-point correlation function:

$$S(\xi, \tau) = \langle \varphi(x+\xi, t+\tau)\varphi(x,t) \rangle = \frac{1}{(2\pi)^2} \int dk \int d\omega\, S(k,\omega) \exp(ik\xi - i\omega\tau), \tag{5.268}$$

where the average is over the ensemble specified below. Utilizing the complete field from Eq. (5.267) with the solution Eq. (5.263), gives for $u > v_d$ and $uv_d < 0$

$$S(k,\omega) = \frac{1}{L} \sum_{n=1}^{N_s} \left\langle \left[12\pi k\rho_s \left(\frac{u_n}{v_d}\right) \operatorname{csch}\left(\frac{\pi k\rho_s}{(1-v_d/u_n)^{1/2}}\right)\right]^2 \delta(\omega - ku_n) \right\rangle, \tag{5.269}$$

and $S(k,\omega) = 0$ for $0 \leq u \leq v_d$. In deriving Eq. (5.269) it is assumed that the solitary wave positions, y_n, are randomly distributed along the length L. Generally, $L = 2\pi r$ the circumference of the confinement device at the radius r which gives the maximum v_d with the axial magnetic field $B\hat{z}$. For a large number of solitary waves, $N_s \gg 1$, the sum in Eq. (5.269) may be converted to an integral over the distribution function, $f_s(u)$, of solitary waves in u space:

$$S(k,\omega) = \frac{1}{L} f_s\left(\frac{\omega}{k}\right) \left\{ 12\pi \rho_s \frac{\omega}{v_d} \operatorname{csch}\left[\pi k \rho_s \left(\frac{\omega}{\omega - kv_d}\right)^{1/2}\right] \right\}^2, \tag{5.270}$$

where $\int_{-\infty}^{\infty} f_s(u)du = N_s$. Before determining $f_s(u)$, we can deduce the qualitative shape of $S(k,\omega)$ directly from Eq. (5.270)) when $k\rho_s \ll 1$.

$$S(k,\omega) \sim \begin{cases} 0, & 0 < \omega < kv_d \\[1ex] \exp\left[-2\pi k\rho_s \left(\frac{kv_d}{\omega - kv_d}\right)^{1/2}\right], & kv_d < \omega \ll kv_d[1+(\pi k\rho_s)^2] \\[1ex] \left(\frac{\omega}{kv_d}\right)^2 \left(1 - \frac{kv_d}{\omega}\right) f_s\left(\frac{\omega}{k}\right), & \omega \gg kv_d[1+(\pi k\rho_s)^2] \text{ or } \omega < 0. \end{cases}$$
(5.271)

Note the $S = 0$ just in the range where small-amplitude excitations contribute. If there is some maximum-amplitude solitary wave, $-|\varphi_{\max}|$, then Eq. (5.263) and (5.270) imply $S(k,\omega) = 0$ for $\omega > (1+\frac{1}{2}|\varphi_{\max}|)\omega_{*e}$.

5.12 Example 5.1 — Elastic Collisions of Two Solitons

A sketch of the contours of $S(k,\omega)$ is given in Fig. 5.7. In addition, the dispersion relation for linear modes is shown by the dashed line. The analysis clearly shows the qualitative difference between the solitary wave spectrum and weak turbulence spectrum concentrated near the linear modes.

(iii) Canonical Distribution Function of Solitary Waves

A quantitative formula for $S(k,\omega)$ requires knowledge of the distribution function, $f_s(u)$. Now we suppose that the solitary waves can be characterized by a Gibbs ensemble. To the extent the solitary waves are solitons they each have an associated conserved quantity. For soliton-bearing equations such as KdV, the inverse spectral transform acts as a canonical transform to action-angle coordinates in which each soliton is represented by a single degree-of-freedom (J, θ) Zakharov and Faddeev (1971), at least for the weakly coupled case. A general theory of action-angle variables for all equations solvable by the inverse scattering transform has been given by Flaschka and Newell (1975). In terms of these coordinates there are N_s conserved quantities, the N_s actions, for a system with N_s solitons. These conserved quantities span the soliton component of the infinite dimensional phase space.

For the true soliton system these degrees-of-freedom are independent, and thus the Gibbs ensemble factors as given by

$$P_n(J_1, J_2, \ldots, J_{N_s}, \theta_1, \ldots, \theta_{N_s}) = \prod_{i=1}^{N_s} P(J_i, \theta_i),$$

$$P(J, \theta) = (1/Z) \exp[-\beta_s E_s(J)], \tag{5.272}$$

where P is the single soliton probability distribution and Z is the partition function (normalization constant). The effective inverse temperature β_s fixes the mean energy in the solitary wave component.

The solitary wave degrees-of-freedom are of course not independent for the RLW equation. Soliton-soliton collisions may in fact provide at least part of the randomization necessary for validity of a statistical description. Numerical computations show that a moderate amplitude solitary waves can survive many collisions with only small modifications of its parameters. Thus our description is meaningful on the timescale with which the solitary waves maintain their integrity, and is not a long-time equilibrium calculation. In the long-time limit the system may reduce to equipartition with energy $1/\beta$ per degree-of-freedom.

Since the inverse spectral transformation probably does not exist for RLW, the canonical transformation to the action-angle coordinates is not possible. Meiss and Horton calculate the probability distribution P using the following procedure. Integration of Eq. (5.272) over θ and a transformation of coordinates from J to E_s gives

$$P(E_s) = \frac{2\pi}{Z} \left| \left(\frac{\partial E_s}{\partial J} \right)^{-1} \right| \exp(-\beta_s E_s). \tag{5.273}$$

Since J is a canonical variable, the derivative $\partial E_s/\partial J$ is the frequency $\dot\theta$. For a soliton, which acts as a free particle

$$\dot\theta = 2\pi(u/L),$$

and therefore

$$P(u) = \frac{L}{Z}\left|\frac{\partial E_s}{\partial u}\frac{1}{u}\right|\exp[-\beta_s E_s(u)]. \tag{5.274}$$

Note that the soliton energy is now expressed as a function of its velocity, which is an easily calculable function. The one-soliton distribution function used in Eq. (5.274) is defined by

$$f_s(u) = N_s P(u). \tag{5.275}$$

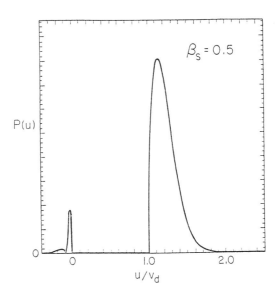

Fig. 5.9. Soliton velocity distribution for the Gibbs ensemble with $T_s = \beta_s^{-1} = 2$ in energy units $[n_e T_e \Delta x\, \rho_s L_c/\eta_e^2]$. The energy density in the soliton gas is $N_s T_s/L_y = n_s T_s$ where N_s is the number of solitons in the system of length L_y. If all the rms wave field $\varphi_0 = e\tilde\Phi/T_e$ were in the soliton component, the soliton density would be $n_s \simeq 0.05\varphi_0^{1/2}/\rho_s$ (solitons per unit length).

Figure 5.9 shows $P(u)$ from Eq. (5.274) with the energy function of Eq. (5.266) and $\beta_s = 1/2$. For low temperatures, $\beta_s \gg 1$, the distribution function simplifies to

5.12 Example 5.1 — Elastic Collisions of Two Solitons

the KdV form since the negative velocities have exponentially small weight:

$$P_{\beta_s \gg 1}(u) \propto \begin{cases} \left(\dfrac{u}{v_d} - 1\right)^{1/2} \exp\left[-12\beta_s \left(\dfrac{u}{v_d} - 1\right)^{3/2}\right], & u > v_d \\ 0, & u < v_d. \end{cases} \quad (5.276)$$

Negative velocity solitary waves become significantly excited for $\beta_s \lesssim 1/2$ with the distribution peaked in the region $-0.096 v_d < u < 0$ and

$$P_{\beta_s \gg 1}(u) \propto \begin{cases} \left(\dfrac{v_d}{-u}\right)^{5/2} \exp\left[-\dfrac{12}{5}\beta_s \left(\dfrac{v_d}{-u}\right)^{1/2}\right], & u < 0 \\ 0, & u > 0. \end{cases} \quad (5.277)$$

The zero of $P(u)$ at $u = -0.096 v_d$ in Fig. 5.5 is due to the Jacobian factor $\partial E_s/\partial u$ in Eq. (5.274).

(iv) Solitary Wave Number Density and Temperature

To utilize the distribution function, Eq. (5.274), it is necessary to know the effective temperature, $T_s = \beta_s^{-1}$, which fixes the mean energy, $\langle E_s \rangle$. The relationship between these quantities is obtained through

$$\langle E_s \rangle = \int_{-\infty}^{\infty} du \, E_s(u) P(u),$$

where $P(u)$ is given in Eq. (5.274). This integral can be done approximately utilizing Eq. (5.276) and Eq. (5.277), yielding

$$\langle E_s \rangle \sim \begin{cases} T_s, & T_s \ll 1 \\ 3T_s, & T_s \gg 1, \end{cases} \quad (5.278)$$

where in the upper (lower) relation only the $u > v_d$ ($u v_d < 0$) solitary waves contribute.

Since the solitary waves represent independent degrees-of-freedom, energy is equipartitioned

$$N_s \langle E_s \rangle \simeq \left(\dfrac{L}{\rho_s}\right) \langle \varphi^2 \rangle = \left(\dfrac{L}{\rho_s}\right) \varphi_0^2, \quad (5.279)$$

where we assume that the fluctuation energy represented by φ_0^2 is entirely due to the solitary waves which have a number density $n_s = N_s/L$.

The total available thermal energy for the drift wave field may be estimated using thermodynamic arguments. There are two energy sources, the diamagnetic kinetic energy and the free energy of expansion, arising from the background temperature and density gradients. When the radial scale of the drift waves is large, $\Delta x \gg \rho_s$, expansion energy dominates and a thermodynamic bound is

$$\varphi_0^2 \lesssim \eta_e^2 (\Delta x/L_n)^2 \simeq \eta_e^2 (\rho_s/L_n). \tag{5.280}$$

Equation (5.262) has been used for Δx, the radial extent of the drift wave solitons. Once the energy available to the field is known, we only need to calculate N_s to obtain the temperature through Eqs. (5.278) and (5.279). This calculation requires knowledge of the number of solitons emerging from a particular initial state, $\varphi(x)$. Since the initial value problem for the RLW equation remains unsolved, we turn again to KdV, recalling that the results will be correct for small T_s.

The inverse scattering transform (Sec. 5.6) allows the determination of the number of solitons emitted by any particular initial state. For moderate amplitude initial states,

$$\varphi_{\max} \gg (\rho_s/L_n)^2,$$

the number of solitons produced is large and a WKB approximation of the inverse problem (Eq. (5.147) can be used to obtain

$$N_s[\varphi] = \frac{1}{\sqrt{6}\pi\rho_s} \int_{\varphi<0} dx [-\varphi(x)]^{1/2}. \tag{5.281}$$

This result is only valid when $\varphi < 0$ for all x: When $\varphi > 0$, non-soliton excitations significantly affect Eq. (5.281) for the general case.

To compute the mean number of solitons, we average Eq. (5.281) using a Gibbs ensemble with the KdV energy

$$E_{\mathrm{KvD}} = \frac{1}{2} \int \varphi^2 \frac{dy}{\rho_s}, \tag{5.282}$$

which is obtained from Eq. (5.155) when $k\rho_s \ll 1$. The mean square potential is fixed to agree with the available energy of Eq. (5.280), $\langle \varphi^2 \rangle = \varphi_0^2$. The mean number of solitons is determined by a functional integral which upon discretization becomes

$$\langle N_s \rangle = \frac{1}{Z} \prod_{i=1}^{n} d\varphi_i N_s[\varphi] \exp\left[-\frac{1}{2}\left(\frac{\varphi_i}{\varphi_0}\right)^2\right], \tag{5.283}$$

where $Z = (2\pi\varphi_0^2)^{n/2}$ is the normalization and $\varphi_i = \varphi(x_i)$. We then obtain

$$n_s = \langle N_s \rangle / L = \alpha(\varphi_0^{1/2}/\rho_s);$$

$$\alpha = \Gamma\left(\frac{3}{4}\right) / (12\sqrt{2}\pi^3)^{1/2} = 0.053. \tag{5.284}$$

Combining Eqs. (5.279) and (5.283) gives the mean energy

5.12 Example 5.1 — Elastic Collisions of Two Solitons

$$\langle E_s \rangle = (1/\alpha)\varphi_0^{3/2}, \tag{5.285}$$

which, in conjunction with Eq. (5.278), gives T_s.
Using the estimate of φ_0 in Eq. (5.280) gives

$$n_s = \frac{\alpha \eta_e^{1/2}}{\rho_s} \left(\frac{\rho_s}{L_n}\right)^{1/4}$$

$$\langle E_s \rangle = \frac{\eta_e^{3/2}}{\alpha} \left(\frac{\rho_s}{L_n}\right)^{3/4}. \tag{5.286}$$

For $\rho_s/L_n \simeq 0.01$ and $\eta_e \simeq 1$, which is appropriate for tokamak experiments, $\langle E_s \rangle = 0.5$ while reversed field pinch experiments (RFP) however, $\rho_s/L_n \simeq 0.1$ and $\langle E_s \rangle = 3.3$. Equation (5.278) then implies $T_s = 1/2$ and $T_s \sim 2 - 3$, respectively. Spectral densities for these temperatures are shown in Fig. 5.10. When $T_s = 1/2$, virtually all the energy is contained in the $u > v_d$ modes as shown in frame (a), while at the higher temperature a substantial fraction of the energy is in the negative frequency modes. Significant excitation of negative frequency modes propagating in the ion diamagnetic direction occurs when there is sufficient thermal energy to overcome the required creation energy, $E_{\min} = 14.01$ shown in Fig. 5.10b.

The frequency shift of the positive spectral peak, $\omega_{\max}$, is given by $\omega_{\max}/kv_d \simeq 1.25$ for $T_s = 0.5$ and 1.5 for $T_s = 3.0$. The frequency shift relative to the linear mode frequency depends upon k, and increases rapidly as $k\rho_s \to 1$. If we use the finite Larmor radius formula for ω_k^ℓ, a frequency shift of $\omega_p/\omega_k^\ell \simeq 5$ is obtained when $k\rho_s \simeq 1$ and $T_s = 0.5$, which is in quantitative agreement with the experiments (Mazzucato, 1976, 1978; Slusher, 1978; TFTR Group, 1977, 1980).

In conclusion, for the root-mean-square fluctuation levels typical of the saturated state of drift wave turbulence, inverse scattering theory for the soliton (KdV) equation is used to show that a large number, $N_s \gg 1$, of solitons can evolve from the drift wave fields. Each drift wave soliton introduces a spatially localized infinite order set of correlations, due to its intrinsic coherence. These correlations are lost in the truncations of statistical turbulence theory, and give rise to new features in the fluctuation spectrum even in the limit of an ideal gas approximation to the many-soliton system. The soliton gas theory is further developed by Tasso (1983, 1987) for the one-dimensional drift wave models using the functional integral analysis.

An approach similar in spirit to the soliton gas model is that of Dupree (1982) who uses phase-space density holes, Bernstein-Greene-Kruskal modes, as the coherent structures in Vlasov-Poisson turbulence. Dupree's hole model of turbulence allows calculation of fluctuation self-trapping which is an effect outside the domain of statistical turbulence theory.

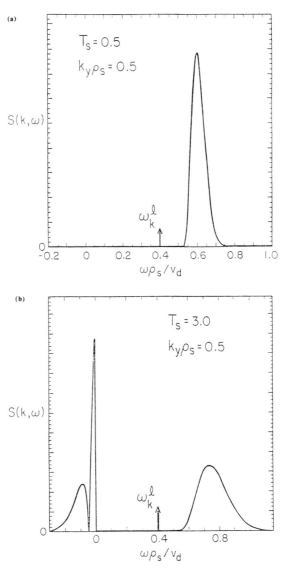

Fig. 5.10. Frequency spectrum $S(k,\omega)$ from the Gibbs ensemble at fixed $k_y \rho_s = 0.5$ (a) The low energy system has solitons propagating at $(3/2)$ the linear phase speed in the electron direction only. (b) The higher energy system has waves at 2 times the linear phase speed and waves propagating in the ion diamagnetic direction.

5.13 Example 5.2 — 3×3 Matrix Formalism of the Inverse Scattering Transformation

To take into account physical effects that depend on higher order dispersion for the soliton propagation in strongly dispersive media we consider the higher order nonlinear Schrödinger equation

$$i\frac{\partial q}{\partial T} + \tfrac{1}{2}\frac{\partial^2 q}{\partial x^2} + |q|^2 q + i\varepsilon\left\{\beta_1 \frac{\partial^3 q}{\partial x^3} + \beta_1 |q|^2 \frac{\partial q}{\partial x} + \beta_3 q \frac{\partial |q|^2}{\partial x}\right\} = 0 \qquad (5.287)$$

which reduces to the modified nonlinear Schrödinger equation (5.205) with values of $\beta_1 = 0$, $\beta_2 = \beta_3$, while it leads to a derivative nonlinear Schrödinger equation (5.212) for the special choices of $\beta_1 = \beta_3 = 0$. Sasa and Satsuma (1991) have shown that Eq. (5.287) can be cast into the 3×3 matrix inverse scattering problem for the specific case of $\beta_1 : \beta_2 : \beta_3 = 1 : 6 : 3$. Firstly, they observe that Eq. (5.287) can be transformed to a complex modified KdV equation

$$\frac{\partial}{\partial t} U + \varepsilon\left\{\frac{\partial^3}{\partial x^3} U + 6|U|^2 \frac{\partial}{\partial x} U + 3U \frac{\partial}{\partial x}|U|^2\right\} = 0 \qquad (5.288)$$

with the variable transformations

$$U(x,t) = q(X,T)\exp\left\{-\frac{i}{6\varepsilon}\left(X - \frac{T}{18\varepsilon}\right)\right\} \qquad (5.289)$$

$$t = T \qquad (5.290)$$

$$x = X - \frac{T}{12\varepsilon}. \qquad (5.291)$$

The 3×3 Lax pairs associated with Eq. (5.288) are given for the three dimensional vector $\Psi^T = (\psi_1, \psi_2, \psi_3)$ as follows:

$$\frac{\partial}{\partial x}\Psi = U\Psi \qquad (5.292)$$

with

$$U = i\lambda\begin{pmatrix} 1 & 0 & 0 \\ 0 & 1 & 0 \\ 0 & 0 & -1 \end{pmatrix} + \begin{pmatrix} 0 & 0 & u \\ 0 & 0 & u^* \\ -u^* & u & 0 \end{pmatrix} \qquad (5.293)$$

and

$$\frac{\partial}{\partial t}\Psi = V\Psi \qquad (5.294)$$

with

$$V = -4i\varepsilon\lambda^3 \begin{pmatrix} 1 & 0 & 0 \\ 0 & 1 & 0 \\ 0 & 0 & -1 \end{pmatrix} + 4\varepsilon(\lambda^2 - |u|^2) \begin{pmatrix} 0 & 0 & u \\ 0 & 0 & u^* \\ -u^* & u & 0 \end{pmatrix} \quad (5.295)$$

$$+ 2i\varepsilon\lambda \begin{pmatrix} |u|^2 & u^2 & u_x \\ u^{*2} & |u|^2 & u_x^* \\ u_x^* & u_x & -2|u|^2 \end{pmatrix} - \varepsilon \begin{pmatrix} 0 & 0 & u_{xx} \\ 0 & 0 & u_{xx}^* \\ u_{xx}^* & -u_{xx} & 0 \end{pmatrix} \quad (5.296)$$

$$+ \varepsilon(u\,u_x^* - u_x u^*) \begin{pmatrix} 1 & 0 & 0 \\ 0 & -1 & 0 \\ 0 & 0 & 0 \end{pmatrix}. \quad (5.297)$$

The compatibility conditions of Eqs. (5.292) and (5.294) with the isospectral condition $\partial\lambda/\partial t = 0$ gives rise to Eq. (5.291).

Following the treatment of Manakov (1974), we can construct the Gel'fand - Levitan-Marchenko equation for the eigenvalue problem (5.292). The one soliton is written down as

$$q(X,T) = \eta \frac{\{2\cosh A + (C-1)\exp(-A)\}}{\cosh(2A - \log|C|) + |C|} \exp(iB) \quad (5.298)$$

where

$$A = \eta\left[X - \left\{\xi - \varepsilon(\eta^2 - 3\xi^2)\right\}T - X^{(0)}\right] \quad (5.299)$$

$$B = \xi\left[X + \left\{(\eta^2 - \xi^2)/(2\varepsilon) + \varepsilon(\xi^2 - 3\eta^2)\right\}T - X^{(0)}\right] \quad (5.300)$$

$$C = 1 - i\eta/\left\{\xi - (6\varepsilon)^{-1}\right\}. \quad (5.301)$$

Here, the real parameters ξ and η are related to the eigenvalue λ by

$$\lambda = (\tfrac{1}{2})\left[-\{\xi - 1/(6\varepsilon)\} + i\eta\right]. \quad (5.302)$$

It is instructive to confirm that in the limit of $\varepsilon \to 0$, (i.e. $C \to 1$). Equation (5.296) reduces to the soliton solution Eq. (5.187). In the opposite limit of $C \to \infty$, (i.e. $\xi \to 1/(6\varepsilon)$), Eq. (5.296) comprise another expression of sech-type soliton. For

5.13 Example 5.2 3 × 3 Matrix Formalism of ...

intermediate values of the parameter C, the shape of the soliton takes two humped shapes, as if it represents the two soliton bound state. These aspects suggest more detailed investigation of the structure of these solitons and their dynamics may be a profitable area for future research.

Chapter 6
Vortex Structures in Hydrodynamic and Vlasov Systems

The theory developed in this Chapter is closely related to the dynamics of fluids on rotating planets and their simulations in the laboratory. Because of the economic importance of this area of geophysical and atmospheric research there is vast literature relevant to the subject of Chapter 6. The reasons why the low frequency transport processes in a magnetized plasma are closely analogous to the dynamics of rotating fluids are made clear in the first section, Sec. 6.1. The most useful source of up-to-date research and further references is to be found in the special issue of CHAOS **4**(2) (1994) edited by M. Nezlin. In this issue one can find the appropriate references to both the plasma physics and geophysical literature for the subject of Chapter 6.

6.1 The Drift Wave–Rossby Wave Analogy

In the dissipationless limit the basic model equation for drift waves is the Hasegawa-Mima equation. The same basic nonlinear wave equation describes the slow, nearly incompressible motions of shallow rotating fluids, where it is known as the Charney equation. Due to this correspondence, the rapidly rotating neutral fluid experiments serve as analog simulations of drift waves, to the degree that the Hasegawa-Mima equation (1979) is an adequate description of the plasma. Thus it is rewarding to develop the theory of drift waves and Rossby waves together and to consider the results of the rotating water tank experiments in view of the analogy.

The analog of the $\mathbf{E} \times \mathbf{B}$ plasma drift velocity in the rotating fluid is the geostrophic flow

$$\mathbf{v}_\perp = (g/f)\hat{\mathbf{z}} \times \nabla H(x, y, t) \tag{6.1}$$

arising from the balance of the Coriolis force $f\mathbf{v} \times \hat{\mathbf{z}}$ with $f \equiv 2\Omega_v = 2\times$ (vertical component of the angular rotation frequency) with the pressure gradient

$$\nabla p = \rho g \nabla H = \rho f \mathbf{v} \times \hat{\mathbf{z}}. \tag{6.2}$$

Here H is the depth of the column of fluid with mass density ρ. Let $\delta H(x,y,t)$ be the variable part of the fluid depth

$$H = H_0(x) + \delta H(x,y,t). \tag{6.3}$$

The analog of the plasma quasineutrality condition $\nabla \cdot \mathbf{j} = 0$ determining the dynamics of the electrostatic plasma potential $\Phi(x,y,t)$ is the height integrated continuity equation for the neutral fluid which gives

$$\frac{dH}{dt} + H\nabla \cdot \mathbf{v}_\perp = 0 \tag{6.4}$$

where the compressible part of the flow velocity $\mathbf{v}^{(2)}$ is given by the balance of the Coriolis force with the inertial acceleration giving

$$\mathbf{v}_\perp^{(2)} = f^{-1}\hat{\mathbf{z}} \times d\mathbf{v}/dt = -\left(\frac{gH_0}{f^2}\right)\frac{d\nabla h}{dt} \tag{6.5}$$

where $h = \delta H/H_0$. The non-geostrophic flow $\mathbf{v}^{(2)}$ is the analog of the plasma polarization current given by the finite inertial acceleration by

$$\mathbf{j}_p(\mathbf{x},t) = \sum_{j=i,e} \frac{c^2 m_j n_j}{B^2} \frac{d\mathbf{E}_\perp}{dt} \simeq -\frac{en_e \rho_s^2}{T_e} \frac{d}{dt} \nabla(e\Phi). \tag{6.6}$$

The scale length for wave dispersion follows from these compressible inertial drift velocities (6.5) and (6.6) and is

ion inertial scale

$$\rho_s = c(m_i T_e)^{1/2}/eB = c_s/\omega_{ci} \tag{6.7}$$

for the plasma drift wave and

Rossby scale length

$$\rho_g = (gH)^{1/2}/f \tag{6.8}$$

for the Rossby wave. Here the ion acoustic speed

$$c_s = \left(\frac{T_e}{m_i}\right)^{1/2} \tag{6.9}$$

is analogous to the gravity wave speed

$$c_g = (gH)^{1/2}. \tag{6.10}$$

6.1 The Drift Wave–Rossby Wave Analogy

In an isothermal fluid the depth H and temperature are related by $gH = kT/m$. For quasi-geostrophic dynamics the Charney-Hasegawa-Mima (CHM) equation follows for x, y in units of the Rossby radius ρ_g, time in units of ρ_g/v_R with

$$v_R = \left(\frac{H_0 g}{f}\right) \partial_x \ell n \left(\frac{f}{H_0}\right) = c_g \left(\frac{\rho_g}{L}\right) \tag{6.11}$$

where $c_g = (gH_0)^{1/2}$, and the amplitude is scaled according to

$$h = \frac{\delta H}{H_0} = \left(\frac{\rho_g}{L}\right) \varphi. \tag{6.12}$$

Now substituting the geostrophic velocity (6.1) rescaled by Eq. (6.12) and the compressible, nongeostrophic correction (6.5) into the conservation Eq. (6.4) and using the definition of the Rossby velocity (6.11) we obtain the Charney equation (1948) describing the Rossby waves in the form

Charney eq.

$$(\rho_g^2 \nabla^2 - 1)\frac{\partial \varphi}{\partial t} + v_R \frac{\partial \varphi}{\partial y} - [\varphi, \rho_g^2 \nabla^2 \varphi] = 0. \tag{6.13}$$

The dimensionless form used in the geophysical literature, including the effect of damping, is written in a coordinate system rotated by $90°$ so that $x \to y$ and $y \to -x$. Following Pedlosky (1987) (his Eq. (3.12.30) and (6.5.21)) and McWilliams and Zabusky (1982) (their Eqs. (1) and (2)) we may write

quasi-geostrophic potential vorticity eq.

$$\frac{\partial}{\partial t} \nabla^2 \phi - \gamma^2 \frac{\partial \phi}{\partial t} + \beta \frac{\partial \phi}{\partial x} + [\phi, \nabla^2 \phi] = k\nabla^4 \phi$$

which becomes the drift wave Eq. (6.15) when $x \to -y$, $y \to x$ (rotation by $-\pi/2$) and $v_R = \beta/\gamma^2 \to v_{de}$. Both the plasma and the neutral fluid anti-cyclones are local high pressure regions.

For the plasma the dynamics of the electrostatic potential $\Phi(\mathbf{x}, t)$ follows from the requirement that $\nabla \cdot \mathbf{j}(\mathbf{x}, t) = 0$. The $\mathbf{E} \times \mathbf{B}$ drift velocity is the same for all the charges and thus produces no electrical current. The current across the magnetic field is given by the finite ion inertial current in Eq. (6.6) called the polarization current. In the case of finite length L_z electrons produce a neutralizing parallel current with

$$\nabla_\parallel j_\parallel \cong -\frac{n_e e^2}{T_e} \left(\frac{\partial}{\partial t} + \mathbf{v}_{de} \cdot \nabla\right) \varphi. \tag{6.14}$$

Combining Eq. (6.14) with the $\nabla \cdot \mathbf{j}_\perp$ from Eq. (6.6) yields the Hasegawa-Mima equation for nonlinear drift waves

Hasegawa-Mima eq.

$$(1 - \rho_s^2 \nabla^2)\frac{\partial \varphi}{\partial t} + v_{de}\frac{\partial \varphi}{\partial y} - [\varphi, \rho_s^2 \nabla^2 \varphi] = 0. \tag{6.15}$$

224   Chapter 6   Vortex Structures in Hydrodynamic and Vlasov Systems

In Sec. 6.4, we derive the H-M equation from systematic reduction of the cold ion fluid equations including the coupling to the parallel compression giving ion acoustic waves.

A summary of the correspondences between the plasma drift wave physics and the Rossby wave physics is given in Table 6.1.

Thus we see that there is a profound analogy between Rossby waves in shallow rotating fluids and drift waves in plasmas, confined transversely by a strong longitudinal magnetic field. In the dissipationless limit the basic model equation for drift waves in plasmas is the Hasegawa-Mima equation (6.15). The same basic nonlinear wave equation (6.13) describes the slow, nearly incompressible motions of shallow rotating fluids. In the case of drift waves the Lorentz force plays a role analogous to that of the Coriolis force on rotating system. The fundamental correspondence follows from low-frequency disturbances where the vector cross-product forms of the accelerations, $\mathbf{v} \times \mathbf{B}$ and $2\mathbf{v} \times \mathbf{\Omega}$, respectively, dominate the dynamics. The drift waves appear as a result of the transverse (relative to the magnetic field) nonuniformity of the electron temperature or plasma density, just like Rossby waves appear owing to the transverse (relative to the local angular rotation velocity $\mathbf{\Omega}$ of the system) nonuniformity of the Coriolis parameter $f = 2\mathbf{\Omega} \cdot \hat{\mathbf{z}}$ or depth of the fluid H. The spatial scale of the dispersion of drift waves is the "Larmor radius, ρ_s, of the ions at the electron temperature," analogous to the Rossby radius ρ_R, and equal to the ratio of the ion acoustic speed c_s to the ion cyclotron frequency ω_{ci}, while the ω_{ci} is analogous to the Coriolis parameter f. Here the ion acoustic speed c_s is analogous to the gravity wave speed c_g. The analogy of the drift velocity v_d is the characteristic Rossby velocity v_R. Based on these facts, it is understandable that the dispersion relation for drift waves is the same as the dispersion relation for Rossby waves.

It follows from the analogy between the two types of waves that, analogously to the hydrodynamic Rossby solitary vortices observed in the rotating water tank experiments, the drift wave solitary vortices should also exist in a magnetized plasma. The detailed correspondences of the analogy are listed in Table 6.1. It should be noted that in plasma physics the direction of inhomogeneity is taken in x, while in geophysics it is taken in y.

The derivation and procedure for solving the dissipative drift wave equation used here is given in Chapter 7. The dissipative drift wave equation differs from the Hasegawa-Mima equation in the presence of electron dissipation. The plasma density fluctuations given by

$$\delta n_e(\mathbf{k}, t) = n_{e0} \left[1 + i\delta_0 k_y (c_0 - k_\perp^2) \right] [e\Phi(\mathbf{k}, t)/T_e] \qquad (6.16)$$

modifies the nonlinear mode coupling in the Hasegawa-Mima equation due to the presence of the $\mathbf{v}_E \cdot \nabla n_e$ nonlinearity in the electron continuity equation. Here δ_0 and c_0 are the appropriate plasma dissipation parameters. It is the $\mathbf{v}_E \cdot \nabla \delta n_e$ nonlinearity from Eq. (6.16) that causes saturation of the drift-wave turbulence at the mixing

Table 6.1. Analogy Between Drift Wave and Rossby Wave

Drift Wave	Rossby Wave
H-M equation: $(1 - \nabla^2)\frac{\partial \varphi}{\partial t} + v_d \frac{\partial \varphi}{\partial y} - [\varphi, \nabla^2 \varphi] = 0$	Charney equation: $(1 - \nabla^2)\frac{\partial h}{\partial t} + v_R \frac{\partial h}{\partial y} - [h, \nabla^2 h] = 0$
Electrostatic potential: $\varphi(x,y,t)$ $\varphi(x,y,t) = (\frac{r_n}{\rho_s})e\Phi(\frac{x}{\rho_s}, \frac{y}{\rho_s}, \frac{c_s}{r_n}t)/T_e$	Variable part of fluid depth: $h(x,y,t)$ $h(x,y,t) = (\frac{L_R}{\rho_R})\delta h(\frac{x}{\rho_R}, \frac{y}{\rho_R}, \frac{c_g}{L_R}t)/H$
Lorentz force: $m_i \omega_{ci} \mathbf{v}_\perp \times \hat{\mathbf{z}}$	Coriolis force: $\rho f \mathbf{v}_\perp \times \hat{\mathbf{z}}$
$\mathbf{E} \times \mathbf{B}$ drift flow: $\mathbf{V}_\perp = \left(\frac{c}{B}\right) \hat{\mathbf{z}} \times \nabla \Phi$	Geostrophic flow: $\mathbf{V}_\perp = \left(\frac{g}{f}\right) \hat{\mathbf{z}} \times \nabla \delta h$
Cyclotron frequency: $\omega_{ci} = \frac{e_i B}{c m_i}$	Coriolis parameter: f
Drift coefficient: $r_n^{-1} = -\frac{\partial}{\partial x} \ell n n_0$	Rossby coefficient: $L_R^{-1} = \frac{\partial}{\partial x} \ell n \left(\frac{f}{H}\right)$
Larmor radius: $\rho_s = \frac{c_s}{\omega_{ci}}$	Rossby radius: $\rho_R = \frac{c_g}{f}$
Ion acoustic speed: $c_s = \left(\frac{T_e}{m_i}\right)^{1/2}$ where T_e is electron temperature.	Gravity wave speed: $c_g = (gH)^{1/2}$ where H is depth of fluid layer.
Drift velocity: $v_d = c_s \rho_s \frac{\partial}{\partial x} \ell n n_0$	Rossby velocity: $v_R = c_g \rho_R \frac{\partial}{\partial x} \ell n\left(\frac{f}{H}\right)$
Dispersion relation: $\omega = \frac{k_y v_d}{1 + k^2 \rho_s^2}$	Dispersion relation: $\omega = \frac{k_y v_R}{1 + k^2 \rho_R^2}$

226   Chapter 6   Vortex Structures in Hydrodynamic and Vlasov Systems

length level (Horton, 1986). Kono and Miyashita (1988) show with simulations of the collisional drift wave that the turbulence evolves through an inverse cascade to a final state containing a large amplitude dipolar vortex structure.

6.2 The Drift Wave Mechanism and Vortex

To illustrate the drift wave mechanism, we consider a dense and collisionless ion column confined in a constant and uniform magnetic field $\mathbf{B} = B_0 \hat{\mathbf{z}}$ as shown in Fig. 6.1. An ion rich plasma column is confined in a constant axial magnetic field $B_0 \hat{\mathbf{z}}$. The ion column causes an electric field $\mathbf{E}$ along the radial direction of the column. The $\mathbf{E} \times \mathbf{B}$ drift causes the column to rotate around its axis. The background plasma has a density gradient along the negative x-direction which causes linear drift waves to propagate along the y-direction.

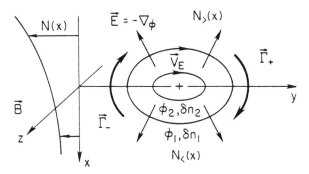

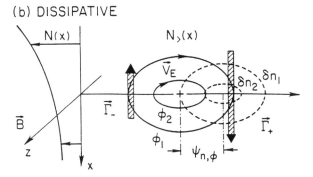

Fig. 6.1. Geometry and the mechanism of the dissipative drift wave instability.

6.2 The Drift Wave Mechanism and Vortex

The positively charged ion column causes a radial electric field $\mathbf{E} = -\nabla\Phi$. The background plasma has a density gradient increasing along the negative x-direction. The radial electric field $\mathbf{E}$ and the magnetic field $\mathbf{B}$ cause the ion column to produce an $\mathbf{E} \times \mathbf{B}$ rotation $\mathbf{V}_{\mathbf{E} \times B} = (c/B_0^2)\mathbf{E} \times \mathbf{B}$ around the axis of the column. The rotation rate can be calculated by $\Omega_E = (c/B_0)k_x k_y \Phi_k$ for a structure characterized by wave numbers (k_x, k_y) and amplitude Φ_k. We call this rotation rate the characteristic nonlinear frequency since it depends on the amplitude Φ_k of the electric potential of the structure of size π/k.

On the other hand, the non-uniform background plasma causes linear drift waves which propagate along the y-direction and have linear drift wave frequency $\omega_k = k_y v_d/(1 + k_\perp^2 \rho_s^2)$. The ratio of the nonlinear frequency given by vortex rotation rate Ω_E to the linear frequency ω_k is defined as the $\mathbf{E} \times \mathbf{B}$ rotation number R_E and is given by

$$R_E = \frac{\Omega_E(k)}{\omega_k} \simeq k_x r_n \left(\frac{e\Phi_k}{T_e}\right). \tag{6.17}$$

Since R_E is the number of rotations that an $\mathbf{E} \times \mathbf{B}$ drifting particle makes in the wave period $T_k = 2\pi/\omega_k$, the value of R_E is a Galilean invariant when ω_k is taken in the plasma rest frame. When $R_E < 1$, the dispersion of the wave packet dominates and only when $R_E \geq 1$ do the vortex structures form. Figure 6.2 is an example comparing the linear $R_E \ll 1$ (left panel) behavior with the strong nonlinear $R_E \sim 10$ propagation (right panel) of the same initial disturbance. The comparison clearly shows the self-binding that occurs in the nonlinear regime. Figure 6.2 shows that in contrast to small amplitude drift waves which spread out and lose strength as they travel as seen in Fig. 6.2a, the solitary drift wave vortices are coherent, self-sustaining structures which retain their strength over long time intervals and distances as they travel.

The most commonly seen vortex structures are monopole and dipole vortices. The monopole vortex represents a net excess in the local charge density and is the natural, finite amplitude solution in the presence of a sheared flow, and the dipole vortex represents a local charge polarization in the local plasma density and is the natural, finite amplitude solution in the absence of a sheared flow. The dipole is seen to be produced in the turbulent wake of a 2-D fluid flowing past an obstacle and is also formed through pairing or coupling of two monopole vortices with opposite rotational directions (Couder and Basdevant, 1986; Horton, 1989). The properties and interactions of both the monopole and dipole vortices have been studied by many authors (Makino et al., 1981; McWilliams and Zabusky, 1982; Horton, 1989). The interactions of the long-lived and particle-trapping vortex structures play an important role in anomalous transport of plasmas (Nycander and Isichenko, 1990).

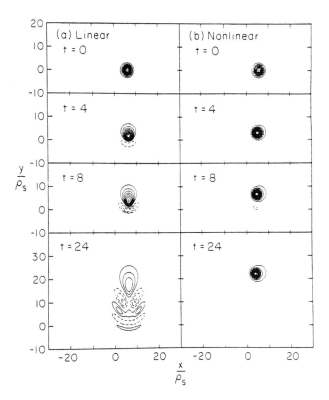

Fig. 6.2. Comparing the propagation of a small amplitude ($R_E \ll 1$) and large amplitude ($R_E = 10$) gaussian potential peak. (a) On the left, the linear pulse spreads due to spectrum of phase velocities from zero to v_{de}. (b) The rapid $\mathbf{E} \times \mathbf{B}$ rotation produces a self-focusing allowing almost dispersionless propagation of a vortex at the speed v_{de}. Time is in units of r_n/c_s. Solid contours are positive potentials and high electron pressure regions anti-cyclones.

6.2.1 Nonlinear drift wave equation and passive scalar equation

The general form of the dissipative drift wave equation may be written in the usual dimensionless variables as

$$(1+\mathcal{L})\frac{\partial \varphi(x,y,t)}{\partial t} + v_d \frac{\partial \varphi}{\partial y} + \left[\frac{\partial \varphi}{\partial x}\frac{\partial}{\partial y}(\mathcal{L}\varphi) - \frac{\partial \varphi}{\partial y}\frac{\partial}{\partial x}(\mathcal{L}\varphi)\right] + \nu \nabla^4 \varphi = 0 \quad (6.18)$$

6.2 The Drift Wave Mechanism and Vortex

with the linear operator $\mathcal{L}$ containing both wave dispersion (Hamiltonian) and the wave dissipation (anti-Hermitian) parts given by

$$\mathcal{L} = -\nabla^2 + \delta_0(c_0 + \nabla^2)\partial_y. \tag{6.19}$$

It is often useful to consider an associated field of dye (or $\mathbf{E} \times \mathbf{B}$ test particles) given by the convection of the passive scalar distribution f

$$\frac{\partial f(x,y,t)}{\partial t} + \frac{\partial \varphi}{\partial x}\frac{\partial f}{\partial y} - \frac{\partial \varphi}{\partial y}\frac{\partial f}{\partial x} - D\nabla^2 f = 0 \tag{6.20}$$

with the background drift vorticity D. The limit of $P = \tilde{\varphi}/D \to \infty$, where P is the dimensionless Peclete number, is of great interest. The derivation of Eqs. (6.18)–(6.19) is given in Horton (1986).

Equations (6.18)–(6.19) are solved using fifth and sixth order Runge-Kutta variable time stepping and a truncated Fourier series transformation $(x,y) \leftrightarrow (k_y, k_y)$ at each time step. Equations (6.18)–(6.19) are written in the usual dimensionless drift wave units of ρ_s for x, y and $r_n/c_s = \rho_s/v_d$ for t and with a constant $k_\| r_n$ assumption. The electrostatic potential Φ is given in terms of $\varphi(x,y,t)$ by $e\Phi/T_e = (\rho_s/r_n)\varphi(x,y,t)$.

The symmetry property of Eq. (6.18) is that for every solution $\varphi(x,y,t)$ there is a reflected solution $\overline{\varphi}$ with $\overline{\varphi} = -\varphi(-x,y,t)$. The nonlinear equation does not have the symmetric solutions $\varphi(-x,y,t) = \varphi(x,y,t)$ of the linear wave equation.

The dimensionless dissipation coefficient ν for the plasma arises from the ion-ion collisional viscosity $\nu = 0.3(r_n\nu_{ii}/c_s)(T_i/T_e)$ and for the neutral fluid $\nu = \nu_{\text{vis}}/\rho_g v_R$ from viscosity. The dimensionless diffusion coefficient for test electrons is $D = (m_e/m_i)(r_n\nu_{ei}/c_s) \sim (m_e/m_i)^{1/2}\nu$ and for test ion is $D \sim \nu$. For typical tokamak plasmas $D \simeq 10^{-4}$ and is the reciprocal of the Peclet number.

The passive transport equation (6.20) is integrated simultaneously with the vorticity equation (6.18) to determine the convection of plasma test particle distribution. Depending on the application, the passive plasma field $f(x,y,t)$ may represent one of a number of physical fields: the distribution of impurities, the temperature field or the poloidal flux function. Here we note that a typical form of the coupling for the back reaction of f onto φ is the addition to the right-hand side of Eq. (6.18) of the term $-g(\partial f/\partial y)$ where $g = v_T^2/R_c$ arises from the curvature $R_c^{-1} = \max|(\mathbf{b}\cdot\nabla)\mathbf{b}|$ of the magnetic field lines. When the back reaction coefficient exceeds a certain critical value the vorticity equation is destabilized, and new types of vortices are formed.

6.2.2 Conservation laws

The drift wave equation (6.18) in the dissipationless limit has three conservation laws: mass, energy, and enstrophy. The equation itself is a statement of local momentum conservation in the plasma flows.

Mass Conservation

Integrating (6.18) over all x, y gives

$$\frac{d}{dt}\varphi_{\mathbf{k}=0} = \frac{d}{dt}\int \varphi \, dx dy = 0 \tag{6.21}$$

which is equivalent to the conservation of mass. To see the equivalence we note that the plasma density is

$$N = N_0\left[1 + \left(\frac{\rho_s}{r_n}\right)\left(\frac{\delta n}{n(x)}\right)\right]$$

where

$$\frac{\delta n}{n} = \varphi + \delta_0(c_0 - \nabla^2)\partial_y\varphi,$$

and thus the total mass M

$$M = \int dx dy \, N = \int dx dy \, \varphi + \text{const.} = \text{const.}$$

Energy Conservation

The local energy density $w(x, y, t)$ is the sum of the electrostatic potential energy $\frac{1}{2}e\delta n_e\Phi \sim \frac{1}{2}\varphi^2 E_{dw}$ and the ion kinetic energy $\frac{1}{2}m_i n_0 v_E^2 \sim \frac{1}{2}(\nabla\varphi)^2 E_{dw}$ where $E_{dw} = N_e T_e\left(\rho_s/r_n\right)^2$. In the dimensionless variables

$$w(x, y, t) = \tfrac{1}{2}\left[\varphi^2 + (\nabla\varphi)^2\right] \tag{6.22}$$

in units of $n_e T_e\left(\frac{\rho_s}{r_n}\right)^2$. The local conservation of energy is derived using the energy flux

$$\mathbf{F}_w = \tfrac{1}{2}\mathbf{v}_d\varphi^2 - \varphi\nabla\frac{\partial\varphi}{\partial t} - \nabla^2\varphi\hat{\mathbf{z}}\times\nabla\left(\frac{\varphi^2}{2}\right) \tag{6.23}$$

in units of $v_{de}T_e\left(\frac{\rho_s}{r_n}\right)^2$, and the conservation law is

$$\frac{\partial w}{\partial t} + \nabla \cdot \mathbf{F}_w = -\nu\varphi\nabla^4\varphi \tag{6.24}$$

derived using Eq. (6.18). Thus,

$$E = \int dx dy \, w = \text{const.}$$

when

$$\oint_c \mathbf{F}_w \cdot d\boldsymbol{\ell} \times \hat{\mathbf{z}} = 0 \quad \text{and} \quad \nu = 0.$$

6.2 The Drift Wave Mechanism and Vortex

Potential Enstrophy Conservation

The potential vorticity for Eq. (6.18) is

$$q(x,y,t) = \nabla^2 \varphi - \varphi + v_d x \tag{6.25}$$

and in the dissipationless (Hasegawa-Mima) limit the dynamical equation (6.18) becomes

$$\frac{dq}{dt} = 0. \tag{6.26}$$

Equation (6.26) is Ertel's theorem for the conservation of potential vorticity. From the convection of q, the conservation law

$$\int q^2 dx dy = \text{const.}$$

and the energy conservation law $E = \int w\, dx dy = \text{const.}$ suggests the definition of the local enstrophy density

$$u(x,y,t) = \frac{1}{2}(\nabla\varphi)^2 + \frac{1}{2}(\nabla^2\varphi)^2 \tag{6.27}$$

and the enstrophy flux

$$\mathbf{F}_u = \mathbf{v}_d(\nabla\varphi)^2 - \nabla\varphi\left(\frac{\partial\varphi}{\partial t} + v_d\frac{\partial\varphi}{\partial y}\right) + \tfrac{1}{2}(\nabla^2\varphi)^2 \hat{\mathbf{z}} \times \nabla\varphi. \tag{6.28}$$

The dynamical equation (6.18) leads to the conservation law

$$\frac{\partial u}{\partial t} + \nabla \cdot \mathbf{F}_u = -\nu \nabla^2 \varphi \nabla^4 \varphi \tag{6.29}$$

with

$$U = \int dx dy\, u = \text{const.}$$

when

$$\oint_c \mathbf{F}_u \cdot d\boldsymbol{\ell} \times \hat{\mathbf{z}} = 0 \quad \text{and} \quad \nu = 0.$$

Momentum Conservation

The equation (6.18) is a constraint (equivalent to the divergence of the current must vanish) derived from the local momentum balance equation. The integration of (6.18) over y leads to a further constraint involving the Reynolds stress

$$\pi(x) = \overline{v_x v_y} \equiv -\int \frac{dy}{L_y} \frac{\partial\varphi}{\partial x}\frac{\partial\varphi}{\partial y} \tag{6.30}$$

giving the flux of y-momentum from x to $x + dx$. The mean flow

$$\bar{v}_y(x,t) \equiv \int \frac{dy}{L_y} \frac{\partial \varphi}{\partial x}(x,y,t) \tag{6.31}$$

satisfies the momentum transport equation

$$\frac{\partial \bar{v}_y}{\partial t} + \frac{\partial \pi}{\partial x} = \nu \frac{\partial^2 \bar{v}_y}{\partial x^2} \tag{6.32}$$

with the kinetic energy transferred from the mean flow $\frac{1}{2}\int \bar{v}_y^2 dx$ to the wave and vortex flow $\varphi(x,y,t)$ given by

$$T = \int \bar{v}_y \frac{\partial \pi}{\partial x} dx. \tag{6.33}$$

In most real confinement systems there is a small radial leakage current $\bar{j}_x$ across the magnetic surfaces. Taking the surface average of $\nabla \cdot \mathbf{j} = 0$ and including this current gives the $\bar{j}_x B_z$ acceleration equation,

$$\frac{\partial \bar{v}_y}{\partial t} + \frac{\partial \pi}{\partial x} = \frac{-j_x B_z}{\rho_m} + \nu \frac{\partial^2 \bar{v}_y}{\partial x^2} \tag{6.34}$$

describing the convective transport of momentum across the magnetic field and the build-up of a shear flow $\bar{v}_y(x,t)$ from the torque $\bar{j}_x B_z$. The shear-flow produced by the Reynolds stress π and/or radial current in Eq. (6.34) gives rise to bifurcations in the confinement regimes described as L-H dynamics in the transport confinement literature. Here ρ_m is the mass density.

The effect of a large scale sheared flow $\bar{v}_y(x)$ with $r_0 \partial \bar{v}_y / \partial x \gtrsim v_{de}$ is the rapid transformation of the dipole vortex into a monopole vortex. The transformation occurs due to the stagnation point introduced on the counter streaming side of the dipole. The sensitivity of the dipole vortices to a large scale ($\gtrsim r_0$) sheared flow may explain the predominance of monopoles in many laboratory experiments.

The conservation of potential vorticity $dq/dt = 0$ in Eq. (6.26) implies the existence of an infinite number of constants of the motion of the form $\int G(q) dx\, dy = $ const which might suggest that the system is integrable. In contrast, the numerical simulations clearly suggest that the presence of the drift wave component makes the vortex collisions inelastic. Mathematical results consistent with the simulations are reported by Zakharov and Schulman (1988) showing that the existence of infinitely many invariants does not imply the integrability of such Hamiltonian field equations. They show that additional constraints on the wave spectrum are required for integrability of the field equations.

6.2.3 Solitary monopolar structures and the trapping condition

The nonuniform background to the plasma causes the linear drift waves which propagate along the y-direction and have linear drift wave frequency $\omega_\mathbf{k} = k_y v_d /$

$(1+k^2\rho_s^2)$. The amount of rotation in a nonlinear structure is measured by the ratio of the vortex rotation rate Ω_E to the linear frequency ω_k given by the rotation number R_E

$$R_E = \frac{\Omega_E(k)}{\omega_k} \simeq k_x r_n \left(\frac{e\Phi_k}{T_e}\right)$$

defined in Eq. (6.17). When $R_E < 1$ the rotation is weak and the dispersion of the wave packet dominates. Only when $R_E \geq 1$ can vortex structures be formed. Fig. 6.2 is an example comparing the linear $R_E \ll 1$ behavior with the strong nonlinear $R_E \sim 10$ propagation of the same size initial disturbance which clearly shows the self-binding effect. Fig. 6.2 shows that in contrast to small amplitude drift waves which spread out and lose strength as they travel, the solitary drift wave vortices are coherent, self-sustaining packets which retain their strength over long time intervals and distances as they travel.

The monopole vortex represents a net excess in the local charge density and is the natural, finite amplitude solution in the presence of a sheared flow. The dipole vortex represents a local charge polarization in the local plasma density and is the natural, finite amplitude solution for the self-focusing of a sinusoidal fluctuation. It is believed, but not well documented, that the interactions of the long-lived and particle-trapping vortex structures play an important role in anomalous transport of plasmas.

In view of the conservation of mass given by Eq. (6.21), we see that the monopole perturbation, such as shown in Fig. 6.2, requires that a net excess mass be involved in forming the perturbation. To avoid this necessity, it is often natural to consider a dipolar perturbation for which $\delta m \propto \int \varphi \, dx \, dy \equiv 0$ by the odd symmetry. It is the dipolar perturbation that is similar to considering a localized, finite amplitude sinusoidal fluctuation of the equilibrium plasma.

6.3 Solitary Dipolar Vortex Solutions

Nonlinearity in the equations for drift waves in a magnetized plasma gives rise to the formation of solitary waves and 2D coherent vortex structures. As in the one-dimensional Korteweg de Vries (KdV) equation, the two-dimensional, drift wave dispersion and nonlinear steepening through mode coupling can balance to form coherent, localized structures. These localized solutions of the drift wave equations are solitary vortices.

Two important physical properties of solitary drift wave vortices are that (i) they propagate with speeds complementary to the speeds of the linear modes, and that (ii) they possess a high degree of stability to perturbations. The first requirement can be understood by considering what would happen if this complementarity condition were not satisfied. If there were linear waves that could propagate at the same speed as the localized structure then such a wave could be linked to the nonlinear structure and bleed-off energy from that structure by the wave energy transport. Thus, to prevent a wave from tapping into the localized, high stored energy density structure formed by the vortex, it is necessary to have the complementary velocity principle. Mathematically, the complementarity of the vortex/soliton speeds arises as a natural consequence of the nonlinear equations and the boundary conditions.

In confinement systems, with sheared magnetic fields the nonlinearity of the wave equation produces an effective potential well that partially reflects and self-focuses the outward propagating wave energy. For small amplitude waves the nonlinear potential well is weak; however, for larger amplitude solitary waves the nonlinear potential well for radial focusing is deep and only an exponentially small outgoing wave tunnels through the barrier to become escaping wave energy. Due to the nonlinear wave barrier, the radiative induced damping of the solitary drift wave is small for vortex structures propagating either faster than the electron diamagnetic drift velocity or in the ion diamagnetic direction.

Many types of plasma equations, including the shear-Alfvén drift modes, FLR-rotating modes, electrostatic and magnetostatic convective cells have now been shown to possess finite amplitude vortex solutions. A review of some of the solutions is given by Shukla et al. (1984). Other recent electromagnetic vortex solutions are given by Liu and Horton (1986) and Mikhailovskii et al. (1984a,b, 1985), a review by Yankov and Petviashvili in the *Reviews of Plasma Physics*, Vol. 14 and *Solitary Waves in Plasmas and in the Atmosphere* by Petviashvili and Pokhotelov (1992).

6.4 Drift Wave-Ion Acoustic Wave Equations

We consider electrostatic electron drift waves and ion acoustic waves in the inhomogeneous plasma slab $n_0(x)$ with a sheared magnetic field

$$\Omega = \left(\frac{e_i B}{m_i c}\right)\left(\hat{\mathbf{z}} + \left(\frac{x}{L_s}\right)\hat{\mathbf{y}}\right). \quad (6.35)$$

The dynamical equations are the pressureless ion fluid and the almost adiabatic electron fluid equations. The condition of quasineutrality relates the ion density to the electron density $n(\mathbf{x}, t)$ which is taken to be

$$\ell n \, \tilde{n} = \ell n \, \frac{n(\mathbf{x},t)}{n_0(x)} = (1 + \mathcal{L}^{\text{ah}})\varphi(\mathbf{x},t) \quad (6.36)$$

where $\varphi = e\Phi/T_e$ is the normalized electrostatic potential, and $\mathcal{L}^{\text{ah}}$ is an anti-Hermitian operator giving the dissipative part of the electron response.

The form of $\hat{\mathcal{L}}^{\text{ah}}$ is given by the electron dissipation in Eq. (6.19), and the nonlinear effect of $\mathcal{L}^{\text{ah}}$ is the subject of Chapter 7 where the transition to temporal chaos is produced by the dissipation in $\mathcal{L}^{\text{ah}}$. In regard to dissipative localized vortex structures in the collisional drift wave, Taniuti (1986) has searched for such nonlinear structures including the effect of magnetic shear. While Taniuti (1986) found

6.4 Drift Wave-Ion Acoustic Wave Equations

evidence for localized solutions he was not able to establish the existence of true long-time stationary, localized waves. The nonlinear structures may be unstable to collapse and radiation in the long-time limit.

The ion fluid equations can be written

$$\frac{d}{dt} \ln \tilde{n} + \frac{v_x}{n_0} \frac{dn_0}{dx} + \nabla \cdot \mathbf{v} = 0 \tag{6.37}$$

$$\frac{d}{dt} \mathbf{v} = -c_s^2 \nabla \varphi + \mathbf{v} \times \mathbf{\Omega}, \tag{6.38}$$

where $c_s^2 = T_e/m_i$. In Eqs. (6.37) and (6.38) the time derivative is the convective derivative

$$\frac{d}{dt} = \frac{\partial}{\partial t} + \mathbf{v} \cdot \nabla. \tag{6.39}$$

An ordering for the drift waves consistent with the experimental measurements of low-frequency fluctuations is given by

$$\varepsilon \equiv \frac{\rho_s}{r_n} \sim \frac{\omega}{\Omega} \sim \frac{v}{c_s} \sim \varphi \ll 1 \tag{6.40}$$

where $\rho_s = c_s/\Omega$, and ω represents a typical frequency. An additional but independent small parameter is $S = r_n/L_s$, measuring the density gradient scale r_n compared with the scale L_s for rotation of the magnetic field vector $\mathbf{B}$ by order unity. The rotation of the magnetic field vector $\mathbf{B}$ couples the ion acoustic oscillations $\omega^2 = k_\parallel^2 c_s^2$ to the drift wave $\omega = \omega_{*e}$ and implies the ordering

$$k_\parallel r_n \sim k_\perp \rho_s \sim \mathcal{O}(1). \tag{6.41}$$

This ordering for $k_\parallel$ is in contrast to that of the Hasegawa and Mima (1977, 1978) ordering where $k_\parallel r_n \sim \mathcal{O}(\varepsilon)$ is introduced to justify the neglect of the parallel ion oscillations.

With the orderings of Eq. (6.40) and (6.41), the natural dimensionless variables are

$$\mathbf{x}_\perp \to \rho_s \mathbf{x}_\perp$$

$$z \to r_n z$$

$$\frac{e\Phi}{T_e} \to \left(\frac{\rho_s}{r_n}\right) \varphi \tag{6.42}$$

$$\mathbf{v}/c_s \to \left(\frac{\rho_s}{r_n}\right) \mathbf{v}$$

$$t \to (r_n/c_s)t,$$

where the variables on the left-hand side of these equations are the physical variables, and those on the right-hand side are dimensionless variables used subsequently.

Resolving the momentum balance in the x-y plane and $\hat{\mathbf{z}}$ direction we can solve Eq. (6.38) for $\mathbf{v}_\perp = (\hat{\mathbf{b}} \times \mathbf{v}) \times \hat{\mathbf{b}}$ and for $v_\parallel$. In the convective derivative in Eq. (6.39), the incompressible $\mathbf{E} \times \mathbf{B}$ flow dominates. As a consequence the $\mathbf{v} \cdot \nabla f$ nonlinearity is written in terms of the Jacobian or Poisson brackets defined by

$$\{a,b\} = \frac{\partial a}{\partial x}\frac{\partial b}{\partial y} - \frac{\partial a}{\partial y}\frac{\partial b}{\partial x} \tag{6.43}$$

with $a = \varphi$ and $b = f$.

With these velocity fields the continuity equation and parallel ion momentum balance equation become to lowest order in ε and S

$$\frac{d}{dt}\left(1 + \mathcal{L} - \nabla_\perp^2\right)\varphi + v_d \frac{\partial \varphi}{\partial y} + \hat{\mathbf{b}} \cdot \nabla v_\parallel = 0. \tag{6.44}$$

$$\frac{dv_\parallel}{dt} = -\hat{\mathbf{b}} \cdot \nabla \varphi \tag{6.45}$$

where

$$\hat{\mathbf{b}} \cdot \nabla = \frac{\partial}{\partial z} + Sx\frac{\partial}{\partial y}. \tag{6.46}$$

The linear dispersion relation from Eqs. (6.44) and (6.45) in the local approximation is given by

$$\omega^2(1 + k_\perp^2 + \mathcal{L}_\mathbf{k}) - \omega k_y v_d - k_\parallel^2 c_s^2 = 0$$

where

$$k_\parallel = k_z + k_y Sx. \tag{6.47}$$

Returning to Eqs. (6.44) and (6.45) and keeping the $\varphi_\mathbf{k}(x)$ dependence on x and using $\nabla^2 = \partial_x^2 - k_y^2$ we obtain the linear eigenmode equation

$$\left[\omega^2(1 + k_y^2 + \mathcal{L}_\mathbf{k} - \partial_x^2) - \omega k_y v_d - k_\parallel^2(x) c_s^2\right]\varphi_\mathbf{k}(x) = 0$$

where $v_d = \rho_s c_s / r_n = 1$ and $\mathcal{L}_\mathbf{k}$ is the $\mathbf{k}$ space representations of the non-adiabatic part of electron operator $\mathcal{L}$ defined in Eq. (6.16).

For systems with sufficiently weak shear $[S < \mathcal{O}(\varepsilon)]$ the coupling of the drift wave to the ion acoustic wave may be neglected. Taking $\hat{\mathbf{b}} \cdot \nabla \sim k_\parallel = k_y Sx \sim 0$ and neglecting the nonadiabatic electron response, $\mathcal{L} = 0$ in Eq. (6.44), the equation reduces to that derived by Hasegawa and Mima Eq. (6.15)

$$\frac{d}{dt}\left(1 - \nabla_\perp^2\right)\varphi + v_d \frac{\partial \varphi}{\partial y} = 0 \tag{6.48}$$

with

$$\frac{d}{dt}\varphi = \frac{\partial \varphi}{\partial t} \quad \text{and} \quad \frac{d}{dt}\nabla_\perp^2 \varphi = \frac{\partial}{\partial t}\nabla_\perp^2 \varphi + \left[\varphi, \nabla^2 \varphi\right].$$

6.4 Drift Wave-Ion Acoustic Wave Equations

Equation (6.48) is the expression of Ertel's theorem (1942) for the conservation of potential vorticity q where

$$\frac{dq}{dt} = 0$$

with

$$q = \nabla^2 \varphi - \varphi + v_d x$$

through $\mathcal{O}(\varepsilon^2)$ as derived below. An equation of this form was first derived by Charney (1948) in a geophysical context for Rossby waves in a rotating neutral field (Pedlosky, 1987). The analogy between Rossby waves and drift waves was first discussed by Hasegawa, Maclennan and Kodama (1979).

6.4.1 Conditions for the conservation of energy and enstrophy

The coupled drift-wave-ion acoustic wave energy density and the enstrophy are defined by

$$E = \tfrac{1}{2} \int d^3 x \left[\varphi^2 + (\nabla_\perp \varphi)^2 + v_z^2 \right] \tag{6.49}$$

$$K = \tfrac{1}{2} \int d^3 x \left[(\nabla_\perp \varphi)^2 + (\nabla_\perp^2 \varphi)^2 \right]. \tag{6.50}$$

These quantities are conserved in the $k_\parallel = 0$ dynamics of Eq. (6.48) but not by the full dynamics of Eqs. (6.44) and (6.45).

The energy and enstrophy changes due to electron dissipation $\mathcal{L}$ and by radiative damping from shear-induced ion acoustic waves are given by

$$\frac{dE}{dt} = - \int d^3 x \varphi \frac{d}{dt} \mathcal{L}\varphi + \int dy dz \varphi \frac{\partial^2 \varphi}{\partial x \partial t} \bigg|_{x=-L}^{x=L}, \tag{6.51}$$

$$\frac{dK}{dt} = \int d^3 x \nabla_\perp^2 \varphi \left(\frac{d}{dt} \mathcal{L}\varphi + \hat{\mathbf{b}} \cdot \nabla v_z \right) + \int dy dz \frac{\partial \varphi}{\partial x} \left(\frac{\partial \varphi}{\partial t} + \frac{\partial \varphi}{\partial y} \right) \bigg|_{x=-L}^{x=L}. \tag{6.52}$$

The fields $\varphi(\mathbf{x}, t)$ and $v_\parallel(\mathbf{x}, t)$ are assumed to decay as $y, z \to \infty$ (or to be periodic). The fields are generally finite at large x due to the magnetic shear. The last term in Eq. (6.51) represents linear radiative damping due to the fields beyond $x > |L|$. We will see in Sec. (6.4.4) that the nonlinearity may be neglected in the radiation region $L \gg r_0$ for a solitary drift wave structure of size r_0. The enstrophy is not conserved due to the coupling of the cross-field motion with the compressible parallel motion.

6.4.2 Ertel's conservation theorem

The ion fluid equation [Eq. (6.38)] with finite pressure terms can be rewritten in terms of the vorticity $\boldsymbol{\omega} = \nabla \times \mathbf{v}$:

$$\frac{d}{dt}(\boldsymbol{\omega} + \boldsymbol{\Omega}) + (\boldsymbol{\omega} + \boldsymbol{\Omega})\nabla \cdot \mathbf{v} = (\boldsymbol{\omega} + \boldsymbol{\Omega}) \cdot \nabla \mathbf{v}, \quad (6.53)$$

provided that $\nabla n \times \nabla p = 0$, which in geophysics is called the barotropic fluid approximation. The η_i-modes and toroidal ion temperature gradient (ITG) modes break the barotropic condition forcing the addition of thermal transport equation.

Using the density equation to eliminate $\nabla \cdot \mathbf{v}$ in (6.53) allows Eq. (6.53) to be rewritten as

$$\frac{d}{dt}\left(\frac{\boldsymbol{\omega} + \boldsymbol{\Omega}}{n}\right) = \left(\frac{\boldsymbol{\omega} + \boldsymbol{\Omega}}{n}\right) \cdot \nabla \mathbf{v} \quad (6.54)$$

where n is the total electron density. Suppose there exists some scalar quantity conserved for each fluid element:

$$\frac{d\lambda}{dt} = 0. \quad (6.55)$$

Then Ertel's theorem states that the potential vorticity

$$\Pi = \frac{\boldsymbol{\omega} + \boldsymbol{\Omega}}{n} \cdot \nabla \lambda, \quad (6.56)$$

is also conserved by the flow. For the case at hand the large toroidal magnetic field $B\hat{\mathbf{z}}$ with constant direction limits consideration to two-dimensional flows in the perpendicular plane ($k_\parallel \to 0$) so that $\nabla \lambda = \hat{\mathbf{z}}$ and

$$\frac{d}{dt}\left(\frac{\omega_z + \Omega}{n}\right) = 0. \quad (6.57)$$

This potential vorticity equation reduces to the Charney equation [Eq. (6.13) and (6.48)] in the limit when the velocity is completely $\mathbf{E} \times \mathbf{B}$ and $n = n_0 \exp(\varphi)$ where $|\varphi| \ll 1$. More generally, Eq. (6.57) includes the effects of ∇B, which is analogous to the β-term of geophysics where Ω is twice the vertical component of the planetary angular rotation frequency, and the effect of higher order nonlinearities. In geophysics, these effects are referred to as non-geostrophic effects with geostrophy defined as the state where the velocity is given entirely by Eq. (6.1) equivalent to the $\mathbf{E} \times \mathbf{B}$ drift. The Charney equation is called the quasi-geostrophic potential vorticity equation (Pedlosky, 1987).

6.4.3 Solitary drift wave vortices: shearless case

Traveling wave solutions of Eq. (6.44) and (6.45) are obtained by letting

$$\eta = y + \alpha z - ut$$

6.4 Drift Wave-Ion Acoustic Wave Equations

$$\varphi(x,y,z,t) = \varphi(x,\eta)$$
$$v_z(x,y,z,t) = v_z(x,\eta) \qquad (6.58)$$

where u is the speed of propagation. Using Eq. (6.58), it is easy to see that

$$v_z(x,\eta) = \frac{\alpha}{u}\varphi(x,\eta) \qquad (6.59)$$

satisfies Eq. (6.45) exactly, since $[\varphi, v_z] = 0$. Substituting Eqs. (6.58) and (6.59) into Eq. (6.44) with $\mathcal{L} = 0$ yields a nonlinear partial differential equation for $\varphi(x,\eta)$ which can be written as a single Poisson bracket relation:

$$\left[\varphi - ux,\; \nabla_\perp^2 \varphi - \varphi + \left(v_d + \frac{\alpha^2}{u}\right)x\right] = 0. \qquad (6.60)$$

This relationship implies that the functions in the bracket are dependent,

$$\nabla_\perp^2 \varphi - \varphi + \left(v_d + \frac{\alpha^2}{u}\right)x = F(\varphi - ux), \qquad (6.61)$$

where $F(X)$ is an arbitrary function. We are looking for localized solutions, that is, for each $x, \varphi \to 0$ as $X \to \infty$. Considering the $X \to \infty$ limit of Eq. (6.61) determines the form of $F(X)$:

$$F(X) = -\left(\frac{v_d}{u} + \frac{\alpha^2}{u^2}\right)X, \qquad (6.62)$$

that is, F must be linear for localized solutions. Using Eq. (6.62) for F in Eq. (6.61) gives

$$\nabla_\perp^2 \varphi + \left(\frac{v_d}{u} - 1 + \frac{\alpha^2}{u^2}\right)\varphi = 0. \qquad (6.63)$$

Isolation of the solution requires that the speed of the vortex satisfy

$$u^2 - uv_d - \alpha^2 > 0 \qquad (6.64)$$

so that φ has positive curvature $\nabla^2\varphi/\varphi > 0$.

It is important to note that the linear modes $\exp(i\mathbf{k}\cdot\mathbf{x} - i\omega t)$ have phase velocities ω/k_y which obey the opposite condition from the inequality in Eq. (6.64) due to the requirement that $\nabla^2\varphi/\varphi = -k_\perp^2 < 0$ for oscillatory solutions. Figure 6.3 shows the allowed regions for the localized, nonlinear solutions called solitary waves. The solutions with $u > v_d$ are nonlinear electron drift waves, while solutions with $u < 0$

are nonlinear ion acoustic waves retarded by the density gradient v_d and rotating in the ion diamagnetic direction.

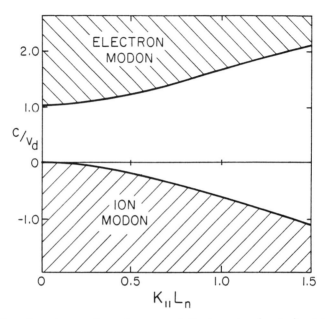

Fig. 6.3. Complementary regions of waves and dipole vortex (modon) solutions of the coupled drift wave-ion acoustic wave Eqs. (6.44) and (6.45).

6.4.4 Larichev-Reznik dipole vortices

Equation (6.63) has solutions which are a sum of modified Bessel functions $\sum K_m(kr)\cos(m\theta)$ with 2^m multipole structure. At present we consider only the lowest mode in this sum consistent with inhomogeneity $v_d x$. The exterior dipolar vortex solution of Eq. (6.63) is for $r > r_0$

$$\varphi = AK_1(kr)\cos\theta$$

with

$$r^2 \equiv x^2 + \eta^2 \quad \text{and} \quad \cos\theta \equiv \frac{\eta}{r} = \frac{y + \alpha z - ut}{r}$$

and

$$k(\alpha, u) = \left(1 - \frac{v_d}{u} - \frac{\alpha^2}{u^2}\right)^{1/2}. \tag{6.65}$$

6.4 Drift Wave-Ion Acoustic Wave Equations

When the trapping condition $\varphi_{\max} > u\Delta x \cong ur_0$ is satisfied where $\Delta x \simeq r_0$ is the radius of the vortex, there are contours of the function $\psi = \varphi - ux$ which do not extend to $x, y \to \infty$. On these closed contours the function $F(X)$ of Eq. (6.61) cannot be determined by the argument leading to Eq. (6.62). For the single mode solution, closed contours occur within the circle of radius $r = r_0$ defined by $\varphi - ux = 0$. Within this circle the function F may be chosen arbitrarily subject to the requirement that φ be twice differentiable across the boundary.

The simplest choice of $F(X)$ in the interior is

$$F^{\text{int}}(X) = -(p^2 + 1)X \qquad (6.66)$$

where $p^2 > 0$ for regularity at $r \to 0$. The interior solutions are then $J_m(pr)\cos(m\theta)$. Continuity of φ and $\nabla_\perp^2 \varphi$ at $r = r_0$ determines the dipole solution

$$\varphi(r,\theta) = \begin{cases} \dfrac{ur_0 K_1(kr)}{K_1(kr_0)} & \text{for } r > r_0 \\[2mm] \left[ur\left(1 + \dfrac{k^2}{p^2}\right) - \dfrac{ur_0 k^2 J_1(pr)}{p^2 J_1(pr_0)} \right]\cos\theta & \text{for } r < r_0 \end{cases}. \qquad (6.67)$$

Requiring that the azimuthal flow velocity $v_\theta = -\partial_r \varphi$ be continuous across r_0 gives a unique relation between p and k through

$$\frac{K_2(kr_0)}{kr_0 K_1(kr_0)} = -\frac{J_2(pr_0)}{pr_0 J_1(pr_0)} \equiv \delta(kr_0). \qquad (6.68)$$

Equation (6.68) has an infinite set of roots $p_n r_0(kr_0), n = 1, 2, \ldots,$. As $kr_0 \to 0$ the roots $p_n r_0$ become the nth zero of $J_1(\gamma_n) = 0$ while as $kr_0 \to \infty$ they become the nth zero of $J_2(\gamma_n) = 0$. Thus, the variation of a particular $p_n r_0$ over the entire range of kr_0 is of order $\pi/2$.

In the dipole vortex solution, Eqs. (6.67) and (6.68), the parameters $\{r_0, u, \alpha\}$ or $\{A, u, \alpha\}$ may be chosen independently subject to the inequality in Eq. (6.64). All other parameters in Eq. (6.67) are determined for a specific branch of Eq. (6.68).

6.4.5 Maximum amplitude and size relations

Defining the actual size of the vortex r_m, as the radius at which $\varphi(r)$ is maximum $\partial \varphi/\partial r(r = r_m) = 0$. In the limits of large and small kr_0, r_m takes the values

$$\frac{r_m}{r_0} = \begin{cases} 0.55 & kr_0 \to 0 \\ 0.32 & kr_0 \to \infty \end{cases}. \qquad (6.69)$$

The peak amplitude $\varphi_{dp} = \varphi(r_m)$ varies as

$$\varphi_{dp} = \varphi(r_m) \simeq \begin{cases} 1.28 r_0 u & kr_0 \ll 1 \\ 0.076 r_0^3 (u - v_d) & kr_0 \gg 1 \end{cases}. \quad (6.70)$$

Note that for a given radius r_0 the vortex with $u > v_d$ requires a finite amplitude field for creation $\varphi_{dp} > 1.28 r_0 v_d$. For the condition $kr_0 \ll 1$ to be satisfied the speed u must be close to v_d so that Eq. (6.70) gives $\varphi_{dp} \simeq 1.28 r_0 v_d$ which is an amplitude just above the mixing length amplitude.

6.4.6 Energy and enstrophy of the dipole vortex

The dipole vortex energy and enstrophy integrals (Eqs. (6.49) and (6.50)) can also be computed from Eq. (6.67) and (6.68). For $k_\parallel = \alpha = 0$ we obtain

$$E = L_c \frac{\pi r_0^4 u^2}{4} \left[1 + \left(\frac{\beta}{\gamma}\right)^2 \right] \left\{ (\delta\beta)^2 + (4\delta + 1) \left[\frac{3}{2} \left[\left(\frac{\beta}{\gamma}\right)^2 + \frac{1}{2} - \frac{v_d}{u} \right] \right] \right\}$$

$$K = L_c \frac{\pi r_0^2 u^2}{4} \beta^2 \left[1 + \left(\frac{\beta}{\gamma}\right)^2 \right] \left[\left(\frac{\beta\gamma\delta}{r_0}\right)^2 + 4\delta + 1 \right], \quad (6.71)$$

where $\beta = kr_0, \gamma = pr_0$ and L_c is the effective length of the physical system containing the two-dimensional structure. From Eqs. (6.42) and (6.49) the units of the E and K integrals are $n_e T_e \rho_s^2 r_n$ and $n_e T_e r_n$, respectively.

These $E(u)$ and $K(u)$ formulas are shown in Fig. 6.4. The asymptotic forms for large and small β are

$$E = \begin{cases} 2\pi r_0^2 u^2 L_c \left[1 + \frac{r_0^2}{4} (0.3616 - \ell n k r_0) \right] & kr_0 \ll 1 \\ \dfrac{\pi r_0^8}{4\gamma^4} L_c (u - v_d)^2 \left[\left(\dfrac{\gamma}{r_0}\right)^2 + \dfrac{3}{2} \right] & kr_0 \gg 1 \end{cases} \quad (6.72)$$

$$K = \begin{cases} \pi r_0^2 u^2 L_c \left[2 + \left(\dfrac{\gamma}{r_0}\right)^2 \right] & kr_0 \ll 1 \\ \dfrac{\pi r_0^6}{4\gamma^2} L_c (u - v_d)^2 & kr_0 \gg 1 \end{cases}. \quad (6.73)$$

6.4 Drift Wave-Ion Acoustic Wave Equations

The energy of the Eulerian vortex dipole ($kr_0 \to 0$) is infinite because of the slow fall-off of $\varphi \propto p\cos\theta/r$ as $r \to \infty$. This property implies that there is a dipole vortex with $u > v_d$ with minimum energy for a given r_0. As an example when $r_0 = 1[\rho_s]$, $E_{\min} = 10.2 L_c$ at $u = 1.04 v_d$ while at $r_0 = 10.0[\rho_s]$, $E_{\min} = 2.2(10^4) L_c$ at $u = 1.008 v_d$. Thus, the large radius dipoles store large energies and propagates close to the drift speed.

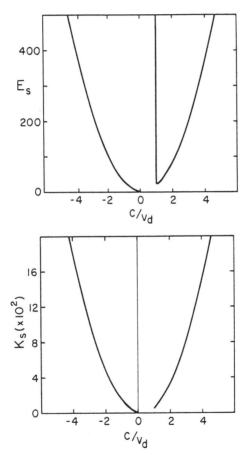

Fig. 6.4. Energy E and enstrophy K_s of the dipole vortex of radius $r_0 = \rho_s$ as a function of the speed c ($= u$) relative to the drift-Rossby wave speed.

6.5 Experimental Features and Computer Simulations of the Dipole Vortices

6.5.1 Plasma drift wave experiments

Numerous observations of steady state, low frequency fluctuations with frequencies, wavelengths, and amplitudes characterized by $\omega \sim k_y v_d$, $k_\perp \rho_s \lesssim 1$, $k_\parallel r_n \lesssim 1$, and $\varphi = e\Phi/T_e \sim \varepsilon$ where $\varepsilon = \rho_s/r_n \ll 1$ are consistent with the nonlinear partial differential equation describing drift waves (Hendel et al., 1968; Brossier et al., 1973; Mazzucato, 1976, 1982; Brower et al. 1987). The drift-wave equations are derived from the two component fluid equations assuming negligible ion pressure, weak electron non-adiabaticity and quasineutrality. The plasma geometry considered is the nonuniform slab with a sheared magnetic field defined by $S = r_n/L_s$.

For weak shear, $S < \varepsilon$, the coupling to ion acoustic waves is negligible, and the reduced drift wave equation admits localized, two-dimensional solutions. These solitary drift waves are equivalent to Rossby wave modons in geophysics, and they are governed by the dimensionless partial differential equation first derived by Charney (1948) for Rossby waves and later obtained for nonlinear drift waves by Hasegawa and Mima (1977, 1978). These solitary drift waves propagate in regions complementary to the $\omega(k_y, k_\parallel)$ regions of the linear oscillations as shown in the phase velocity diagram in Fig. 6.3.

The charge separations in these solitary vortices have an electric dipole core with radius r_0 and an exponentially-screened exterior decaying as $\exp(-kr)$, where $k = (1 - v_d/u)^{1/2}$ in units of ρ_s^{-1}. When kr_0 is small, the peak amplitude $e\Phi_m/T_e \simeq 1.28 r_0/r_n$ occurs at $r_m \simeq r_0/2$. The energy, E_s, of a solitary drift wave is of order $r_0^2 L_c n_e T_e (r_0/r_n)^2$. For speeds $u > v_d$ there is a minimum energy for excitation given by $E_s \sim 4\pi r_0^2 L_c n_e T_e (\rho_s/r_n)^2$ for $r_0 \lesssim \rho_s$ occurring at $kr_0 \sim 10^{-3} - 10^{-2}$. At $kr_0 \to 0$ where $u = v_d$, the solitary drift wave becomes identical to a dipole vortex solution of the Euler equation. In this case, the exponential screening is replaced by the unscreened dipole potential $\varphi = p\cos\theta/r$, and the energy becomes infinite. Solitary vortices traveling in the ion-diamagnetic direction ($uv_d < 0$) have energies which increase monotonically with increasing $-u/v_d$.

Estimating the maximum density of solitary vortices to be $n_s = 1/\pi r_0^2 L_c$, where the packing fraction is unity, gives an estimate of the maximum energy of a dipole vortex gas system. For densities larger than this, interactions distort the vortices in time of order r_0/v_d. At the critical density, the energy $n_s E_s$ in solitary waves as a fraction of the plasma thermal energy, $n_e T_e$ is $n_s E_s/n_e T_e \simeq (r_0/r_n)^2 \sim (\rho_s/r_n)^2$. This bound on the drift wave energy density is consistent with the free energy estimates of thermodynamics.

In the presence of strong shear $S \gtrsim \varepsilon$, the electron drift wave is intrinsically coupled to the ion acoustic oscillations due to the dispersion in $\alpha = k_\parallel/k_y = Sx$ with distance x from the $k_\parallel = 0$ rational surfaces. The field of the solitary wave is a quasi-bound state of a Schrödinger equation with a finite potential barrier. Well outside the core $r = r_0$ of the drift wave dipole vortex the solutions are outgoing coupled drift-ion acoustic waves. These outgoing waves radiate energy from the core to regions of ion-Landau damping as in the Pearlstein-Berk linear eigenmodes (Pearlstein and Berk, 1969).

In the radiation region, the sheared-field dipole vortices have an exponentially small amplitude determined by the tunneling of waves through the nonlinear potential barrier. For speeds such that $|u/v_d - 1| > S$, the outer dipole vortex scale, $1/k$, and the Pearlstein-Berk turning point, x_T, are well separated. The rate of loss of energy due to radiation gives the damping rate $\gamma_s \sim (c_s/L_s)\exp[-L_s/r_n|u/v_d - 1|]$. For speeds such that $|u/v_d - 1| \leq S$ the turning point is close to the outer scale, and we expect the radiative damping rate to be qualitatively the same as for the linear eigenmodes which have $0 < u < v_d$ and $\gamma \sim c_s/L_s$.

The analysis of the nonlinear drift wave-ion acoustic wave equations are an important aspect of the nonlinear dynamics of drift waves. The solitary drift waves are coherent, localized structures with plasma trapped in the core of radius r_0, energies proportional to $r_0^2 L_c n_e T_e$, and exponentially small exterior fields. The structures exist both in the shearless and sheared magnetic fields. In all cases the dynamics of the solitary wave components provides $\mathbf{k}\omega$ spectral components that are complementary to those of the linear modes. The solitary drift-ion acoustic waves are an important component of the long-time limit turbulent fluctuations in confined magnetized plasmas. These coherent structures store fluctuation energy from the turbulent inverse cascade of smaller scale fluctuations to the large scales.

Numerical experiments by Makino et al. (1981) confirm that the dipole vortex solution with $p = p_1(kr_0)$ is a stable, though the dipole vortices do not have the extreme stability of a KdV soliton. McWilliams et al. (1981, 1982) subjected the dipole vortex to random initial perturbations and found that it survived if the rms strength of the perturbation was less than 10-20% of the vortex amplitude. For larger perturbations the vortex was destroyed by the large scale shearing of its flow contours. External sheared flows produce an X-point where there is a stagnation of the flow on the counter flowing side of the dipole. This opens up the interior contours on that X-point side releasing the trapped fluid and its vorticity into the external sheared flow. The result is the evolution to a monopole vortex known to occur in sheared flows.

The dipole vortices propagating parallel to v_d ($uv_d > 0$) are also subject to a rotational or tilt instability. The theoretical demonstration of the instability is difficult and is still an active area of research. The tilt instability is seen most clearly in the simulations of Jovanović and Horton (1993), while the theoretical studies of Muzylev and Reznik (1992) and Nycander (1992) give the mathematical aspects of the stability analysis of these two dimensional structures. Physically, the tilting instability occurs most easily from adding a monopolar perturbation $\varepsilon\varphi_0(r/r_0) \sim \varepsilon\exp(-r^2/r_0^2)$ of radius r_0 comparable to that of the dipole. This makes one pole of the dipole stronger, which then produces a rotation of the entire structure similar to the case of two unequal point vortices of opposite sign.

Experiments on the collisions of dipole vortices show that for some parameter ranges the coaxial collisions with zero impact parameter b are nearly elastic (Makino et al., 1981; McWilliams et al., 1982). Numerical simulations of collisions with the impact parameter $b \cong r_0$ result in reactions with a different final configurations of monopoles and dipoles (Horton, 1989). Internal degrees-of-freedom are excited by the collisions, and fission, or splitting up, of the dipoles occurs as well as coalescence of like-signed lobes. The result depends on the kinematics of the collision.

Fleirl et al. (1980) have shown that the solution in Eq. (6.67) can be generalized by the addition of an azimuthally-symmetric "rider" field or monopole component $K_0(kr)$ for $r > r_0$ and $J_0(pr)$ for $r \leq r_0$. The radial structure of this field is determined by matching conditions at $r = r_0$ as before, but the amplitude of the symmetric rider can be freely chosen. Thus, it is possible for the azimuthally-symmetric component $\varphi_0(r)$ to mask the $\cos\theta$ nature of the underlying dipole vortex. It is the dipolar part of the solution that corresponds to the solitary drift wave or Rossby solution. They are driven by v_{de} or the β-term.

Numerical simulations and rotating water tank experiments starting with a symmetric monopole vortex show that the v_d or geophysical β-term drive a $\varphi_1(r)\cos\theta$ dipole component to the vortex. The dipole charge separation directly produces the nonlinear $\mathbf{E} \times \mathbf{B}$ drift of the solitary wave structure.

6.5.2 Drift wave vortex collisions and coalescence

(i) **Monopoles and Wake Fields**

In the 2D Euler equation the axisymmetric monopole vortex is an exact motionless solution, and the dipole vortex is an exact solution that translates with a constant speed in an arbitrary direction. The Euler equation has no preferred direction and no linear waves. An experimental example of a gas of Euler equation-like dipole vortices is given in the soap film experiments and computer simulations of Couder and Basdevant (1986).

In contrast to the Euler equation, the 2D drift wave equation has a continuous spectrum of linear waves given by

$$\omega_{\mathbf{k}} = \frac{\mathbf{k} \cdot \mathbf{v}_d}{1 + k_\perp^2 \rho_s^2} \tag{6.74}$$

with $\mathbf{v}_d = -(cT_e/eB)\hat{\mathbf{b}}_0 \times \nabla \ell n\, n_e$ giving the preferred direction and velocity of wave propagation. Arbitrary small amplitude $R_E \ll 1$ structures propagate along $x = x_0$, $y \cong y_0 + (\partial\omega/\partial k_y)t$ with the wave dispersion given by

$$\varphi_w(x,y,t) = \sum_{\mathbf{k}} \varphi_{\mathbf{k}} \exp(ik_x x + ik_y y - i\omega_{\mathbf{k}} t) \tag{6.75}$$

6.5 Experimental Features and Computer Simulations of ... 247

where $\varphi_{-\mathbf{k}} = \varphi_{\mathbf{k}}^*$. The central peak of φ_w given by Eq. (6.75) loses strength due to the dispersion of wave energy given by $\Delta x^2 \approx (\partial^2 \omega / \partial \mathbf{k}^2) t \simeq \rho_s^2 \omega_{k_0} t$ which is a strong effect except for very large-scale structures.

The characteristic propagation of a general initial field $\varphi(x, y, 0)$, however, depends sensitively on its amplitude and scale size $1/k_\perp$. The ratio of the nonlinear binding compared with the wave dispersion is given by R_E in Eq. (6.17). For $R_E \lesssim 1$ we observe the dispersion of the wave packet as given by (6.75). For $R_E \gtrsim 1$ the nonlinear self-binding dominates producing a long-lived localized packet. The typical self-binding behavior for a monopolar structure is shown in Fig. 6.2 for $R_E = 10$. This self-binding structure occurs above the transition value $R_E \sim 1$.

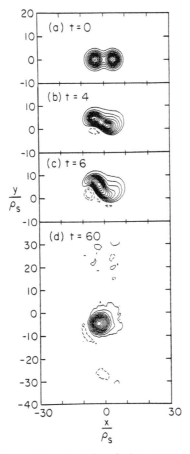

Fig. 6.5. Coalescence of two nearby potential peaks into a strong, nonlinear vortex and a wake of drift waves.

The self-binding dynamics can be further tested by the coalescence of two nearly monopolar structures. The coalescence that occurs when two nearby monopoles are released at $t = 0$ and advanced with the H-M-C equation is shown in Fig. 6.5.

For structures with $R_E \gtrsim 1$, the binding nonlinearity captures most of the wave energy due to the nonlinear feedback that occurs in the structure when the circulation time $2\pi/\Omega_E$ is shorter than the oscillation time $2\pi/\omega_k$. The typical $R_E \gtrsim 1$ structures are described as long lived monopole and dipole vortices with localized fields of the form $\varphi_v(x, y - ut) = \varphi_v(r, \theta)$ where $r = [x^2 + (y - ut)^2]^{1/2}$ and $r\cos\theta = x$. For $R_E \gtrsim 1$ the computed fields are better described by the combined field $\varphi = \varphi_v + \varphi_w$ of vortices φ_v and waves φ_w than by either individual component.

The steady-state symmetric monopole $\varphi(r)$, while long-lived, is not an exact solution of Eq. (6.48) due to the linear terms which drive up an asymmetric x dependence to φ. The dipole solution has the correct x asymmetry to balance the linear wave terms and is an exact, long-time solution to the Hamiltonian wave equation. The positive potential monopolar-like solution that develops from $\varphi(r)$ has a shelf on the positive x (low density-shallow depth) side with a smaller flow velocity and a steeper gradient, higher flow velocity on the high density side. The negative monopolar-like solution has the shelf and steeper gradient sides reversed. The reversal follows from the symmetry property of Eq. (6.48) with the reflected solution $\overline{\varphi} = -\varphi(-x, y, t)$.

The nonlinear conversion by localization of the extrema of the wave field into a vortex field is an important mechanism for minimizing the effects of wave diffraction and wave scattering by inhomogeneities in the system. The generic nonuniformity in toroidal confinement systems is the twisting of the magnetic field lines called magnetic shear. For drift waves in a sheared magnetic field the local radial dependence of the parallel gradient described by $k_\parallel = k_y(x - x_0)/L_s$ is typically the most important inhomogeneity for the wave. Wave packets with $R_E \ll 1$ spread from $x = x_0$ to $x = x_i$ where $k_\parallel = k_y(x_i - x_0)/L_s \equiv \omega_k/v_i$ at which point the wave energy is rapidly absorbed by resonant thermal velocity v_i ions. The distance Δx to the absorption region is $\Delta x_i = L_s(\omega/k_y v_i) \cong \rho_s(L_s/r_n)$. The numerical experiment in Fig. 6.2 contrasts the spreading of the linear and nonlinear disturbance. In Figure 6.2 the results are in accord with the theories (Meiss and Horton, 1983; Su et al., 1991; Horton et al., 1992) that show the nonlinear localization of the vortex energy E_v to $r_0 \lesssim \Delta x_i = \rho_s(L_s/r_n)$ strongly reduces the coupling to the ion acoustic waves with their associated resonant absorption of the energy in the structure. The calculation in Eq. (6.45) gives the decay rate induced by magnetic shear. The nonlinear suppression of shear damping is also investigated in the simulation of Biskamp and Walter (1985) by solving the drift wave-ion acoustic wave equations in the sheared slab model.

From the left column in Fig. 6.2 we see that the long wavelength components of Eq. (6.74) propagate with a speed near v_d but that the small scale components have nearly zero speed. The waves propagate symmetrically in x giving an approximately

6.5 Experimental Features and Computer Simulations of ...

parabolic envelope to the expanding wake $\varphi(x^2, y-v_d t, t)$. In contrast, the nonlinear evolution given in the right-hand column of Fig. 6.2 develops an asymmetrical x dependence. The growth of the asymmetry and the weak wave emission are limited after a few rotation periods with the reformed structure propagating as a vortex $\varphi_v(x, y - ut)$ for as long a time as computed in the dissipationless system.

For the small amplitude $R_E < 1$ structures only the long wavelength head of the structure or wave packet keeps up with the corresponding vortex structure; the rest of the wave energy propagates at much slower speed. Since the passively convected field $f(x, y, t)$, or distribution of test particles, is trapped in the nonlinear vortex, the transport along $\mathbf{v}_d$ of the trapped particles is nearly complete in a nonlinear vortex structure. For the small amplitude wave packet the net transport of the passive particles is essentially zero since they are not trapped.

(ii) Coalescence of Drift Wave Vortices

One of the most important processes in the nonlinear dynamics of vortices is the merging, or coalescence, of like-signed regions of vorticity. In the merging process streamlines reconnect producing a mixing of the transported fields previously trapped in the two separate vortices. Since the merging process itself happens in about one vortex rotation period $2\pi/\Omega_E(k)$, the reconnection quickly mixes the (dye) fields f_1 and f_2 over the new larger vortex. Thus the coalescence of vortices produces a rapid, local transport mechanism. Nezlin (1986, 1993) gives a review of the types of vortices, turbulence, and coalescence phenomena found in the rotating water tank experiments. Antipov et al. (1982) reports the formation of dipole vortices and the anti-cyclonic monopole vortex in the first rotating water tank experiments for Rossby waves.

The coalescence of two-dimensional vortices with cores of constant vorticity for the Euler equation is shown by Overman and Zabusky (1982) to occur for equal strength vortices when the separation d_* of the vortices is less than a critical value of $d_* = 3.2R$ where R is the core radius of the constant vorticity vortex. Asymmetric merging of Euler vortices of unequal strength proceeds as shown in Melander et al. (1987a,b). The merging of vortices is easily observed in the nonlinear stage of the Kelvin-Helmholtz instability where the local growth rate for merging can be predicted by the linear stability analysis of the period doubling perturbation of the periodic vortex chain as given by Pierrehumbert and Widnall (1982) and Horton et al. (1987).

In the case of the drift wave-Rossby wave equation, the coalescence process is observed in the rotating neutral fluid experiments. Rotating water tank experiments by Swinney et al. (1988) and Griffiths and Hopfinger (1986, 1987) show the importance of vortex merging in $\mathbf{E} \times \mathbf{B}$ or geostrophic vortices which are governed by equations close to the drift wave. Swinney et al. (1988) and Sommeria et al. (1991) report that merging of up to five vortices into one vortex takes place when the pumping rate driving the fluctuating flow is sufficiently strong.

250   Chapter 6   Vortex Structures in Hydrodynamic and Vlasov Systems

Driscoll and Fine (1990) use the pure electron plasma to simulate the 2D Euler merging of vortices. They show a sharp transition for separations d_* less than approximately 3 to $4R$ for coalescence. At larger separation the vortices rotate around each other for up to 10^4 times.

In plasma the coalescence of two positive electrostatic potential vortices is demonstrated in the experiments of Pécseli et al. (1984, 1985) In the Q-machine experiment two separated positive vortices larger than the ion gyroradius coalesce in the $\tau_c \sim 100\mu$s which is about one rotation period in the vortex with $v_E \simeq 5 \times 10^3$ cm/s.

For the drift wave coalescence experiment shown in Fig. 6.5 we take two positive, gaussian vortices displaced in radius by $d = 2x_0$ and with $\mathbf{E} \times \mathbf{B}$ rotation frequencies greater than unity. For example, we take

$$\varphi(x,y,t=0) = A_1 \exp\left[-\frac{(x+x_0)^2+y^2}{r_0^2}\right] + A_2 \exp\left[-\frac{(x-x_0)^2+y^2}{r_0^2}\right] \quad (6.76)$$

with $A_1 = A_2 \equiv 50$, $r_0 = 2[\rho_s]$ so that the maximum (isolated) vorticity is $\zeta_0 = -4A/r_0^2 = 50$, and the rotation frequency at the $1/e$ radius of $\varphi_\pm(r_\pm = r_0)$ is $\Omega_E = -2A/r_0^2 e = 9.2$. In the experiment shown in Fig. 6.5 $d = 2x_0 = 8[\rho_s]$, so that the relative strength overlapping of the potentials is $\varphi(0,0,0)/A_1 = 2\exp(-x_0^2/r_0^2) = 0.037$.

Immediately after releasing the field from the initial condition in frame (a), the mutual convection forms the substantial $m = 2$ distortion of the structure shown in frame (b) at $tc_s/r_n = 4$ ($\Omega_E t_b = 37$), and the reconnection of the streamlines in the coalescence is essentially complete in frame (c) at $tc_s/r_n = 6$ ($\Omega_E t_c = 55$) or nine rotations. During the coalescence the circularization of the resulting vortex and the separation from the drift wave wake takes about 5 to 10 rotation periods in this experiment. The resulting monopole vortex has $\Omega_E \lesssim 9.0$ and is shown at $c_s t/r_n = 60$ in frame (d) along with its wake field which has max $|\varphi_{\text{wave}}| \lesssim 0.1$.

The positive $A_{1,2}$ merger shown in Fig. 6.5 corresponds to the anti-cyclonic merger reported experimentally to occur for large relative separations. When the direction of the vortex circulation is reversed with $A_1 = A_2 = -50$, Eq. (6.48) predicts that the merger occurs for exactly the same d/r_0. The pictures obtained with negative $A_{1,2}$ are the x-reflection of the positive $A_{1,2}$ mergers. Thus, for both positive and negative gaussian drift wave vortices the critical separation distance d^* is about $d^*/r_0 \simeq 5.5/2 \simeq 2.8$. Increasing $A = \varphi_{\text{max}}$ increases d^* slowly as $d^* \ell n(A/r_0 v_d)$. To break the symmetry between the positive (anti-cyclonic) and negative (cyclonic) merging rates requires the presence of the KdV hydrodynamic nonlinearity of Chapter 5.

(iii) Dipole Vortex Formation From Pairing

Two oppositely charged drift wave vortices can also merge to form a dipole vortex through a process we may call pairing or coupling. Since drift waves are emitted during the pairing, the inelastic merging process is analogous to the atomic

6.5 Experimental Features and Computer Simulations of ...

process of radiative recombination. The presence of the drift wave component of the excitation spectrum is particularly important in the pairing process since it is through the shedding of energy and enstrophy in the radiated wake that the vortices can be modified to produce a bound state characteristic of the dipole vortex state. In the absence of drift wave radiation the invariants of the equation of motion would exclude the formation of the dipole vortex.

Why don't the nearby positive and negative charged regions continue to pull together and thereby completely eliminate the local charge separation? The conservation of energy-enstrophy invariants is sufficient to prevent the complete overlapping and thus neutralization of the charge separation or vorticity ($\nabla^2\varphi$) contained in the neighboring lobes of the dipole vortex. Since the dipole vortex state is known to be an exceptionally stable state, the overall process is well described as a radiative recombination collision with the cross-section σ for pairing comparable to the vortex diameter. (Recall that cross-section in two-dimensions is linear in size.)

For a drift wave pairing or recombination experiment we use Eq. (6.76) with $A_2 = -A_1 = 50$ and $r_0 = 2.0$ giving $\zeta_0 = .50$ and $\Omega_E(r_\pm = r_0) = 9.2$. In the experiment shown we choose $d = 2x_0 = 14[\rho_s]$ so that the initial overlap at $x = 0, y = 0$ is $\varphi_1(0,0,0) = A\exp\left(-(x_0/2)^2\right) = 4.8 \times 10^{-4}$. The experiment is shown in Fig. 6.6 with a fixed contour interval of $\Delta\varphi = 4.0$.

After launching the initial condition shown in Fig. 6.6(a) the mutual convection advances the vortex cores with a speed $u \gtrsim v_d$ while leaving the wake of oscillations of amplitude of order unity shown in Fig. 6.6(b) at $tc_s/r_n = 20$ ($\Omega_E t = 184$ or ~ 20 rotations). The pulling together of the vortex cores is a slow process compared with that in the corresponding like-signed merging. At time $tc_s/r_n = 20$ the core separation is $d(t = 20) \simeq 8$, and at $tc_s/r_n = 60$, shown in Fig. 6.6(c), the pulling together is nearly complete with $d(t = 60) = 6.0$. The final separation of the dipole centers is $d \simeq 6[\rho_s]$ — one half the initial separation giving an average attracting velocity of $\dot{x}_{12} \sim -6/60 = -0.1[v_d]$ during the drift along the magnetic surface with $\dot{y} \simeq 60/60 \sim 1.0[v_d]$ during the pairing process.

Reversing the signs of the initial monopoles ($A_1 = -A_2 = 50$) produces a different result. With the negative potential on the left and the positive potential on the right the monopoles do not pull together to form a dipole pair. Repeating the experiment in Fig. 6.6 with this sign reversal we find that at $t = 60r_n/c_s$ the monopoles have slightly separated to $d \cong 15\rho_s$ from the initial $d_0 = 12\rho_s$. The physical difference between the two configurations is given by the magnitude of the potential vorticity q. With the polarity leading to coupling the $\varphi - \nabla^2\varphi$ contribution partially cancels the ambient density gradient $v_d x$, whereas with the repelling polarity the solitary wave adds to the density gradient. Another way of stating the difference is that when the shelves of the monopoles extend toward each other the coupling process is inhibited.

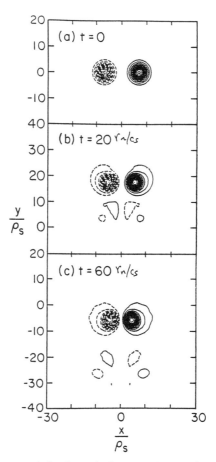

Fig. 6.6. Pairing of two oppositely charged plasma regions to form a dipole vortex with a wake of plasma waves. The box is of length $20\pi\rho_s = 62.8\rho_s$ and periodic in y. Time is in units of r_n/c_s, and the $\mathbf{E} \times \mathbf{B}$ rotation frequency at the center of the vortices is approximately $10(c_s\rho_s/r_0^2)$ with $r_0 = 10\rho_s$.

(iv) **Stability of the Dipole Vortex**

The dipole vortex is an exact solitary wave solution to the Hamiltonian or dissipationless drift wave equation in the limit of constant equilibrium gradients. The solitary wave properties of the dipole solutions have been the subject of numerous theoretical, numerical, and experimental studies. The dipole vortices have been created and studied in several neutral fluid experiments.

6.5 Experimental Features and Computer Simulations of ...

A demonstration of the stability and robustness of the dipoles is given by their ability to reconstruct their initial form after the strong distortions that occur during the zero impact parameter $b = 0$ head-on and overtaking collisions first shown by Makino et al. (1981) and investigated in detail by McWilliams and Zabusky (1982).

McWilliams and Zabusky state that the zero impact parameter head-on collision "is remarkably non-destructive." They report that the longer the overlap or collision time the stronger the merging tendency with the extreme example of a smaller diameter dipole vortex being absorbed by the larger diameter dipole during an overtaking collision. Horton (1989) emphasizes the influence of the impact parameter b on the collisions, and the fact that the Bessel function solutions are not unique to the solitary vortex properties of the nonlinear system. Horton shows that inelastic collisions occur for $b \sim r_0$ with like-signed regions of vorticity merging to form new vortex states under collisions.

The $J_1(pr) - K_1(kr)$ solutions (6.29)–(6.30) of the Rossby-Drift Wave equation are not the only kind of solitary solutions. They assume linearity of the $q = \nabla^2 \varphi - \varphi + x$ versus $\varphi - ux$ functional and a circular boundary $r = a$. Good dipole solitary waves are created from the initial condition

$$\varphi(x, y, 0) = Ar \cos \theta \, e^{-kr} \tag{6.77}$$

with $k = (1 - v_d/u)^{1/2}$ and $A \gtrsim 3v_d/k$. The curvature of this function is continuous and changes sign at $r_1 = 3/k$ which defines the effective radius a of the corresponding Bessel function dipole vortex. The amplitude dependence of the speed of the dipole vortex generated from Eq. (6.77) is weaker than that of the Bessel function vortex.

Investigations of the stability of the dipole vortices indicate that there are two dangerous perturbations to the lifetime of a dipole. One is an ambient shear flow with $\delta\varphi = v'_E x^2/2$ and the second is tilting of the dipole. The sheared flow places a stagnation point ($\times$ point) in the flow at a distance $\Delta x = u/v'_E$ on the side of the dipole with opposing flows. When this $\times$ point is within the distance $u/v'_E < \max(\pi/k, 2r_0)$ of the vortex center, the lobe of the vortex with the stagnation point opens up dumping its vorticity into the counterstreaming plasma. The remaining lobe forms a shear flow driven monopole vortex which can be expressed in terms of $J_0(kr)$ and $J_2(kr) \cos(2\theta)$ (Horton-Tajima-Kamimura, 1987)).

The second form of the instability results from strengthening one lobe of the dipole relative to the second lobe. The instability is studied in Jovanović and Horton (1993) both numerically and with Lyapunov stability theory. The study appears to agree with the analytic study of Muzylev and Reznik (1992).

To consider whether the dipole vortices are attracted to the circular shape used in constructing the Bessel function solutions, experiments with $r(x, y)$ replaced with $r = (x^2 + e^2 y^2)^{1/2}$ were run with $e = 0.5$ and $e = 2.0$ by Horton (1989). In both cases an approximately circular dipole propagates out of the initial data and leaves a wake behind. The elliptical perturbation with the long axis parallel to wave propagation direction produces less distortion than the distortion with the elongation perpendicular to the direction of wave propagation.

6.5.3 Anomalous transport during inelastic vortex collisions

In the first studies of the dipole vortex collisions the elastic nature of the collisions was emphasized. Subsequent studies with less symmetry in the initial states show that the collisions give rise to anomalous transport across the magnetic field. This transport is strongest for inelastic collisions due to the permanent reconnection of the streamlines that takes place during the inelastic collisions.

The symmetry present in the zero impact parameter ($b = 0$) collisions keeps the net anomalous transport small or vanishing. Large net radial transport due to collisions occurs when the impact parameter b is comparable to the vortex radius r_0. In collisions with $b \sim r_0$ the symmetry of the vortex field is broken, and like regions of vorticity merge permanently into a new final state. During the merging the charged particles, initially trapped in separate vortices, mix over the core of the new vortex. Large net values of radial flux $\Gamma = \langle \delta f v_x \rangle \approx -r_0 u \left(d\overline{f}/dx \right)$ are observed during this type of collision (Horton, 1989; Nycander and Isichenko, 1990).

As an example we consider the collision of the two vortices shown in Fig. 6.7. Both initial dipoles, with opposite polarity, have the radius $r_0 = 6\rho_s$ for their circular boundaries. The impact parameter defined by the relative displacement of the dipole centers in the x-direction is taken as $b = 5\rho_s$. The amplitude and structure of the dipoles follows from Eqs. (6.67)–(6.68) for the chosen speeds $u_1 = 2v_d$ and $u_2 = -v_d$. The lobes with negative-φ and positive vorticity-$\nabla^2\varphi$ regions meet in the collision and merge into a single strong vortex as shown in Figs. 6.7. The merging event is similar to that shown in Fig. 6.3 with a strong $m = 2$ distortion followed by re-circularization.

During the collision the far left-hand lobe of positive φ and negative vorticity is rotated under the newly formed, larger positive vortex ($\nabla^2\varphi > 0$) and couples to it while growing in strength. This process forms a new dipole vortex with approximately the same radius for the circular boundary $a' \simeq 6\rho_s$ and strength as the original $u_2 = -v_d$ dipole during the time interval $t = 32 - 40$ and shown at $t = 80$.

The newly created dipole ($t > 30$) contains a mixture of plasma from the original dipoles and is created with its axis rotated by about 45° from the direction of the magnetic surface. The passively convected dye trapped in the new vortex is transported across the magnetic surface by $\Delta x \simeq 2r_0 \simeq 10\rho_s$ in a time of order $\Delta t \simeq 20 r_n/c_s$ as seen in Figs. 6.7. The newly formed dipole propagates with $\dot{y} \simeq -0.5 v_d$, and its axis slowly rotates with $\dot\theta \simeq -0.03\,\mathrm{rad/s}$ continuing to rotate past the usual aligned position ($t = 80$). The monopole propagates with speed $\dot{y} \cong 1.0 v_d$ and contains plasma that moved from $x \cong -a$ to $x \cong 0$ during the course of the collision.

6.5 Experimental Features and Computer Simulations of ... 255

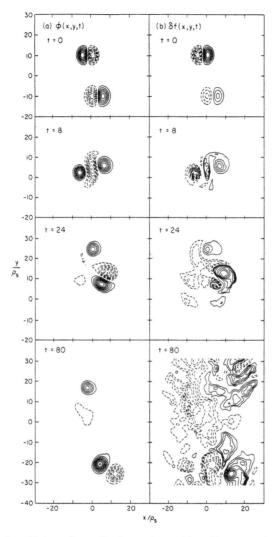

Fig. 6.7. Inelastic collision of two dipole vortices with radius $r_0 = 6\rho_s$ and separation $b = 5\rho_s$. The upper dipole is computed with $u = -v_{de}$ and the lower dipole with $u = 2v_{de}$. (a) The two negative potential regions of low density merge between $t = 8 - 24$ forming a new dipole, a monopole while radiating drift waves. (b) Excess density or temperature (δf) of a trapped, passively convected scalar that is mixed and transported during the inelastic collision. There is large pulse of $\langle v_x \delta f \rangle$ during the time of the collision. Note the polarity of δf depends on the product of the sign of φ and u.

Some aspects of the vortex collisions can be predicted by a point vortex representation of the flow. Kono and Miyashita (1989) show how to compute the space-time variation of the strength of the point vortex cores as they convect and interact in the plasma. Benzi et al. (1988) in a study of decaying 2D decaying turbulence directly compare the point vortex dynamics with the motion of the vortices in the decaying turbulence finding close agreement.

The evolution of the perturbation δf to the distribution f of test particles (Eq. (6.50)) in the vortex field is shown in Fig. 6.7(b). The passive scalar is assumed to have a uniform background gradient $d\overline{f}/dx \equiv -\overline{f}/L_f = $ const. across the vortex $L_f \gg r_0$. The evolution of δf is computed from

$$\left(\partial_t + \mathbf{v}_E \cdot \nabla - D\nabla^2\right)\delta f = \frac{\partial \varphi}{\partial y}\frac{\partial \overline{f}}{\partial x} \tag{6.78}$$

and shown in Figs. 6.7(b) for $D = 0.05$.

Integrating Eq. (6.78) over y gives the net convective flux $\Gamma(x)$ across the magnetic surface at x with

$$\frac{\partial \overline{f}(x)}{\partial t} + \frac{\partial}{\partial x}\left(\Gamma(x) - D\frac{\partial \overline{f}}{\partial x}\right) = 0 \tag{6.79}$$

where $\Gamma(x) = \overline{nv_x} = -L_y^{-1}\int dy \partial_y \varphi \delta f$. The total or global flux Γ_T is given by

$$\Gamma_T = \int \frac{dx}{L_x}\Gamma(x) = -\int \frac{dxdy}{L_xL_y}\frac{\partial \varphi}{\partial y}\delta f = \sum_{\mathbf{k}} ik_y\varphi^*_{\mathbf{k}}\delta f_{\mathbf{k}} \tag{6.80}$$

and is evaluated as a function of time during the collision. The approximate formula for Γ_T following from Eqs. (6.78) and (6.80) is

$$\Gamma_T \cong -\sum_{\mathbf{k}\omega} \frac{k_y^2|\phi_{\mathbf{k}\omega}|^2}{-i\left(\omega - \Omega_E(\mathbf{k})\right) + \mathbf{k}_\perp^2 D}\frac{\partial \overline{f}}{\partial x} \tag{6.81}$$

where $\Omega_E(\mathbf{k})$ is the vortex rotation frequency used as a local estimate of $\mathbf{v}_E \cdot \nabla \delta f_{\mathbf{k}} \simeq i\Omega_E(\mathbf{k})\delta f_k$.

The exact calculation of Γ requires the solution of the second order partial differential Eq. (6.78) for δf. The solution for Γ_T in the presence of a single sinusoidal wave is given in Rosenbluth et al. (1987). For large Peclet number $P_E = \ell v_E/D$ the problem is difficult due to the formation of boundary layers of width $\Delta x = (D/\Omega_E)^{1/2}$ along the separatrices between the vortices. The effective diffusivity from the transport across the boundary layer is $D^* = 1.4(Dv_E r_0)^{1/2}$. This enhancement of the diffusivity D by the 2D incompressible flow is reported for Couette flow by Solomon et al. (1993).

For isolated vortices the convection of δf evolves to $\mathbf{v}_E \cdot \nabla \delta f = 0$ giving rise to the stationary trapped dye distribution

$$\delta f(x, y - ut) = \frac{\varphi}{u}\frac{\partial \overline{f}}{\partial x} \tag{6.82}$$

with $\varphi(x, y - ut)$ localized for isolated dipole vortices. This stationary distribution gives perfect transport of the trapped plasma along the magnetic surface with speed u and zero net transport across the magnetic surface since

$$\Gamma(x) = \langle nv_x \rangle = -\frac{1}{u}\frac{\partial \overline{f}}{\partial x} \int_1^2 \frac{dy}{L_y} \varphi \frac{\partial \varphi}{\partial y} = -\frac{1}{u}\frac{\partial \overline{F}}{px}\frac{1}{L_y}\left(\varphi_2^2 - \varphi_1^2\right) = 0 \qquad (6.83)$$

between any points 1,2 where $\varphi_1^2 = \varphi_2^2$. In the presence of the small background diffusivity $D(P_E = \ell v_E/D \gg 1)$ there is a residual diffusion varying as $D_x = (Dav_d)^{1/2}$ due to the sharp boundary layer gradients occurring between the trapped and passing particles. Here $P_E = \ell v_E/D$ is called the Peclet number of the medium.

The distribution computed from Eq. (6.82) is shown in Fig. 6.7(b) at $t = 0$ for $\partial \overline{f}/\partial x < 0$ and describes an excess of density $\delta f > 0$ (solid lines) in the upper positive vortex ($\nabla^2 \varphi > 0$) coming down ($u_2 < 0$) and a density deficit in the lower positive vortex going up ($u_1 > 0$).

During the collision the regions of positive and negative δf are brought together creating sharp gradients ($\gg |\delta f|/r_0$) in the merged vortex as shown at $t = 8r_n/c_s$. After one vortex turnover time these regions are mixed with further turnovers producing a spiraled filamentary structure of positive and negative δf. With the estimate that the gradients of δf build up as $k_\perp^2 \simeq \Omega_E^2 t^2/r_0^2$ the time for the background diffusivity to stop the filamentation is $t_d = (r_0^2/D\Omega_E^2)^{1/3}$, whereupon the distribution again evolves to the uniformly mixed state with $\mathbf{v}_E \cdot \nabla \delta f \simeq 0$ and with δf given again by Eq. (6.82).

Although this description of the effect of the collision on the test particle distribution is qualitative, it shows an important transport mechanism. Some fraction of the trapped particles is detrapped during the collision. The essential features of the collisional interaction, however, is a complete mixing of the trapped plasma over the vortex size r_0 after each collision. Assuming the time between collisions τ_v is long compared with interaction time Δt_{col}, we can construct a model for the quasi-coherent intermittent transport of plasma induced by the vortex collisions.

6.6 Intermittent Transport from Vortex Collisions

The vortex collision experiments in Sec. 6.5 show a fast, transient plasma transport over the vortex radius $r_0 \gg \rho_s$ during a vortex collision. It remains, however, to determine the net plasma transport than can occur due to the vortex-vortex interactions.

For a gas of N_v vortices with line density $n_v = N_v/L_x L_y$ where L_x and L_y are the dimensions of the periodic 2D box, we define the average distance r_{ij} between the vortices by $\pi r_{ij}^2 N_v = L_x L_y$ or $r_{ij} = (\pi n_v)^{-1/2}$. For N_v vortices with average radius r_0 the packing fraction f_p is defined by

$$f_p = \frac{N_v \pi r_0^2}{L_x L_y} = \pi n_v r_0^2 = \frac{r_0^2}{r_{ij}^2} \quad (6.84)$$

with the restriction that $f_p < 1$. The fractional area perpendicular to $\mathbf{B}$ covered by vortices is f_p.

For an electrostatic potential field

$$\varphi(x,y,t) = \sum_{i=1}^{N_v} \varphi_i^v(x - x_i, y - y_i - u_i t) + \sum_{\mathbf{k}} \varphi_{\mathbf{k}}^w(t) e^{i\mathbf{k} \cdot \mathbf{x}} \quad (6.85)$$

composed of vortices and waves, the energy density W in the drift wave field φ compared with free energy density $W_f = nT_e(\rho_s/r_n)^2$ has a vortex and wave component given by

$$\frac{W}{W_f} = f_p \overline{A_v^2} \left(1 + \frac{\rho_s^2}{r_0^2}\right) + (\tilde{\varphi}^w)^2 \left(1 + \overline{k}_\perp^2 \rho_s^2\right) \quad (6.86)$$

where $\overline{A_v^2}$ is the average (dimensionless) vortex amplitude

$$\overline{A_v^2} = \frac{1}{N_v} \sum_{i=1}^{N_v} A_i^2, \quad (6.87)$$

and $\tilde{\varphi}^w$ is the root-mean-square wave amplitude

$$\tilde{\varphi}^w = \left(\sum_{\mathbf{k}} |\varphi_{\mathbf{k}}^w|^2\right)^{1/2}. \quad (6.88)$$

In obtaining Eq. (6.86) the waves and vortices are assumed uncorrelated $\langle \varphi_{\mathbf{k}}^w \varphi_i^v \rangle = 0$.

The ratio of the energy in the vortices to the waves can be estimated for a spectrum of waves saturated at the mixing length level $\tilde{\varphi}^w = \alpha^{1/2}/\overline{k}_x$ (corresponding to $e\Phi/T_e \sim \alpha^{1/2}/\overline{k}_x r_n$) and noting that the dimensionless vortex amplitude satisfies $A_v \gtrsim u r_0$ for self-binding. We use the dimensionless constant $\alpha < 1$ to describe the saturation level of those fluctuations not strong enough to form vortices (Horton and Estes, 1979). From these amplitudes the ratio of the energy densities in vortices to waves is given by

$$\frac{E^v}{E^w} \simeq f_p \overline{k}_x^2 r_0^2. \quad (6.89)$$

Since both simulations and experiments show that large scale fluctuations coalesce into vortices, we may expect that $\overline{k}_x^2 r_0^2 \gg 1$ for the wave spectrum. Thus, for a packing fraction $f_p \gtrsim 1/\overline{k}_x^2 r_0^2$ the energy in the vortices is equal to or greater than the waves. The division of the fluctuation spectrum into waves and vortices is discussed in Sec. 6.7 and is based on the onset of self-binding that occurs for $R_E \gtrsim 1 - 2$.

6.6 Intermittent Transport from Vortex Collisions

From the vortex collision experiments in Sec. 6.5.3 we find that the cross-section $\sigma(b)$ for strong inelastic collisions is peaked at $b \simeq r_0$ where $\sigma_{\max} \simeq 2r_0$ and is negligible for $b \gtrsim 3r_0$. This range is comparable with the coalescence range reported by Overman and Zabusky (1982) and Griffiths and Hopfinger (1986, 1987) for neutral fluids.

The mean-free-path λ_{mfp} between vortex collisions is defined by

$$n_v \lambda_{\mathrm{mfp}} \sigma = 1 \tag{6.90}$$

and with $\sigma = 2r_0$, where r_0 is the average vortex radius, and the mean-free-path is $\lambda_{\mathrm{mfp}} = 1/2r_0 n_v$. The time τ between collisions is

$$\tau u n_v \sigma = 1,$$

and the vortex-vortex collision frequency is

$$\nu_v = \frac{1}{\tau} = n_v u \sigma \simeq 2 n_v r_0 v_d \tag{6.91}$$

taking the average vortex speed as $u \gtrsim v_d$. The numerical experiments indicate that the vortex-vortex collisions lead to an effective diffusion D_v of the vortices given by

$$D_v = \nu_v r_0^2 \equiv n_v r_0^2 v_d \sigma \tag{6.92}$$

which increases as r_0^3 at fixed n_v and v_d. The plasma trapped in a vortex is $\pi r_0^2 n_e$ per unit length, and the total trapped fraction of the plasma density is $n_e \pi r_0^2 N_v / L_x L_y = f_p n_e$. Thus the net plasma diffusivity D_p from vortex trapped plasma is related to the vortex diffusion D_v by

$$D_p = f_p D_v = f_p^2 \left(\frac{u\sigma}{\pi} \right) \cong f_p^2 r_0 v_d \tag{6.93}$$

where $r_0 v_d = (r_0 / r_n)(cT_e / eB)$.

Adding the transport from the uncorrelated wave field $\tilde{\varphi}^w$ we obtain that the total plasma diffusivity is given by

$$D = f_p^2 r_0 v_d + \alpha \bar{\lambda}_x v_{de} \tag{6.94}$$

where the vortex induced diffusion arises from structures with $R_E \gtrsim 1$ or $c\Phi_v / B > r_0 v_d$ and the wave diffusion from $c\varphi^w / B \simeq \lambda_x v_d \alpha^{1/2}$ with α the empirical mixing length saturation constant describing the broadband fluctuations. In dimensional units the diffusion is given by

$$D_p = f_p^2 \frac{r_0}{r_n} \frac{cT_e}{eB} + \alpha \frac{\bar{\lambda}_x}{r_n} \frac{cT_e}{eB}. \tag{6.95}$$

The strong packing fraction square f_p^2 dependence of the vortex diffusion component in Eq. (6.95) arises from the increases of the vortex-vortex interaction rate with f_p and the increasing fraction of plasma trapped in the vortices with f_p. For $f_p \lesssim 1$ and $r_0 \lesssim r_n$ the transport in Eq. (6.95) is the Bohm diffusion law. The characteristic of vortex transport is that it is intermittent in space and time with large pulses occurring in the transport flux.

The packing fraction f_p and vortex size r_0 are associated with the rate of buildup of wave energy at large scales due to mode coupling or the inverse cascade of fluctuation energy from active regions of scale ρ_s or smaller to large scales of order r_0. The rate of damping of the large scale motion and the effect of magnetic shear may both influence the final value of r_0. For sufficiently energetic systems the mean value of r_0 may be as large as r_n which could account for the large amplitude, perhaps coherent structures, observed in the edge regions of confined plasmas.

In neutral fluids the experiments of Couder and Basdevant (1986) indicate the possibility of high packing fractions under certain conditions.

The statistical distribution of 2D neutral fluid vortices is given in Benzi *et al.* (1988) along with estimates of the vortex merging cross-section σ_i and a balance equation for the number of vortices of each size scale from computer simulations.

6.7 Interaction Energies and Kurtosis in Distributions of Vortices and Waves

Here we compute the three invariant moments of the dynamics for a distribution of vortices and incoherent waves. The vortices φ_i^v are parameterized $i = 1$ to N_v with amplitude A_i and radius $r_i \equiv 1/\gamma_i$. The waves are given by $\varphi_{\mathbf{k}}^w$ and are taken to be the Fourier modes of the difference field $\varphi - \varphi^v = \varphi - \sum_{i=1}^{N_v} \varphi_i^v$. At a given time the total field is represented by

$$\varphi = \sum_{i=1}^{N_v} \varphi_i^v(x - x_i, y - y_i - u_i t) + \sum_{\mathbf{k}} \varphi_{\mathbf{k}}^w(t) e^{i \mathbf{k} \cdot \mathbf{x}} \quad (6.96)$$

To make the calculation analytic we assume gaussian shapes $A_i \exp(-r^2/r_i^2)$ for the vortices and assume the waves $\varphi_{-\mathbf{k}}^w = (\varphi_{\mathbf{k}}^w)^*$ are uncorrelated with φ_i^v. That is $\langle \varphi_i^v \varphi_{\mathbf{k}}^w \rangle = 0$

We calculate the three moments defined in Eqs. (6.21), (6.22), and (6.27) as follows:

Mass Moment

$$M = \int dx dy \varphi = \pi \sum_{i=1}^{N_v} A_i r_i^2 + L_x L_y \varphi_{\mathbf{k}=0}^w \quad (6.97)$$

6.7 Interaction Energies and Kurtosis in Distribution of Vortices and Waves

Energy Moment

$$E = \sum_{i=1}^{N_v} E_i + \sum_{i<j}^{N_v} E_{ij} + L_x L_y \sum_{\mathbf{k}}(1+k^2)|\varphi_{\mathbf{k}}^w|^2 \quad (6.98)$$

where the vortex energy is (with $\gamma_i^2 = 1/r_i^2$)

$$E_i = \frac{\pi A_i^2}{4\gamma_i^2}(1+2\gamma_i^2) = \frac{\pi}{2}A_i^2\left(1+\frac{1}{2}r_i^2\right) \quad (6.99)$$

and E_{ij} is the vortex interaction energy. For the simple case $\gamma_i = \gamma_j = \gamma$, the interaction energy E_{ij} is the function of the separation

$$r_{ij}^2 = (x_i - x_j)^2 + (y_i - y_j)^2$$

given by

$$E_{ij} = \frac{\pi}{2}\frac{A_i A_j}{\gamma^2}\left(1+2\gamma^2-\gamma^4 r_{ij}^2\right)\exp\left(-\frac{\gamma^2 r_{ij}^2}{2}\right). \quad (6.100)$$

The interaction energy has local extrema at

$$r_{ij} = 0 \quad \text{and} \quad r_{ij} = \frac{(1+4\gamma^2)^{1/2}}{\gamma^2} \quad (6.101)$$

and vanishes ($E_{ij} = 0$) at

$$r_{ij} = \infty \quad \text{and} \quad r_{ij} = \frac{(1+2\gamma^2)^{1/2}}{\gamma^2}. \quad (6.102)$$

Enstrophy Moment

$$U = \sum_{i=1}^{N_v} U_i + \sum_{i<j}^{N_v} U_{ij} + L_x L_y \sum_{\mathbf{k}} k^2(1+k^2)|\varphi_{\mathbf{k}}^w|^2 \quad (6.103)$$

where

$$U_i = \frac{\pi A_i^2}{2}(1+4\gamma_i^2) \quad (6.104)$$

and the interaction enstrophy is only simple when $\gamma = \gamma_i = \gamma_j$ where

$$U_{ij} = \pi A_i A_j \left[1 + 4\gamma^2 - \left(\frac{\gamma^2}{2}+4\gamma^4\right)r_{ij}^2 + \frac{1}{2}\gamma^6 r_{ij}^4\right]\exp\left(-\frac{\gamma^2 r_{ij}^2}{2}\right). \quad (6.105)$$

262   Chapter 6   Vortex Structures in Hydrodynamic and Vlasov Systems

Packing Fraction

The vortex packing fraction f_p for this model field is defined by

$$f_p = \sum_{i=1}^{N_v} \frac{\pi r_i^2}{L_x L_y} = \frac{N_v \pi r_0^2}{L_x L_y} = \pi r_0^2 n_v \qquad (6.106)$$

where we define the average vortex radius r_0 by $r_0 = \left(\frac{1}{N_v} \sum_{i=1}^{N_v} r_i^2\right)^{1/2}$ and the line density of vortices by $n_v = N_v / L_x L_y$.

The line mass density is then

$$m = \frac{M}{L_x L_y} = f_p \overline{A} + \varphi_{\mathbf{k}=0}^w$$

where $\overline{A} = \sum A_i r_i^2 / \sum r_i^2$ and $\varphi_{\mathbf{k}=0}^w = 0$ by definition of the wave component.

The line energy and enstrophy density are complicated functions that depend on the pair correlation $P_v(r_{ij})$ function of the vortices. For a dilute, uniform distribution where $\langle r_{ij}^2 \rangle = L_x L_y / \pi N_v = (\pi n_v)^{-1}$ the interactions E_{ij}, U_{ij} are weak. In this limit we have

$$w = \frac{E}{L_x L_y} \cong f_p A^2 \left(1 + \frac{2}{r_0^2}\right) + \sum_{\mathbf{k}} (1 + k^2)|\varphi_{\mathbf{k}}^w|^2 \qquad (6.107)$$

and

$$u = \frac{U}{L_x L_y} \cong \frac{f_p A^2}{r_0^2}\left(1 + \frac{4}{r_0^2}\right) + \sum_{\mathbf{k}} k^2(1 + k^2)|\varphi_k^w|^2 \qquad (6.108)$$

where A^2 is an average of A_i^2 over the distribution of vortices.

The presence of the vortices can strongly change the skewness and kurtosis of the fluctuation field. For small f_p the second, third, and fourth order momentums are

$$\langle \varphi^2 \rangle = \sum_{\mathbf{k}} |\varphi_{\mathbf{k}}^w|^2 + f_p \langle A^2 \rangle$$

$$\langle \varphi^3 \rangle \cong \sum_{\mathbf{k}=\mathbf{k}_1+\mathbf{k}_2} \varphi_{-\mathbf{k}} \varphi_{\mathbf{k}_1} \varphi_{\mathbf{k}_2} + 3 f_p \langle A \rangle \sum_{\mathbf{k}} |\varphi_{\mathbf{k}}^w|^2 + f_p \langle A^3 \rangle \qquad (6.109)$$

$$\langle \varphi^4 \rangle \cong 3 \left(\sum_{\mathbf{k}} |\varphi_{\mathbf{k}_1}^w|^2\right)^2 + 6 f_p \langle A^2 \rangle \sum_{\mathbf{k}} |\varphi_{\mathbf{k}}^w|^2 + f_p \langle A^4 \rangle \qquad (6.110)$$

where $\langle A^n \rangle = \frac{1}{N_v} \sum^{N_v} A_i^n$. Thus, the kurtosis is

$$K = \frac{f_p A^4 + 6 f_p \langle A^2 \rangle \langle \varphi^2 \rangle + 3 \langle \varphi^2 \rangle^2}{(f_p A^2 + \langle \varphi^2 \rangle)^2}. \qquad (6.111)$$

For $f_p A^2 > \langle \varphi^2 \rangle$ the kurtosis is determined by the packing function and becomes

$$K \simeq \frac{1}{f_p} = \frac{\langle r_{ij}^2 \rangle}{r_0^2} \qquad (6.112)$$

which can be very large when describing the spotiness of the vortex turbulence. For $f_p A^2 < \langle \varphi^2 \rangle$ we have $K \cong 3$ typical of gaussian random wave turbulence.

Spotty and patchy distributions of vorticity are often produced in the 2D simulations of fluids and plasmas. In such regimes the vorticity is intermittent and characterized by a high kurtosis.

Assuming that the decomposition in Eq. (6.96) with a quasi-normal distribution for φ^w is valid and that suitable ordinary differential equations could be derived, the original problem of $2N\mathbf{k}$ space equations would be reduced to $2N_v$ equations for $x_i(t)$, $y_i(t)$ coordinates of the vortices and $2N_w$ equations for $\varphi_\mathbf{k}^w$. The reduction would be significant when N_w, $N_v \ll N$. In the limit of $N_w = 0$ the dynamics of the point vortices (Kono and Horton, 1991) and modulated point vortices (Zabusky and McWilliams, 1982) can describe the collisions between distributed vortices. In general, however, in the drift-wave-vortex collisions the numbers N_v and N_w change during a collision, and the prospect for a quantitative reduced description with small N_w and N_v seems rather dim. A rate equation for dN_v/dt based on coalescence simulations is proposed in Benzi et al. (1988).

For the one-dimensional model the vortices are replaced by the solitons discussed in Sec. 5.11. In this limit, it is possible to make the soliton vortex division of the field in Eq. (6.96) in a rigorous, self-consistent manner. For the KdV model the division is given by the inverse scattering transform (IST) in Sec. 5.8. Tasso (1983a, 1987) has introduced an alternative method using Gibbs statistics and a functional integration technique to address these issues. Using this theory that does not depend upon IST, Tasso and Lerbinger (1983b) obtain the wavenumber spectrum $I(k) = I_0/(k^2 + k_0^2)$ with constant I_0 and k_0^2 for the one-dimensional drift-wave problem.

6.8 Fluctuation Spectrum from a Gas of Dipole Vortices

Electromagnetic scattering experiments in confinement devices measure the electron density correlations from fluctuations through the dynamical form factor $S(\mathbf{k}, \omega)$. The observed fluctuation spectra are approximately isotropic in $\mathbf{k}_\perp$. Some experiments show a peak in $k_\perp$ near $k_\perp \rho_s \lesssim 0.3$. For each $\mathbf{k}_\perp$ the frequency distribution is peaked at $\omega_m \gtrsim \omega_{*e}$ with a width $\Delta\omega \gtrsim \omega_m$. The question arises as to how these characteristics compare with fluctuations produced by an ensemble of dipole drift wave vortices.

The fluctuation spectrum of a uniform distribution of $df_{dp}(r_0, u) = f_{dp}(r_0, u)\, dr_0\, du$ dipole vortices with given r_0, u is given by

$$S(\mathbf{k}, \omega)\delta(\mathbf{k} - \mathbf{k}')\delta(\omega - \omega') = \frac{df_{dp}}{(2\pi)^4} \left\langle \varphi_{dp}(\mathbf{k}, \omega)\varphi_{dp}^*(\mathbf{k}', \omega') \right\rangle$$

where $\varphi_{dp}(\mathbf{k}, \omega) = \int d^3x\, dt\, \exp(-i\mathbf{k} \cdot \mathbf{x} + i\omega t)\varphi_{dp}(\mathbf{x}, t)$, and the average is over the initial position $\mathbf{x}_0$ of the vortices. The Fourier transform of Eq. (6.67) gives

$$S(\mathbf{k}, \omega) = \frac{(2\pi)^3 df_{dp}}{V} \left[ur_0\beta^2(\beta^2 + \gamma^2) \frac{k_x}{k_\perp^2} \frac{J_2(k_\perp r_0) + k_\perp r_0 \delta_k J_1(k_\perp r_0)}{(\beta^2 + k_\perp^2 r_0^2)(\gamma^2 - k_\perp^2 r_0^2)} \right]^2$$

$$\times \delta(\omega - k_y u)\delta(k_z - \alpha k_y), \qquad (6.113)$$

where $\beta \equiv kr_0 \equiv r_0(1 - v_d/u - \alpha^2/u^2)^{1/2}$ and $\gamma = pr_0$, and δ is given in Eq. (6.68). The spectrum $S(\mathbf{k})$ is proportional to $k_x^2 \widehat{S}(\beta)$ where $\widehat{S}(k_\perp)$ is isotropic in $\mathbf{k}_\perp$ The spectrum vanishes at $k_x = 0$ due to the antisymmetry of the dipole field. Since the vortex propagates at a fixed speed, only frequencies satisfying the dispersion relation $\omega = k_y u$ are excited. Figure 6.8 shows $S(k_\perp) = \int dk_z d\omega\, S(\mathbf{k}, \omega)$ for $k_y = 0$.

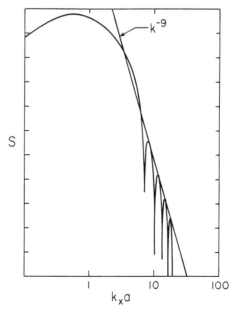

Fig. 6.8. The wavenumber spectrum for an isolated drift wave-Rossby wave dipole vortex with parameters $k_y = 0$, $r_0 = \rho_s$, and $u/v_d = 3/2$.

In this graph the spectrum is peaked at $k_\perp = k_x = 0.54 r_0^{-1}$ and decays as k^{-9} for $k_\perp \gg r_0^{-1}$ with the oscillations due to the Bessel functions in Eq. (6.67). Background wave turbulence will support the wave spectrum with a lower spectral index at $k_\perp \rho_s \gtrsim 1$.

A general state consisting of a superposition of vortices with various sizes and speeds has a spectrum given by Eq. (6.113) with $df_{dp}(u, r_0)$ integrated over the number distribution of vortices with parameters (r_0, u). This fluctuation spectrum will have a frequency width proportional to the spread in velocity distribution Δu in $f_{dp}(u, r_0)$ of the dipole vortices.

6.9 Monopolar Vortices Produced by Sheared Flows

The presence of a nonuniformity in the diamagnetic drift speed $v_d \to v_d(x)$ leads to an additional nonlinearity of the KdV type encountered in Chapter 5. We may consider this nonlinearity in the form of a structural perturbation of the original Hasegawa-Mima equation. We show that this simple structural change results in the splitting of the dipole vortices and that it allows the creation of monopolar vortices.

Returning to the derivation of Eq. (6.44) we see when the convective derivative of $\ell n \, \tilde{n}$ in Eq. (6.37) is evaluated by taking into account the $T_e(x)$ dependence in the $\tilde{n} = n_0(x)[e\Phi/T_e(x)]$ that the KdV nonlinearity proportional to $\alpha = \rho_s \partial_x \ell n \, T_e$ is generated. The nonlinear equation that replaces Eq. (6.44) is

$$(1 - \nabla^2)\frac{\partial \varphi}{\partial t} + v_d \frac{\partial \varphi}{\partial y} + \alpha v_d \varphi \frac{\partial \varphi}{\partial y} + [\nabla^2 \varphi, \varphi] = 0. \tag{6.114}$$

The long wavelength speed of propagation in Eq. (6.114) is $u = v_d(1 + \alpha \varphi)$ so that α gives a faster speed for $\alpha \varphi > 0$ and a slower speed for $\alpha \varphi < 0$ relative to the drift velocity v_d.

6.9.1 Monopole solutions due to the combined action of the Poisson bracket and KdV nonlinearities

In order to interpret the results for the splitting of the dipoles due to the KdV term we first present exact monopole solutions to Eq. (6.114). Assuming a travelling steady-state solution $\varphi = \varphi(x, y - ut)$, Eq. (6.114) becomes

$$-u(1 - \nabla^2)\frac{\partial \varphi}{\partial y} + v_d \frac{\partial \varphi}{\partial y} + \alpha \varphi \frac{\partial \varphi}{\partial y} + [\nabla^2 \varphi, \varphi] = 0. \tag{6.115}$$

For axisymmetric monopoles, $\varphi(r, \theta) \to \varphi(r)$ and $[\varphi(r), \nabla^2 \varphi(r)] = 0$, leaving

$$\frac{1}{r}\frac{d}{dr}\left(r\frac{d\varphi}{dr}\right) = k^2 \varphi - \frac{\alpha}{2u}\varphi^2 \tag{6.116}$$

where
$$k^2 = \left(1 - \frac{v_d}{u}\right). \tag{6.117}$$

Defining $\varphi = \varphi_m \psi(\xi)$, where $\xi = kr$, Eq. (6.116) becomes the following in terms of ψ:

$$\frac{d^2\psi}{d\xi^2} + \frac{1}{\xi}\frac{d\psi}{d\xi} = \psi - \gamma\psi^2. \tag{6.118}$$

Here we have set

$$\varphi_m = \frac{3\gamma u k^2}{\alpha} \tag{6.119}$$

where γ is a constant that remains to be determined. For a given amplitude φ_m Eqs. (6.117) and (6.119) determine the speed of propagation by

$$u = v_d + \frac{\alpha\varphi_m}{3\gamma}.$$

We compare Eq. (6.114) to the analogous equation for one-dimensional solitary drift waves that are solutions of the Petviashvili Eq. (5.261) (Meiss and Horton, 1982; Morrison et al., 1984),

$$\phi_t - \phi_{yyt} + v_d\phi_y + \alpha\phi\phi_y = 0. \tag{6.120}$$

This equation is the one-dimensional restriction of Eq. (6.114) and is called the regularized long-wave (RLW) equation in geophysical hydrodynamics. Inserting $\varphi = \varphi_m \psi\left(k(y - ut)\right)$ yields

$$\frac{d^2\psi}{d\xi^2} = \psi - \psi^2 \tag{6.121}$$

where $\xi \equiv k(y - ut)$, $k^2 = (1 - v_d/u)$, and $\varphi_m = 12uk^2/\alpha$. Equation (6.121) possesses the well-known soliton solution

$$\varphi = \varphi_m \operatorname{sech}^2\xi.$$

We now describe the 2D solution of Eq. (6.118) which is more strongly localized than the 1D solution of (6.121). The monopole speed-width relation $k = k(u)$, Eq. (6.117), is identical with that for the one-dimensional case. However, the amplitude relation of Eq. (6.119) differs by the presence of the factor γ, which accounts for peaking in 2D relative to 1D. Equation (6.118), together with the boundary conditions

$$\frac{d\psi}{d\xi}(\xi = 0) = 0 \quad , \quad \lim_{\xi \to \infty} \frac{1}{\psi}\frac{d\psi}{d\xi} = -2, \tag{6.122}$$

and the condition $\psi(\xi = 0) = 1$, defines a nonlinear eigenvalue problem for γ. Physically, the eigenvalue is the amplitude of the monopole. Numerically we obtain

6.9 Monopolar Vortices Produced by Sheared Flows

$\gamma = 1.5946$; the shape of the eigenfunction defined by this procedure is shown in Fig. 6.9.

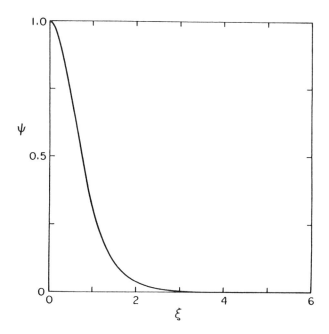

Fig. 6.9. The electrostatic potential (stream function) for the 2D axisymmetric monopole vortex satisfying Eq. (6.118).

The above calculation is only valid for $k^2 > 0$, which implies $u > v_d$ or $uv_d < 0$. In light of Eq. (6.119) positive velocity ($uv_d > 0$) monopoles have $\varphi_m > 0$. These are called anticyclones since they are regions of high density and pressure. The negative ($uv_d < 0$) velocity monopoles have negative φ_m and are called cyclones and have low pressure centers. In the case $k^2 < 0$ (where $0 < u < v_d$) the two-dimensional Eq. (6.116) does not possess exponentially localized solutions. When $k^2 < 0$ the wave structure propagates energy with speed u where $0 < u < v_d$, and in two-dimensions there is a weakly localized oscillatory (outgoing radiation) solution where the amplitude decays as $\xi^{-1/2}$ to conserve the flux of wave energy in two-dimension. This difference in k^2 and u/v_d is the mechanism by which the monopoles become detached from the wave turbulence.

In Fig. 6.10 we plot solutions to Eq. (6.118). Curve 1 has $\psi(0) = 1.85$. Curve 2 has $\psi(0) = \gamma$ as in Fig. 6.9; it is homoclinic to zero: $\psi(kr = \infty) = 0$. Curves 3 and 4 have $\psi(0) = 1.3$ and $\psi(0) = 0.1$, respectively. These, like all solutions with $\gamma > \psi(0) > 0$, are homoclinic to 2/3. Curve 5 has $\psi(0) = -0.1$. All of these solutions have $d\psi(\xi = 0)/d\xi = 0$. For $\psi(\xi = 0) > \gamma$ the solution diverges to $-\infty$. Solutions with $0 < \psi(\xi = 0) \lesssim \gamma$ are homoclinic to $\psi = 2/3$. The nonlinear solutions homoclinic to 2/3 are the radiation solutions mentioned above and describe finite amplitude cylindrically symmetric waves propagating with speed u such that $0 < u < v_d$, that is, $k^2 < 0$. Mathematically, these finite amplitude wave solutions can be "pulled down" by the following symmetry relation:

$$\overline{\varphi}(-k^2; r, \theta) = \varphi(k^2; r, \theta) - \frac{8uk^2}{\alpha}. \tag{6.123}$$

If φ solves $\nabla^2 \varphi - 4k^2 \varphi + \alpha \varphi^2/2u = 0$, then $\overline{\varphi}$ solves $\nabla^2 \overline{\varphi} + 4k^2 \overline{\varphi} + \alpha \overline{\varphi}^2/2u = 0$. Thus, Fig. 6.10 compactly displays both the localized solitary wave and the finite amplitude radiation solutions (Su et al., 1992).

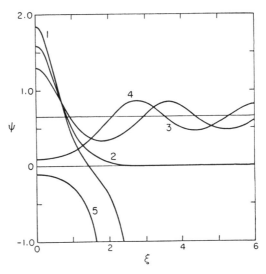

Fig. 6.10. The complete set of 2D axisymmetric solutions for the nonlinear wave equation (6.118).

A symmetry of the Charney-Hasegawa-Mima equation is as follows: For every solution of Eq. (6.114) with $\alpha = 0$, $\varphi(x, y, t)$ there is a solution — $\varphi(-x, y, t)$ (i.e., in the $\alpha = 0$ limit there is the symmetry $\varphi(x, y, t) \to -\varphi(-x, y, t)$.) This symmetry is lost when the finite scalar nonlinearity α is taken into account. This antisymmetry in x is the symmetry property possessed by the nonlinear dipole vortex solutions.

When the scalar nonlinearity is added there is no symmetry for a given α, but for the two different equations with α and $-\alpha$ there is the symmetry relation

$$\varphi(-\alpha; x, y, t) = -\varphi(\alpha, -x, y, t). \tag{6.124}$$

6.9 Monopolar Vortices Produced by Sheared Flows

Thus a small, finite α lifts a degeneracy in the $\alpha = 0$ equation. Now we consider the effect of the KdV nonlinearity on the dipole vortices.

6.9.2 Dipole vortex splitting into monopoles

The dipole vortex solution of Sec. 6.4.4 possessed by Eq. (6.114) in the absence of scalar nonlinearity ($\alpha = 0$) is given by

$$\varphi = \begin{cases} \left[-\dfrac{k^2 r_0}{p^2 r} \dfrac{J_1(pr)}{J_1(pr_0)} + \left(1 + \dfrac{k^2}{p^2}\right) \right] ur \cos\theta & (r < r_0) \\ ur_0 \dfrac{K_1(kr)}{K_1(kr_0)} \cos\theta & (r > r_0) \end{cases} \quad (6.125)$$

where $r^2 = x^2 + (y - ut)^2$, $u = v_d/(1 - k^2)$, $x = r\cos\theta$, and $y = r\sin\theta$. The parameters p and k are related by

$$\dfrac{1}{kr_0} \dfrac{K_2(kr_0)}{K_1(kr_0)} = -\dfrac{1}{pr_0} \dfrac{J_2(pr_0)}{J_1(pr_0)},$$

which follows from continuity of the flow velocity $\mathbf{v} = \hat{\mathbf{z}} \times \boldsymbol{\nabla}\varphi$ across $r = r_0$.

The presence of the shear measured by α in the diamagnetic velocity gives the amplitude and width-speed relations of Sec. 6.9.1. for the monopoles. From Eqs. (6.117) and (6.119) we obtain the following speed for anticyclones:

$$u = v_d + \dfrac{\alpha|\varphi_m|}{3\gamma}, \quad (6.126)$$

while for cyclones, $\varphi_m < 0$, and the speed is given by

$$u = v_d - \dfrac{\alpha|\varphi_m|}{3\gamma} \quad (6.127)$$

where exponential localization ($k^2 > 0$) requires $\alpha|\varphi_m| > 3\gamma v_d$. The anticyclone and cyclone propagate with different speeds, the relative speed being

$$u_+ - u_- = \dfrac{\alpha|\varphi_m^+|}{(3\gamma)} + \dfrac{\alpha|\varphi_m^-|}{(3\gamma)}. \quad (6.128)$$

This result explains why the initial dipole vortex pair will split apart into two isolated monopoles with opposite signs. The time scale for the breakup of the dipole vortex can be estimated from

$$\Delta t \sim \dfrac{3r_0 \gamma}{(2\alpha|\varphi_m|)}. \quad (6.129)$$

This relation for the breakup time is confirmed in the simulations of Su et al. (1992).

270   Chapter 6   Vortex Structures in Hydrodynamic and Vlasov Systems

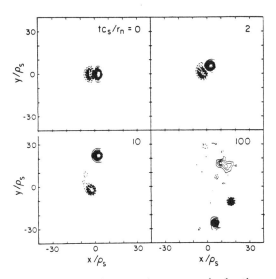

Fig. 6.11. The splitting of the dipole vortex into monopoles by the scalar nonlinearity. The initial condition is a dipole vortex with $r_0 = 6\rho_s$ and $u = 2v_d$ while the strength of the scalar nonlinearity parameter is $\alpha = 0.1$. The contour interval is $\Delta\varphi = 4.0[\rho_s/L_n]$.

Figure 6.11 shows the streamlines $\varphi(x,y,t) = $ const. at times $tc_s/r_n = 0, 2, 10$, and 100. Observe that at $tc_s/r_n = 10$, the initial dipole vortex splits into the two completely isolated monopoles, and the two monopoles approach circular shape. We also observed that the amplitudes of the vortices increase slightly, but the radius of both the vortices does not change much until $t \simeq 30$. The observed speed of the anticyclone is $u \simeq 2.0v_d$, and that of the cyclone is $u \simeq -0.3v_d$. The amplitude of the anticyclone is $\varphi_m^+ \simeq 52$, and that of the cyclone is $\varphi_m^- \simeq -62$.

Moreover, when we turn off the vector nonlinearity and re-do the same experiment, we observe that the radius of the cyclone decreases significantly very rapidly, and its amplitude increases a significant amount and then forms a very localized monopole vortex or solitary wave, while the radius of the anticyclone doesn't change appreciably, and its amplitude increases slightly.

Now we consider whether the anticyclone and cyclones can naturally evolve from the one-dimensional solitary waves described by Eq. (6.121).

We know that Eq. (6.120) has the one-dimensional solitary wave ($k_x \approx 0$) solution given by Petviashvili (1977),

$$\varphi = \frac{3}{\alpha}(u - v_d)\text{sech}^2\left[\frac{1}{2}\left(1 - \frac{v_d}{u}\right)^{1/2}(y - ut)\right], \qquad (6.130)$$

where u is the speed of the solitary wave.

6.9 Monopolar Vortices Produced by Sheared Flows

However, the one-dimensional solitary wave is unstable to a finite k_x filamentation instability shown as follows. Taking Eq. (6.130) as the initial condition, we find that computer solution of Eq. (6.114) evolves to the two-dimensional, nearly circularly symmetric monopole vortices. The results are shown in Fig. 6.12.

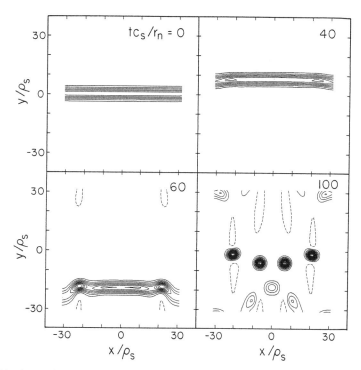

Fig. 6.12. A one-dimensional solitary drift wave at $t = 0$ is subjected to a weak finite k_x modulation. The modulation is unstable growing to produce reconnection of the flow lines ($t = 60$) into four strong monopoles ($t = 100$). Here $\alpha = 0.1$ and initially $u = 2v_d$.

The solution (6.130) is possible only when $u > v_d$ or $v_d u < 0$. For $u > v_d$, we find that the positive solitary wave eventually evolves into two-dimensional anticyclones, as shown in Fig. 6.12. For $u < 0$, we also find that the negative solitary wave evolves into two-dimensional cyclones with $k_x \sim k_y$. From Eq. (6.130), compared with Eq. (6.119), we can see that $\varphi_m < 0$ occurs for $\alpha|\varphi_m| > 3\gamma/v_d$ where $u < 0$.

Now we ask if the cyclone will disappear eventually; in other words can the cyclone be the solution of Eq. (6.114).

From Sec. 6.9.1 the solitary wave solution of Eq. (6.114) does exist and has the form,

$$\varphi = \left[\frac{3\gamma}{\alpha}(u - v_d)\right]\psi(kr), \quad (6.131)$$

according to Eqs. (6.116) and (6.119), where $r = [x^2 + (y - ut)^2]^{1/2}$. So for $u > v_d$, we get a solitary solution for positive amplitude, and the amplitude can be small by having $u \geq v_d$. For $uv_d < 0$ we get for negative amplitude and positive k^2 provided $\alpha|\varphi_m| > 3\gamma v_d$. Thus, the cyclone can only occur if there is sufficient energy in the initial structure.

This means that both the cyclones and anticyclones can exist in the system after the initial dipole vortex pair breaks up if the amplitude φ_m of the dipole pair is large enough so that $\alpha|\varphi_m| > 3\gamma v_d$ where $\gamma = 1.5946$ has been determined in Sec. 6.9.1. This condition requires a substantial amplitude and energy for the solitary wave to form the cyclone. If one uses the approximate formula $\varphi_{dp} = 1.28 v_d r_0$ for the amplitude of the dipole vortex, the condition for the onset of a cyclone can be re-expressed as the condition $r_0 > 3.74/\alpha$.

6.10 Discussion and Conclusions

The nonlinear drift wave-Rossby wave studies show that for a potential fluctuation with an amplitude large enough that the $\mathbf{E} \times \mathbf{B}$ rotation rate $\Omega_E(\mathbf{k})$ around the fluctuation of scale $\mathbf{k}_\perp$ is greater than the linear wave oscillation frequency $\omega_*(\mathbf{k})$ the convective nonlinearity of the plasma vorticity produces a self-binding in the fluctuation. The self-binding allows the fluctuation to propagate over long distances without the loss of energy that occurs in the equivalent small amplitude disturbance. The amplitudes required for the onset of self-binding are shown to be just above the equivalent mixing length levels where the role of the mean radial wavelength $\overline{\lambda}_x$ in the mixing length formula is replaced with the vortex diameter $2r_0$.

The monopolar potential structures from a local charge excess in the plasma or a local high or low pressure in the Rossby wave equation trap any passively convected field and coherently transport the field for long distances. An initial monopole develops an antisymmetric dipolar component due to the v_d term and propagates over long distances without appreciable loss of energy.

The importance and prevalence of 2D vortical flows arises from the self-binding or self-focusing nature of these finite amplitude ($R_E \gtrsim 1$) structures. The self-focusing of the structure allows it to avoid the destructive effects of system inhomogeneities leading to diffraction and energy absorption such as by Landau damping, or for Rossby vortices by bottom topography, which are so effective in limiting the lifetime of small amplitude fluctuations. Further research is required to determine the limits of the 2D approximation to the 3D systems and the limits on the inhomogeneity parameters for the validity of the 2D vortex description.

6.10 Discussion and Conclusions

In this chapter we have shown that the nonuniformity of the drift speed over the radius of the dipole introduces the scalar nonlinearity whose effect is to split the dipole apart. For plasmas the magnitude of the scalar nonlinearity is associated with a gradient of the plasma electron temperature (Petviashvili, 1977, 1980), η_e, and is described by the parameter $\alpha = \eta_e \rho_s / r_n$. In the Rossby waves the scalar nonlinearity arises from the variation of the depth of the fluid with wave amplitude, as in the classical KdV equation for shallow water waves. The lifetime τ of dipole of radius r_0 is estimated from $\tau \sim r_0/\alpha\varphi_m$, where φ_m is a measure of vortex amplitude. We showed that there is a critical amplitude for the formation of the cyclone. Above the critical amplitude we have shown that both the anticyclone ($\varphi > 0$) and cyclone ($\varphi < 0$) exist and do not disappear after the dipole vortex pair breaks up.

The question of the existence of the cyclone solution is important for the interpretation of the rotating water tank experiments. In the early experiments of Antipov et al. (1982, 1983) only anticyclones were reported to be long-lived vortices. In the work of Antonova et al. (1983) with a larger tank it is reported that both cyclones and anticyclones are formed. We suggest that cyclones, although requiring sufficient amplitude to form, are a natural solution of the nonlinear drift wave-Rossby wave equation. The cyclones may be especially important for anomalous transport because of their low propagation velocities.

Chapter 7

Statistical Properties and Correlation Functions for Drift Waves

The subject of plasma turbulence is a large and complex field. Here we restrict the Chapter to that part of the subject that follows naturally from adding forcing from instabilities and dissipation from wave-particle interactions to the drift wave problem developed in Chapter 6. The introduction of growth and damping to the drift wave-Rossby wave spectrum is of key importance to both the problem of plasma confinement in tokamaks and to the description of Rossby wave turbulence. The techniques developed here are general ones, applying to many mode coupling problems. As an example, the problems of ionospheric turbulence (Sudan and Keskinen, 1977), as well as numerous other forms of plasma turbulence, are treated by the techniques presented in this Chapter.

Early works on plasma turbulence that may be helpful to the reader are the *Turbulence in Plasmas*, Tsytovich (1977) and *Nonlinear Plasma Physics*, Davidson (1972). Of course, some, but certainly not all, aspects of plasma turbulence are related to those of fluid turbulence. There are many texts available on the subject of fluid turbulence. The most complete compendium of results is the two volumes of Monin and Yaglom (1971) covering theory and experiments. A good source of original works is CHAOS by Hao (1984).

7.1 Introductory Remarks

The statistical properties of fluctuations generated from the dissipative, two-dimensional, one-field equation for nonlinear drift waves show both coherent structures and incoherent turbulent fluctuations. Which of these characteristics dominates depends on the system parameters $\{\mu\}$. The two-point correlation functions $\langle \varphi(\mathbf{r}_1 t_1) \varphi(\mathbf{r}_2 t_2) \rangle$ and the probability distribution functions of the amplitude and phase of the complex valued Fourier modes $\varphi_{\mathbf{k}}(t)$ can be used to characterize the fluctuations. Formulas derived from the renormalized, Markovianized spectral equations are used to interpret some features of the turbulent fluctuations in laboratory plasmas. The large-scale coherent fluctuation structures are better interpreted in terms of the dipole and monopole vortices of Chapter 6.

We give the conditions required of the dissipative mode coupling system for the existence of an attractor in the phase space of the system. The attractor is chaotic and appears to produce statistically unique steady-state solutions. That is, initial data with either smaller or larger fluctuations are pulled into the same turbulent steady state. The attractor for flow in the phase space is illustrated schematically in Fig. 7.1.

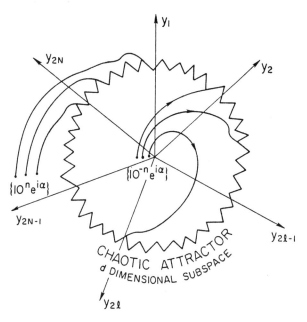

Fig. 7.1. Representation of the chaotic attractor embedded in the high dimensional phase space $\{y_\ell\}$ giving that state $\varphi_{k_\ell}(t) = y_{2l-1}(t) + i\, y_{2\ell}(t)$ of the truncated $\ell = 1, 2, \ldots, N$ system. Due to the properties of the mode coupling system specified in Sec. 7.2 there exists a sub-manifold to which the solutions $\mathbf{Y}(t \to \infty)$ are attracted.

Studies show that the statistics of the fluctuations on the attractor are a strong function of the growth and damping parameters in the equation. Strong growth rates lead to states with nearly Gaussian statistics and exponentially decaying two-point correlation functions. Weak growth rates produce fluctuations with non-Gaussian statistics and long-range correlations. Stable systems, which eventually decay to

7.1 Introductory Remarks

zero energy, but are initialized from states with high levels of turbulence, develop large scale, strongly correlated vortical structures with lifetimes one to two orders of magnitude longer than the correlation times of the growth-rate driven turbulence. These long-lived coherent structures that emerge in the decaying turbulence are readily interpreted in terms of the exponentially shielded vortex solutions of the corresponding dissipationless partial differential equation. Examples of these long-lived fluctuation structures produced by the dissipative drift turbulence described in this chapter are shown in Fig. 7.2. Even in the case of an externally driven system vortex structures can be dominant. Recent simulations by the Orszag group with high $k_f \rho_s$ — forcing show the formation of a quasi-crystallization of the vortex gas in the limit $v_E \gg v_d$ (Kukharkin et al., 1995).

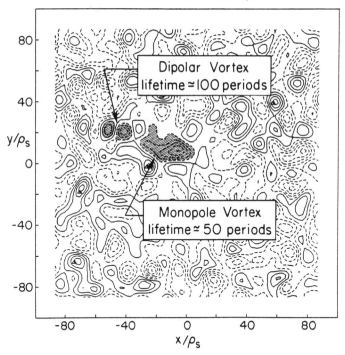

Fig. 7.2. Spatial pattern of the turbulent flows given by the contours of constant potential in the turbulent steady state. Positive values of potential are solid lines and negative values are dashed lines. Characteristic monopole and dipole structures are labelled.

7.2 Nonlinear Drift Wave Equation

A dissipative nonlinear one-field equation for describing drift-wave turbulence has been extensively studied in fusion transport studies by Horton (1986), Terry and Horton (1982), Sudan and Pfirsch (1985), Waltz (1983), Terry and Diamond (1985), Balk, et al., (1990, 1991) and others. The basic simplification in the nonlinear drift-wave model is that for sufficiently small ρ_i/L_n the dominant nonlinear process in drift waves is the convection of the density,

$$\mathbf{v}_E \cdot \nabla n, \qquad (7.1)$$

and the vorticity,

$$\mathbf{v}_E \cdot \nabla \omega, \qquad (7.2)$$

by the $\mathbf{E} \times \mathbf{B}$ velocity

$$\mathbf{v}_E = \frac{c\mathbf{E} \times \mathbf{B}}{B^2}. \qquad (7.3)$$

Here

$$\omega = \hat{\mathbf{z}} \cdot \nabla \times \mathbf{v}_E = \frac{c\nabla^2 \Phi}{B}. \qquad (7.4)$$

A derivation of the drift-wave equation is given in Sec. 6.4.

For a 2D model of drift waves we characterize the mode structure along $\mathbf{B}$ by an average wavenumber $\langle k_\parallel^2 \rangle^{1/2}$. The electron response in the model is approximated by

$$\frac{n_\mathbf{k}}{n_0} = (1 - i\delta_\mathbf{k})\left(\frac{e\Phi_\mathbf{k}}{T_e}\right) \qquad (7.5)$$

valid for $\langle k_\parallel^2 \rangle^{1/2} v_e > \omega$. In toroidal geometry a typical value for $\langle k_\parallel^2 \rangle^{1/2}$ is $1/(qR)$ and in linear systems simply π/L_z where L_z is the length of the device parallel to $\mathbf{B}$. The non-adiabatic, dissipative response $\delta_\mathbf{k}$ gives rise to instability with growth rate

$$\gamma_\mathbf{k} \simeq \omega_{*e}\delta_\mathbf{k} \qquad (7.6)$$

and to mode coupling with

$$\mathbf{v}_E \cdot \nabla n \simeq \frac{cen_0}{T_e}\left(\frac{\mathbf{k}_1 \times \mathbf{k} \cdot \hat{\mathbf{z}}}{B}\right)\delta_{\mathbf{k}_1}\Phi_{\mathbf{k}_1}\Phi_{\mathbf{k}-\mathbf{k}_1} \sim \frac{\partial n}{\partial t} = \gamma_\mathbf{k} n_\mathbf{k}. \qquad (7.7)$$

Thus, the mode coupling balances the growth rate, $\gamma_\mathbf{k}$, at the mixing length level of saturation

$$\frac{\delta n_k}{n_0} \sim \frac{e\Phi_k}{T_e} \sim \frac{1}{k_x L_n} \qquad (7.8)$$

approximately independent of $\delta_\mathbf{k}$. In obtaining (7.8) we use Eqs. (7.5)–(7.7).

The standard dimensionless drift-wave units are $x, y \to \rho_s x, \rho_s y$ and $t \to tL_n/c_s$ and the electrostatic potential Φ is given by $e\Phi/T_e = (\rho_s/L_n)\varphi(x,y,t)$. Here

7.2 Nonlinear Drift Wave Equation

$\rho_s = c(m_i T_e)^{1/2}/eB$, $L_n^{-1} = -\partial \ln n_0/\partial r$ and $c_s = (T_e/m_i)^{1/2}$. In these units the diamagnetic drift velocity $v_d = 1$ (although v_d is left in the equations below to identify the drift term) and the $\mathbf{E} \times \mathbf{B}$ convective derivative is given by

$$\partial_t f + \mathbf{v}_E \cdot \nabla f = \{\partial_t f + [\varphi, f]\}\left[\frac{c_s}{L_n}\right]$$

where $[\varphi, f] = \hat{\mathbf{z}} \cdot \nabla \varphi \times \nabla f = \partial_x \varphi \partial_y f - \partial_y \varphi \partial_x f$.

7.2.1 Nonlinear drift wave equation in configuration space

The drift-wave model is defined in terms of the linear operator (Terry and Horton, 1982, 1983) $\mathcal{L}$ which divides into the hermitian part $\mathcal{L}_h$ giving the wave dispersion and into the anti-hermitian part $\mathcal{L}_{ah}$ giving the wave dissipation:

$$\mathcal{L} = \mathcal{L}_h + \mathcal{L}_{ah} = -\nabla^2 + \delta_0(c_0 + \nabla^2)\frac{\partial}{\partial y} \quad (7.9)$$

with Fourier transform

$$\mathcal{L}e^{i\mathbf{k}\cdot\mathbf{x}} = (\chi'_\mathbf{k} + i\chi''_\mathbf{k})e^{i\mathbf{k}\cdot x} = \chi_\mathbf{k} e^{i\mathbf{k}\cdot\mathbf{x}} \quad (7.10)$$

$$\chi'_\mathbf{k} = k^2 \quad \text{and} \quad \chi''_\mathbf{k} = \delta_0 k_y(c_0 - k^2).$$

The complex drift-wave frequency for $\varphi_\mathbf{k} \exp(-i\omega t)$ oscillations is

$$\omega = \frac{k_y v_d}{(1 + \chi_\mathbf{k})} \quad (7.11)$$

where $\chi_\mathbf{k} = \chi'_\mathbf{k} + \chi''_\mathbf{k}$. The drift-wave parameters δ_0 and c_0 are defined further in terms of the system parameters in Sec. 7.2.1.2.

The two-dimensional drift-wave equation is

$$(1 + \mathcal{L})\frac{\partial \varphi(x, y, t)}{\partial t} + [\varphi, \mathcal{L}\varphi] + v_d \frac{\partial \varphi}{\partial y} + \hat{\gamma}_i \varphi = 0 \quad (7.12)$$

where $[\varphi, \mathcal{L}\varphi]$ gives the $\mathbf{E} \times \mathbf{B}$ convection of the density and vorticity. Equation (7.12) is the condition that $\nabla \cdot \mathbf{j} = 0$ for the quasineutral plasma. At sufficiently large space scales Eq. (7.12) gives convection of the initial disturbance $\varphi(x, y, 0)$ to $\varphi(x, y - v_d t, 0)$. At small scales, first dissipation and then dispersion and nonlinearity determine the dynamics.

7.2.1.1 Ion and electron dissipation parameters

In general both short and long wavelength drift waves are damped by either (i) ion-ion or ion-neutral collisions, (ii) ion Landau-damping (iii) coupling to damped modes such as ion-acoustic waves ($k_\| v_e > \omega$) or convective cells ($k_\| v_e < \omega$) and (iv) the spatial propagation of wave energy out of the active region. These processes are represented by $\hat\gamma_i$ in Eq. (7.12) with Fourier transform

$$\gamma_i(\mathbf{k}) = \gamma_0 \exp\left(-\frac{k_y^2}{k_0^2}\right) + \nu_0 k^4 \qquad (7.13)$$

using the parameters γ_0, k_0 to control the low $k \leq k_0$ damping and the parameter ν_0 to control the high-k damping. The exponential dependence in (7.13) arises from the evaluation of the ion Landau damping $\exp(-\omega^2/k_\|^2 v_i^2)$ at $\omega = \omega_{*e}$. An example for the value of these parameters is given by the low-k coupling to damped ion-acoustic waves at $\omega_{*e} = \left\langle k_\|^2 \right\rangle^{1/2} c_s$ giving $k_0 = \left\langle k_\|^2 \right\rangle^{1/2} L_n (T_i/T_e)^{1/2}, \gamma_0 = (\pi/2)^{1/2}(T_e/T_i)^{3/2}$ and the high-k ion-ion collisional viscosity giving $\nu_0 = 0.3(\nu_{ii} L_n/c_s)(T_i/T_e)^2$.

An example for the electron dissipation parameters are the three regimes in a low-beta tokamak with safety factor $q = rB/RB_\theta$ for which $\left\langle k_\|^2 \right\rangle^{1/2} L_n = L_n/qR = \epsilon_n/q$. (i) The collisional drift waves $\nu_e > \left\langle k_\|^2 \right\rangle^{1/2} v_e$ have $c_0 = 3\eta_e/2$ and $\delta_0 = (m_e/m_i)^{1/2}\nu_e/\left\langle k_\|^2 \right\rangle L_n v_e$; (ii) the collisionless or plateau regime drift waves with $\epsilon^{3/2} \left\langle k_\|^2 \right\rangle^{1/2} v_e < \nu_e < \left\langle k_\|^2 \right\rangle^{1/2} v_e$ have $c_0 = \eta_e/2$ and $\delta_0 = (\pi m_e/2m_i)^{1/2}/\left\langle k_\|^2 \right\rangle^{1/2} L_n$; (iii) the trapped electron modes have $c_0 = -\eta_e$ and $\delta_0 = \epsilon^{3/2}(c_s/L_n \nu_e)$ for $\nu_e < \epsilon^{3/2}\left\langle k_\|^2 \right\rangle^{1/2} v_e$. An analytic response function $\delta_\mathbf{k}(\eta_e, \epsilon, \nu_e/\left\langle k_\|^2 \right\rangle^{1/2} v_e, \left\langle k_\|^2 \right\rangle^{1/2} L_n)$ from kinetic theory may be used to describe these regimes more accurately with continuous transitions.

Turbulence studies in the collisional regime using two-field equations coupling the potential $e\varphi/T_e$ and density fluctuations $\delta n/n$ fluctuations are reported by Hasegawa and Wakatani (1983) and by Terry and Diamond (1985). The dispersion relation is quadratic in ω with an unstable drift-wave branch and a damped mode. The one-field Eq. (7.12) with the values of δ_0 and c_0 given by part (i) in the preceding paragraph describe the dynamics along the unstable branch while losing the coupling to the damped branch. Coupling to the damped branch by the two-field equations may potentially lower the amplitude of the turbulence and the anomalous flux from that predicted by the one-field description used here. Details on the relationship between these two models are given in Sec. 7.5.

7.2.1.2 Dimensionless system parameters $\{\mu\}$

Equations (7.10) and (7.12) define the drift-wave model with five dimensionless parameters $\mu = \{\delta_0, c_0, \nu_0, \gamma_0, k_0\}$. Sub-models are $\mu_3 = \{\delta_0, c_0, \nu_0, 0, 0\}$ with only high-k ion damping, and $\mu_2 = \{\delta_0, c_0 > 0, 0, 0, 0\}$ with only electron dissipation ($c_0 > 0$). This two-parameter $\{\delta_0, c_0\}$ model was used by Terry and Horton (1982, 1983) to study drift wave turbulence.

7.2 Nonlinear Drift Wave Equation

A typical distribution of the linear frequencies $\omega^\ell(k_x, k_y)$ and growth-damping rate $\gamma^\ell(k_x, k_y)$ for the parameters $\{\mu\} = \{\delta_0 = 0.25, c_0 = -0.25, \nu_0 = 0.005\}$ is shown in Fig. 7.3. The frequency units are c_s/L_n and wavenumber units ρ_s^{-1}.

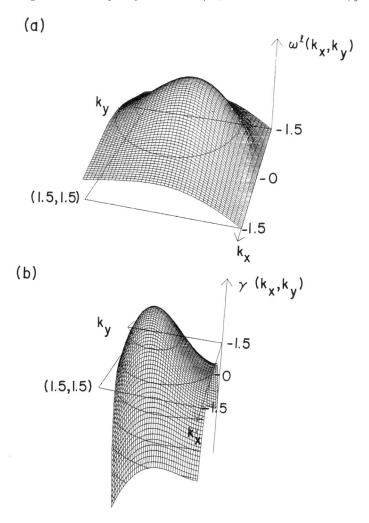

Fig. 7.3. A typical distribution of the drift wave angular frequency $\omega_{\mathbf{k}}^\ell$ and the growth-damping rate $\gamma_{\mathbf{k}}^\ell$ on the grid used in simulations which is $85k_1 \times 85k_1$ with $k_i \rho_s = 0.1$. The maximum and minimum growth and damping rates are $\gamma_{\max} = 0.15(c_s/L_n)$ and $\gamma_{\min} = -0.42(c_s/L_n)$.

The present study is for the statistical properties of the solutions of the mode coupling equations (7.12) obtained from a finite N truncation the original nonlinear partial differential equation (pde). For a given truncation with N **k**-vectors and a $2N$ dimensional phase space the solutions of the mode coupling equations approximate the solutions of the associated pde (7.12) only on some limited subset D_{pde} of the parameter values $\mu \subset D_{\text{pde}}(\Delta k, N)$. Determining the subset D_{pde} of parameters is a difficult problem beyond the scope of the present discussion. The present study adopts the finite N mode coupling equation as the basic dynamical equation considering the properties of its solutions. The validity of statistical turbulence theory for predicting the properties of the solution of the mode coupling equation is independent of the question of the relationship of the solutions of the truncated equations to the solutions of the continuum pde.

7.2.2 Nonlinear drift wave equation in k space

Turbulence theory for drift waves is formulated in **k** space defined by the Fourier transform

$$\varphi(x, y, t) = \sum_{\mathbf{k}} \varphi(\mathbf{k}, t) \exp(i\mathbf{k} \cdot \mathbf{x}) \tag{7.14}$$

with $\varphi(-\mathbf{k}, t) = \varphi^*(\mathbf{k}, t)$ where each **k** vector in (7.14) introduces one degree-of-freedom in the dynamics. Truncating the representation (7.14) with N **k**-vectors and computing their dynamics with Eq. (7.12) gives a reduced description of the system that conserves the quadratic invariants of the convective nonlinearity $[\varphi, \mathcal{L}\varphi]$. The dynamical equations are

$$(1 + \chi_{\mathbf{k}}) \frac{d\varphi(\mathbf{k}, t)}{dt} = [-ik_y v_d - \gamma_i(\mathbf{k})] \varphi(\mathbf{k}, t)$$

$$+ \tfrac{1}{2} \sum_{\mathbf{k}_1 + \mathbf{k}_2 = \mathbf{k}} \hat{\mathbf{z}} \cdot \mathbf{k}_1 \times \mathbf{k}_2 (\chi_{\mathbf{k}_2} - \chi_{\mathbf{k}_1}) \varphi(\mathbf{k}_1, t) \varphi(\mathbf{k}_2, t). \tag{7.15}$$

For small amplitudes, $\sum_k |k \chi_k \varphi_k|^2 < 1$, Eq. (7.15) is linear and diagonal. The modes evolve with $\varphi_{\mathbf{k}}(t) = \varphi_{\mathbf{k}}(0) \exp(-i\omega_{\mathbf{k}}^l t + \gamma_{\mathbf{k}}^l t)$ where

$$\omega_{\mathbf{k}}^l + i\gamma_{\mathbf{k}}^l = \frac{k_y v_d - i\gamma_0 \exp(-k_y^2/k_0^2) - i\nu_0 k^4}{1 + k^2 + i\delta_0 k_y(c_0 - k^2)}. \tag{7.16}$$

The evolution from Eq. (7.15) is conventionally calculated by using both **k**-space and **x**-space at each time step. The convolution sum is calculated in **x**-space. The use of the fast Fourier transform (FFT) software makes these calculations possible. Aliasing errors are avoided by limiting the physical $\varphi_{\mathbf{k}}$ array to 2/3 the length of the FFT array. The results given here are for a 85^2 physical **k** space obtained from a 128^2 FFT array. The system then has $N = 3612$ degrees-of-freedom with $2N$ first order real differential equations.

7.2.3 Volume contraction in the phase space

We use the index $l = 1, 2, 3 \ldots N$ to count the $\mathbf{k}$ vectors and define the $2N$ dimensional state vector $\mathbf{Y}(t) = \{y_i(t)\}_{i=1}^{2N}$ in the phase space Γ^{2N} defined by

$$\varphi_{\mathbf{k}_l}(t) = y_{2l-1}(t) + i\, y_{2l}(t) \tag{7.17}$$

for $l = 1, 2, \ldots N$. With this normalization the square of the radius vector in the phase space is twice the energy of the state $2E(t) = \|\mathbf{Y}\|^2$. The finite N truncation (7.17) of the dynamics in Eq. (7.15) preserves the nonlinear constraints

$$\int d\mathbf{x}[\varphi, \mathcal{L}\varphi]\varphi = 0 \tag{7.18}$$

and

$$\int d\mathbf{x}[\varphi, \mathcal{L}\varphi]\mathcal{L}\varphi = 0 \tag{7.19}$$

since

$$\sum_{\mathbf{k}} \sum_{\mathbf{k}_1 + \mathbf{k}_2 = \mathbf{k}} \hat{\mathbf{z}} \cdot \mathbf{k}_1 \times \mathbf{k}_2 (\chi_2 - \chi_1) \varphi_{\mathbf{k}_1} \varphi_{\mathbf{k}_2} (\varphi_{-\mathbf{k}}, \chi_{-\mathbf{k}} \varphi_{-\mathbf{k}}) = 0 \tag{7.20}$$

in each triad $\mathbf{k}_1 + \mathbf{k}_2 = \mathbf{k}$. For proving relations such as (7.20) it is often convenient to introduce $\mathbf{k}_3 = -\mathbf{k}$ so that $\mathbf{k}_1 + \mathbf{k}_2 + \mathbf{k}_3 = 0$.

We define the probability density $\rho^{2N}(\mathbf{Y}, t)$ in the phase space Γ^{2N} where $\rho^{2N} dy_1 dy_2 \ldots dy_{2N}$ gives the probable number of systems in the state $d\mathbf{Y}$ at $\mathbf{Y}, t$. Conservation of the number of systems gives

$$\frac{\partial \rho^{2N}}{\partial t} + \sum_{i=1}^{2N} \frac{dy_i}{dt} \frac{\partial \rho^{2N}}{\partial y_i} = -\rho^{2N} \sum_{i=1}^{2N} \frac{\partial}{\partial y_i}\left(\frac{dy_i}{dt}\right) = -2\gamma_T(\mu)\rho^{2N} \tag{7.21}$$

where the rate of volume contraction is uniform throughout the phase space and is given by the parameters μ through

$$2\gamma_T(\mu) = \sum_{i=1}^{2N} \frac{\partial}{\partial y_i}\left(\frac{dy_i}{dt}\right) = \sum_{l=1}^{N} 2\gamma_{\mathbf{k}_l}^l. \tag{7.22}$$

For system parameters such that $\gamma_T(\mu) < 0$ the dynamics contracts every volume Ω_{2N} to zero as $t \to \infty$ at the rate $\Omega_{2N} \exp(2\gamma_T t)$. Therefore, the long time limit of the phase space trajectories define a subspace or manifold with dimension $d < 2N$. The attracting manifold is found to be chaotic by numerically showing that neighboring trajectories on the attractor $\mathbf{Y}$ and $\mathbf{Y} + \delta \mathbf{Y}$ exponentially diverge $\|\delta \mathbf{Y}\| \sim \|\delta \mathbf{Y}(0)\| \exp(\lambda t)$. The chaotic attractor is a subspace with a of fractal dimension. We do not attempt to discuss the dimension d of the chaotic attractor which is a strong function of the parameters μ and is typically large enough (> 10) to make its actual value of little practical use.

7.2.3.1 Conditions for nonlinear saturation

Volume contraction alone does not guarantee nonlinear saturation since the manifold defined by $Y(t \to \infty)$ could be open to infinity. The existence of an attracting manifold with bounded energy values $\|Y\|^2$ follows from three properties of the system:

(1) energy conservation of the mode coupling,

(2) volume contraction of the flow and

(3) that the mode coupling drives the system to equipartition.

Conditions (1) and (2) are established analytically, whereas condition (3) must be established by numerical experiments. We observe that the equipartition condition (3) is also a necessary condition for the applicability of a modal kinetic equation derived in the random phase approximation (RPA) or from renormalized (RN) turbulence theory since these statistical theories yield kinetic equations with the property that for sufficiently high total energy the kinetic equation drives the system to equipartition.

For states with small $\|Y\|^2$ the state vector $Y(t)$ moves out from the origin on the unstable manifold defined by $\gamma_{\mathbf{k}}^l > 0$. For states with sufficiently large $\|Y\|$ the nonlinear interactions dominate the linear terms and by property (3) the system evolves to equipartition. The rate ν_{eq} of approach to equipartition increases as $\nu_{\text{eq}} = \text{const}\sqrt{E(t)}$ and is fast compared to the $\max |\gamma_{\mathbf{k}}^l|$ for large $2E = \|Y\|^2$. Using properties (1) and (3), we have for sufficiently large $\|Y\|$ that

$$\frac{dE}{dt} = 2\sum_{i=1}^{2N} \gamma_{\mathbf{k}}^l E_{\mathbf{k}} \cong 2\left(\sum_{i=1}^{2N} \gamma_{\mathbf{k}}^l\right) \frac{E}{N} \equiv 2\langle\gamma\rangle E(t). \qquad (7.23)$$

Thus, when the flow is volume contracting ($\langle\gamma\rangle = \gamma_T(\mu)/N < 0$), property (2), all states with sufficiently large $\|Y\|$ decay back toward the origin in phase space at the rate $\langle\gamma\rangle$. Thus, when conditions (1)–(3) are satisfied all solutions are attracted into a subspace or manifold of phase space on which $E(t)$ has bounded variations. For a dissipationless system there is no such contraction in the flow and modes tend toward phase-locked states with increasing amplitudes and narrower spatial localization as the energy is increased.

In the dissipative system it is easy to observe this contracting behavior with computer experiments by taking random initial states with either large or small amplitudes. The experiments also show that when more than a few modes are unstable ($\gamma_{\mathbf{k}}^l > 0$) the motion is chaotic in time as discussed in detail for $N = 3$ by Terry and Horton (1982) and for $N = 10$ in Terry and Horton (1983). The temporal chaos is due to the divergence of neighboring trajectories as measured by the positive Lyapunov exponents on the attracting manifold.

7.2 Nonlinear Drift Wave Equation

7.2.3.2 Conservation laws and equipartition

In the dissipationless limit $\delta_0 = \gamma_0 = \nu_0 = 0$ Eq. (7.12) reduces to the Hasegawa-Mima equation (Hasegawa and Mima, 1977, 1978) which conserves the total energy

$$E(t) = \frac{1}{2} \int \frac{d\mathbf{x}}{V} [\varphi^2 + (\nabla\varphi)^2] = \frac{1}{2} \sum_{\mathbf{k}} (1+k^2)|\varphi_{\mathbf{k}}|^2 = \sum_{l=1}^{N} E(\mathbf{k}_l, t)$$

where $E(\mathbf{k}_l) = \frac{1}{2}(1+k^2)|\varphi_{\mathbf{k}_l}|^2$ and the potential enstrophy

$$U(t) = \frac{1}{2} \int \frac{d\mathbf{x}}{V} [(\nabla\varphi)^2 + (\nabla^2\varphi)^2] = \sum_{l=1}^{N} k_l^2 E(\mathbf{k}_l, t).$$

The physical consequences of the conservation laws are discussed in Sec. 6.4. The dissipationless motion occurs on the $2N-2$ dimensional subspace, or manifold $\Gamma_{E,U}$ defined by

$$\rho^{2N} = \delta(E(Y) - E(t_0))\delta(U(Y) - U(t_0)) \quad \text{in } \Gamma^{2N}.$$

Numerical experiments for drift wave equation, following those of Kells and Orszag (1978) for the 2D-Euler equation with $N = 10$, show that the isolated, truncated system is ergodic, mixing and goes to statistical equilibrium. The distribution of $|\varphi_{\mathbf{k}}|^2$ in the statistical equilibrium is given approximately by the Gibbs canonical ensemble

$$\rho^{2N} = \exp(-\mu E - \beta U) \tag{7.24}$$

giving $E_{\text{eq}}(\mathbf{k}) = \text{const}/(\mu + \beta k^2)$ with μ and β determined by constant $E(t_0)$ and $U(t_0)$. Fyfe and Montgomery (1979) show the same result for a system with 32^2 modes for the dissipationless ($\delta_0 = \nu_0 = \gamma_0 = 0$), waveless ($v_d = 0$) limit of Eq. (7.12), that is $dq/dt = 0$ with $q = (1 - \nabla^2)\varphi(x, y, t)$.

Equipartition of energy occurs when the mode coupling conserves energy but not enstrophy. Taking $\gamma_i(\mathbf{k}) = v_d = 0$ in Eq. (7.12) but retaining the dissipative contribution of $\chi_{\mathbf{k}}''$ in the mode coupling gives an equation that conserves only the invariant

$$P(t) = \frac{1}{2} \int \frac{d\mathbf{x}}{V} [(1+\mathcal{L})\varphi]^2 = \frac{1}{2} \sum_{\mathbf{k}} |1 + \chi_{\mathbf{k}}|^2 |\varphi_{\mathbf{k}}|^2. \tag{7.25}$$

This invariant is a generalization of the plasma wave-momentum density

$$P_y = \sum \left(\frac{k_y}{\omega_{\mathbf{k}}}\right) W(\mathbf{k}) = \frac{1}{2} \sum_{\mathbf{k}} (1+k_\perp^2)^2 |\varphi_{\mathbf{k}}| = E + U. \tag{7.26}$$

Numerical experiments show that this system evolves to

$$E(\mathbf{k}, t \to \infty) = E_{\text{eq}}(\mathbf{k}) = \frac{1}{N} E(t_0). \tag{7.27}$$

The tendency of the mode coupling to drive the system toward equipartition is an important property of the system as emphasized by Kraichnan and Montgomery (1980). Unfortunately, it appears beyond the scope of analytic theory to determine which nonlinear pde's will establish an equipartition in the long-time limit. An example of the evolution to the equipartition state (7.27) is given in Fig. 6 of Terry and Horton (1983).

7.2.4 Statistical turbulence theory

The basis for a reduced theoretical description of drift-wave turbulence is that the modes $\varphi_{\mathbf{k}}(t)$ are sufficiently random for an expansion about Gaussian statistics to be a valid. Statistical turbulence theories differ principally in how they compute the three mode correlation function $\langle \varphi_{k_1} \varphi_{k_2} \varphi_{k_3} \rangle$. The four mode correlation function $\langle \varphi_1 \varphi_2 \varphi_3 \varphi_4 \rangle$ is truncated in terms of lower order correlation functions. Tractable theories neglect non-Gaussian statistics in the four-mode correlation function by adopting the

quasinormal approximation

$$\langle \varphi_{k_1} \varphi_{k_2} \varphi_{k_3} \varphi_{k_4} \rangle \cong \langle \varphi_{k_1} \varphi_{k_2} \rangle \langle \varphi_{k_3} \varphi_{k_4} \rangle + \langle \varphi_{k_1} \varphi_{k_3} \rangle \langle \varphi_{k_2} \varphi_{k_4} \rangle + \langle \varphi_{k_1} \varphi_{k_4} \rangle \langle \varphi_{k_2} \varphi_{k_3} \rangle . \quad (7.28)$$

With this quasinormal approximation the chain of coupled correlations is broken and we have *closure*. The correlation function $\langle \varphi_{k_1} \varphi_{k_2} \rangle$ can then be computed with the resulting closed turbulence theory.

Turbulence theories differ in how they approximate (or neglect) higher-order correlations. Renormalized turbulence theory (RNT) and the so-called Direct Interaction Approximation (DIA) sum certain "diagonal contributions" to all orders in $\langle \varphi_{k_1} \varphi_{k_2} \rangle$.

However, it was shown by Ogura (1963) that the quasinormal approximation can erroneously predict negative energies. This closure has also been shown to violate the near-equilibrium statistics by incorrectly predicting a time-reversible behavior (Orszag, 1970). The origin of these difficulties is that the quasinormal closure is not *realizable*. A closure is said to be realizable if there exists an underlying probability density function for the predicted statistics (Orszag, 1970). Fortunately, physically realizable closures can be constructed from the quasinormal approximation by also renormalizing the linear propagator in the equation that describes the evolution of the correlation function (Orszag, 1970; Longcope and Sudan, 1991; Yakhot and Orszag, 1986).

The complete sets of two-mode correlation functions $\langle \varphi_{k_1} \varphi_{k_2} \rangle$ determines the two-point field correlation function through

$$\langle \varphi(\mathbf{x},t)\varphi(\mathbf{x}+\mathbf{r},t+\tau) \rangle = \sum_{\mathbf{k}_1,\mathbf{k}_2} \langle \varphi_{\mathbf{k}_1}(t)\varphi_{\mathbf{k}_2}(t+\tau) \rangle \exp[i(\mathbf{k}_1+\mathbf{k}_2)\cdot\mathbf{x}+i\mathbf{k}_2\cdot\mathbf{r}]. \quad (7.29)$$

7.2 Nonlinear Drift Wave Equation

For spatially homogeneous and temporally stationary turbulence

$$\langle \varphi_{\mathbf{k}_1}(t)\varphi_{\mathbf{k}_2}(t+\tau)\rangle = \delta_{\mathbf{k}_2,-\mathbf{k}_1} I_{\mathbf{k}_2}(\tau). \tag{7.30}$$

With condition (7.30) satisfied the normalized two-point correlation function $C(\mathbf{r}, \tau)$ is

$$\langle \varphi(\mathbf{x},t)\varphi(\mathbf{x}+\mathbf{r},t+\tau)\rangle = \langle \varphi^2 \rangle C(\mathbf{r},\tau) = \sum_{\mathbf{k}} I_{\mathbf{k}}(\tau)e^{i\mathbf{k}\cdot\mathbf{r}} \tag{7.31}$$

An example of $C(\mathbf{r}, \tau = 0)$ for a typical strong turbulence state produced by Eq. (7.12) is shown in Fig. 7.4. Note the weak oscillations of the correlation function in the direction of wave propagation.

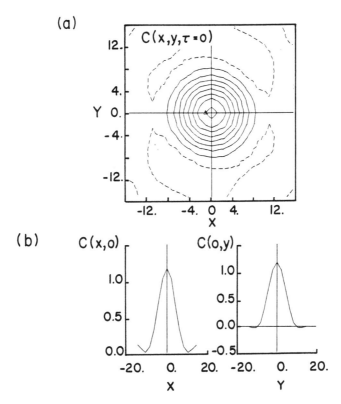

Fig. 7.4. The two-space point, one-time correlation function for the random spatial pattern of the electrostatic potential $\phi(x,y,t)$ produced in the saturated state. (a) contours of constant $C(x,y,\tau=0)$ with x and y in units of ρ_s. The turbulence is weakly anisotropic.

For turbulence theory to be applicable the correlation function must be sufficiently localized with $C(\mathbf{r}, \tau) \simeq 0$ for $|\mathbf{r}| > l_c$ or $\tau > \tau_c$. The localization condition of $C(\mathbf{r}, \tau)$ rules out the solitons and coherent vortices. Solitons and other coherent structures given in Chapter 6, have long correlation lengths and lifetimes. In particular the two-time correlations must $I_\mathbf{k}(\tau) \to 0$ for $|\tau| \to \infty$ sufficiently rapidly to yield time local balance equations for $E(\mathbf{k}, t)$. For exponentially decaying two-time correlations

Markovian approximation

$$I_\mathbf{k}(\tau) \sim I_\mathbf{k}(0) \exp(-\nu_\mathbf{k} |\tau|) \tag{7.32}$$

the time history integrals in renormalized turbulence theory are performed analytically to yield a Markovian or time-local theory for the dynamics of the correlation functions. All time-history integrals are replaced by propagators containing $\{\nu_\mathbf{k}\}$.

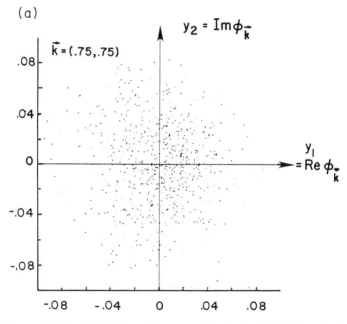

Fig. 7.5. The probability distribution of the potential of a single high $\mathbf{k}$ mode in the drift wave turbulence. (a) distribution in the phase space and (b) the probability density and first four moments of the real part of $\phi_\mathbf{k}(t)$.

7.2 Nonlinear Drift Wave Equation

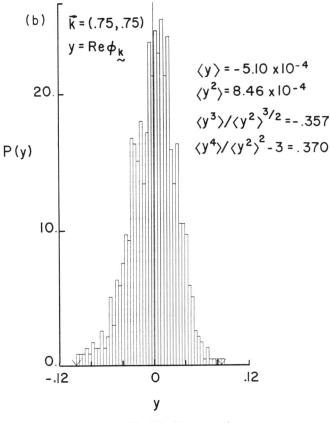

Fig. 7.5. (*Continued*)

7.2.4.1 Example of mode statistics

As an example, consider the turbulence generated by the trapped electron mode by taking $c_0 = -0.25, \gamma_0 = 0, \nu_0 = 0.005$ and $\delta_0 = 0.25$. The linear frequency and growth rate from Eq. (7.16) are shown in Fig. 7.3. Figure 7.5 shows the probability density obtained from $\varphi(\mathbf{x}, t)$ in the single mode phase space $y_1 = \text{Re}\,\varphi_{\mathbf{k}}(t)$ and $y_2 = \text{Im}\,\varphi_{\mathbf{k}}(t)$ for a typical mode in the spectrum. The density $P(a_k, \theta_k)$ is symmetric in phase angle $\theta_{\mathbf{k}} = \tan^{-1}(y_2/y_1)$ and Gaussian in amplitude $P(a_k, \theta_k) = a_k \exp(-a_k^2/2\,\langle y_1^2\rangle)/(2\pi\,\langle y_1^2\rangle)$. As a further measure of the Gaussianity of the modal statistics the probability density $P(y) = \frac{1}{T}\int_0^T dt\,\delta(y - y_1(t))$ and its moments are constructed in Fig. 7.5b. Again the deviation from Gaussianity appears weak for this $\{\mu\}$ parameter set.

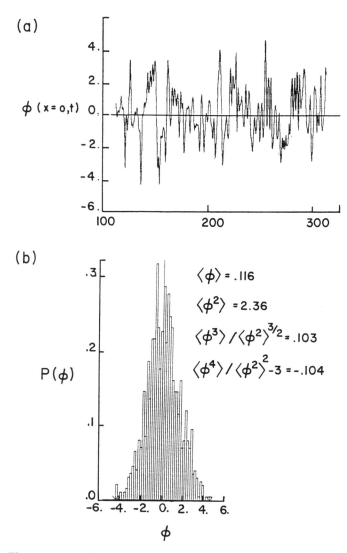

Fig. 7.6. The stationary, chaotic time series obtained for the potential at a single space point x from the steady-state turbulence produced by the nonlinear drift wave equation (7.12). (a) the time series $\phi(\mathbf{x}, t)$ at fixed x and (b) the probability distribution of the amplitude from (a), as defined by Eq. (7.33), along with its first four moments.

7.2.4.2 Statistics of the field measured at a single space point

The simplest measure of the statistics of the field is to construct the probability distribution $P(\varphi)$ of the field at a given space point defined by

$$P(\varphi) = \frac{1}{T} \int_0^T dt\, \delta(\varphi - \varphi(\mathbf{x},t)) = \int_{-\infty}^{+\infty} \frac{d\eta}{2\pi} e^{i\eta\varphi} \int_0^T \frac{dt}{T} e^{-i\eta\varphi(\mathbf{x},t)}$$

$$= \int_{-\infty}^{+\infty} \frac{d\eta}{2\pi} e^{i\eta\varphi} \left\langle e^{-i\eta\varphi(\mathbf{x},t)} \right\rangle. \tag{7.33}$$

For a sinusoidal wave of amplitude a the distribution is $P(\varphi) = \pi^{-1}(a^2 - \varphi^2)^{-1/2}$ and for a $\varphi(\mathbf{x},t)$ given by the sum of a large number N of independent random modes $P(\varphi) = \exp(-\varphi^2/\varphi_0^2)/(2\pi\varphi_0^2)^{1/2}$ with $\varphi_0^2 = N^{-1}\sum_l \langle \varphi_l^2 \rangle$ by the central limit theorem.

The probability distribution for the turbulent state in Fig. 7.6a is shown in Fig. 7.6b along with its first four moments. The relatively small skewness $s = \langle \varphi^3 \rangle / \langle \varphi^2 \rangle^{3/2}$ and the closeness of the kurtosis $\langle \varphi^4 \rangle / \langle \varphi^2 \rangle^2$ to the Gaussian value of 3 suggests that in fact the correlations between the modes are weak for this μ parameter set. The time series in the turbulent steady state is shown in Fig. 7.6a. The probability distribution $P(\varphi)$ of the field value φ in range $d\varphi$ is constructed from the time series by binning the sampled values $\varphi_n = \varphi(n\Delta t; \mathbf{x}_0)$ into suitably chosen sets of bins of size $\Delta\varphi$ according to the size of the ensemble and φ_0.

7.2.5 Two-time correlation functions

Homogeneous, stationary turbulence is characterized by the lack of correlations between modes $\varphi_{\mathbf{k}_1}(t_1)$ and $\varphi_{\mathbf{k}_2}(t_2)$ for either $\mathbf{k}_1 + \mathbf{k}_2 \neq 0$ or $\mathbf{k}_1 + \mathbf{k}_2 = 0$ and $|t_1 - t_2| \gg \tau_c(k_1)$.

For an ergodic system, the time average over a finite sample time $T \gg \tau_c(k)$, where $\tau_c(k) = 1/\nu_k$ is the correlation time, is equivalent to an ensemble average over an effective number $N_{\text{eff}} = \nu_k T$ of independent experiments. For the finite sample time, the condition for the two-time correlation function $I_{12}(\tau) = \frac{1}{T}\int_0^T \varphi_{\mathbf{k}_1}^*(t)\varphi_{\mathbf{k}_2}(t+\tau)dt$ to show zero correlation between $\varphi_{\mathbf{k}_1}$ and $\varphi_{\mathbf{k}_2}$, is that the random function $I_{12}(\tau)$ must satisfy

$$\langle I_{12}(\tau) \rangle = 0 \quad \text{and} \quad \langle I_{12}^2 \rangle = \frac{I_{k_1}(0)I_{k_2}(0)}{(\nu_{k_1} + \nu_{k_2})T} \tag{7.34}$$

for $\mathbf{k}_1 + \mathbf{k}_2 \neq 0$. Horton (1986) has tested Eq. (7.34) by varying T for sample $\varphi_{\mathbf{k}_1}(t_1)$ and $\varphi_{\mathbf{k}_2}(t_2)$ in the turbulent steady state. The T scaling tests follow Eq. (7.34) showing that $\varphi_{\mathbf{k}_1}(t_1)$ and $\varphi_{\mathbf{k}_2}(t_2)$ are uncorrelated for $\mathbf{k}_1 + \mathbf{k}_2 \neq 0$ and for $\mathbf{k}_1 + \mathbf{k}_2 = 0$ when $|t_1 - t_2| \gg \tau_c(\mathbf{k}_1)$.

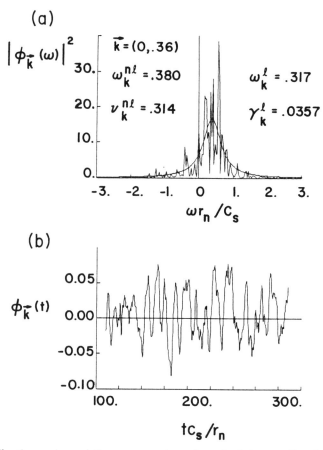

Fig. 7.7. The time series and the power spectrum for a single intermediate k mode. The mode $k \rho_s = (0.0, 0.36)$ is well below the maximum growth rate. The power spectrum $\langle |\phi_\mathbf{k}(\omega)^2| \rangle$ is fit to the Lorentzian with the parameters $\{\omega_\mathbf{k}^{n\ell}, \nu_\mathbf{k}^{n\ell}\}$ with the result shown as the solid line curve in frame (a). The time series is shown in frame (b).

7.2.6 Properties of correlation functions for $\mathbf{k}_1 + \mathbf{k}_2 = 0$

Now we consider the time dependence of the correlation function $I_\mathbf{k}(\tau) = T^{-1} \int_0^T \varphi_\mathbf{k}^*(t) \varphi_\mathbf{k}(t+\tau) dt$. The real and imaginary parts of $\ln I_\mathbf{k}(\tau)$ yield the nonlinear frequency and decorrelation rate. For small τ the magnitude of $I_\mathbf{k}(\tau)$ decreases as $I_\mathbf{k}(0)(1 - \tau^2/\tau_1^2)$. After a short roll-over period $\tau \lesssim \tau_1$ the function decays exponentially with

$$\mathrm{Re}\left[\ln I_\mathbf{k}(\tau)\right] = -\nu_\mathbf{k}|\tau|$$

7.2 Nonlinear Drift Wave Equation

for $\tau > \tau_1$. The imaginary part of $\ln I_\mathbf{k}(\tau)$ also varies linearly in time. We conclude that a good parameterization of the function is

$$I_\mathbf{k}(\tau) = I_\mathbf{k}(0)\exp(-i\omega_\mathbf{k}\tau - \nu_\mathbf{k}|\tau|) \tag{7.35}$$

valid for $|\tau| \gtrsim \tau_1 \simeq 1/\omega_{\max} \simeq 0.1[L_n/c_s]$ and $\nu_\mathbf{k}|\tau| < \ln(\nu_\mathbf{k}T)$. The rollover time τ_1 reflects the highest frequency in spectrum with significant power.

A second test of the parameterization of $I_\mathbf{k}(\tau)$ is made from the frequency spectrum of $\varphi_\mathbf{k}(t)$

$$\varphi_\mathbf{k}(\omega) = \int_0^T \varphi_\mathbf{k}(t)\exp(i\omega t)\frac{dt}{T}. \tag{7.36}$$

The spectral density implied by Eq. (7.32) is

$$I_\mathbf{k}(\omega) = \left\langle |\varphi_\mathbf{k}(\omega)|^2 \right\rangle = \int_0^T \int_0^T \frac{dt_1 dt_2}{T^2}\left\langle \varphi_\mathbf{k}^*(t_1)\varphi_\mathbf{k}(t_2) \right\rangle \exp[i\omega(t_2-t_1)]$$

$$= \frac{2\nu_\mathbf{k} I_\mathbf{k}(0)}{(\omega-\omega_\mathbf{k})^2 + \nu_\mathbf{k}^2}, \tag{7.37}$$

that is $I_\mathbf{k}(\omega)$ is the Lorentzian in $\omega - \omega_\mathbf{k}$ with width $\nu_\mathbf{k}$.

In Fig. 7.7 we show the Lorentzian parameters $I_\mathbf{k}(0), \omega_\mathbf{k}, \nu_\mathbf{k}$ obtained from a nonlinear regression fit of Eq. (7.37) to the measured $|\varphi_\mathbf{k}(\omega)|^2$. An alternate parameterization of $I_\mathbf{k}(\omega)$ by a Gaussian gives poorer results as indicated by the linear decrease of $\text{Re}[\ln I_\mathbf{k}(\tau)]$ with τ. A linear increase of $I_\mathbf{k}(\tau)$ for $\tau \to 0$ implies the high-frequency behavior of $I_\mathbf{k}(\omega)$ decreases as $1/\omega^2$.

7.2.7 Space correlation function and the wavenumber spectrum

Even for system parameters $\{\mu\}$ giving small growth rates in a restricted region of $\mathbf{k}$ space the spectrum $E(\mathbf{k})$ extends over many modes with the range Δk giving rise to a spatial correlation distance $l_c = 1/\Delta k$. In Fig. 7.4 we show level contours of the space correlation function $C(\mathbf{r}, \tau = 0)$ given by

$$\left\langle \varphi^2 \right\rangle C(\mathbf{r}) = \sum_\mathbf{k} I_\mathbf{k} \exp(i\mathbf{k}\cdot\mathbf{r}) \tag{7.38}$$

where

$$I_\mathbf{k} = I_\mathbf{k}(0) = \frac{1}{T}\int_0^T |\varphi_\mathbf{k}(t)|^2 dt/T$$

for the example presented here. For the case of weakly anisotropic spectrum Eq. (7.38) reduces to

$$\left\langle \varphi^2 \right\rangle C(r) = \int_0^\infty dk\, k\, I(k) J_0(kr).$$

The small r decay of the correlation function is consistent with the power law spectrum

$$I(k) = \langle I_{\mathbf{k}} \rangle \cong 1/k^m \qquad (7.39)$$

corresponds to

$$C(r \to 0) \simeq \int^{\infty} dk k^{1-m} J_0(kr) \sim 1 - r^{m-2}. \qquad (7.40)$$

A typical spectral index for 2D turbulence is $m = 4$ for which $C(r \to 0) \sim 1 - \alpha_2 r^2$. At large r the decay is approximately exponential

$$C(r > l_c) \simeq \exp\left(\frac{-r}{l_c}\right)$$

corresponding to

$$\langle I(\mathbf{k}) \rangle \simeq I_0/(1 + k^2 l_c^2)^{\frac{3}{2}} \sim I_0(1 - 3k^2 l_c^2/2) \qquad (7.41)$$

for $k^2 l_c^2 \ll 1$ and $\langle I(\mathbf{k}) \rangle \simeq I_0/(kl_c)^3$ for $k^2 l_c^2 \gg 1$.

7.2.8 Effect of dissipation on space-time evolution

In Fig. 7.2 we show a typical flow pattern associated with the electrostatic potential $\varphi(x, y, t)$ which is the stream function for the $\mathbf{E} \times \mathbf{B}$ convection of the plasma. A time series of such figures shows that the maxima and minima form and disappear on the scales l_c and τ_c in accord with physical interpretation of these correlation scales. The level contours of $\varphi(x, y, t)$ and $\nabla^2 \varphi(x, y, t)$ form and dissolve, reconnecting to form another similar pattern on these space-time scales in a random manner. The higher maxima and lower minima keep their identity for longer periods of time.

There is a competition between the tendency for the fluctuations to self-organize into coherent structures and the linear dissipative driving mechanism $\gamma_k \sim \delta_k \omega_*$ to stir or pump the fluctuations into a chaotic state with short-range correlations as shown in Fig. 7.2.

The ability of the field dynamics to self-organize into large-scale vortex motion is made clear by reducing or by switching off the instability driving mechanism ($\delta_0 = 0$ for $t > t_1$) in the turbulent steady. The small scale fluctuations rapidly decay from the viscous ν_0 $k^4/(1+k^2)$ damping leaving the intermediate and large scale, long-lived fluctuations which then self-organize into a vortex turbulence. McWilliams has studied the emergence of coherent vortices in decaying turbulence in detail (1984, 1990) and proposed the power law for their rate of decay through mergers (Sec. 6.4) as $N_0(t) \sim t^{-\zeta}$ with $\zeta \sim 3/4$. Experiments by Driscoll and Fine (1995) using the pure electron plasma to simulate 2D Euler hydrodynamics appear to confirm this rate of merging.

The stream function contours in the weakly decaying turbulent state show large-scale structures with $l_c \sim 10 - 20 \, \rho_s$ which propagate with velocities $\dot{x} \simeq 0$, $\dot{y} = u \simeq$

v_d maintaining their identities for times of order $\tau_v \sim 100\, \rho_s/v_d$. These coherent structures are of the type of vortices investigated in Chapter 6.

Due to their long lifetime and relative coherence we interpret these structures as the monopolar and dipolar drift-wave vortices. For example the four negative contours shaded in Fig. 7.2 translate together with $u \gtrsim v_d$ for a time $\tau_v \sim 50$ $\mathbf{E} \times \mathbf{B}$ rotation periods and act as a coherent monopolar vortex, and the pair of positive and negative contours shaded in Fig. 7.2 form a dipolar vortex. We note that the exponential shielding $K_0(kr) \simeq r^{-1/2} \exp(-kr)$ with $k = (1 - v_d/u)^{1/2}[\rho_s^{-1}]$ of both monopolar and dipolar vortices is responsible for their close packing compared with the long-range Euler vortices ($v_d = 0$) with $K_0 \to -ln(r)$ potential ($k = 0$) shown in Figs. 3 and 4 of McWilliams (1984). If the structures shown in Fig. 7.2 were linear structures they would spread by dispersion on time scales small compared with their observed lifetimes.

7.3 Renormalized, Markovianized Spectral Equations

The quadratic nonlinearity of Eq. (7.12) couples the correlation function $\left\langle \varphi_{\mathbf{k}}^*(t)\varphi_{\mathbf{k}}(t) \right\rangle$ to the three mode correlation function $\left\langle \varphi_{\mathbf{k}}^*(t)\varphi_{\mathbf{k}_1}(t_1)\varphi_{\mathbf{k}-\mathbf{k}_1}(t_1) \right\rangle$. In weak turbulence theory this correlation is computed perturbatively by inverting the *linear* operator $1/\epsilon_k^l$ where from Eqs. (7.15) and (7.16) the operator is

$$\epsilon_{\mathbf{k}}^l(\omega) = (1 + \chi_{\mathbf{k}})(\omega - \omega_{\mathbf{k}}^l - i\gamma_{\mathbf{k}}^l). \tag{7.42}$$

In the next order the quasinormal correlation approximation formula (7.28) are used to express the four-mode correlation function in terms of sum of quadratic products of $\langle |\varphi_{\mathbf{k}}(t)|^2 \rangle$. The expansion parameter in this iteration of Eq. (7.15) is R_k^0 where

$$R_k^0 = \left| \frac{1}{\epsilon_{\mathbf{k}}^l(\omega)} \sum_{\mathbf{k}_1} (\hat{\mathbf{z}} \cdot \mathbf{k} \times \mathbf{k}_1)(\chi_{\mathbf{k}-\mathbf{k}_1} - \chi_{\mathbf{k}_1})\varphi_{\mathbf{k}_1} \right|. \tag{7.43}$$

For $\omega \simeq \omega_{\mathbf{k}}^l$, even for small rms fluctuation levels $\widetilde{\varphi} = (\sum_{\mathbf{k}} |\varphi_{\mathbf{k}}|^2)^{1/2} \ll 1$, this perturbation expansion fails. Using Eqs. (7.10), (7.11), and (7.42) we find that $R_k^0 \sim 1$ at the mixing length level given by $k^2 \widetilde{\varphi} = \Omega_E \sim |k_y v_d|$.

In renormalized turbulence theory the three mode correlation function is computed to all orders in $\langle |\varphi_{\mathbf{k}}(t)|^2 \rangle$ for a certain class of secular terms (Horton and Choi, 1979). The terms retained at each order of the summation are restricted first to those surviving the average over quasi-Gaussian statistics, and subsequently to those with a sufficiently high degree of secularity in the fluctuation propagator $1/\epsilon_k^l$; i.e., those for which $R_k^0 \to \infty$ with highest exponent $(R_k^0)^l$. Summation of the selected (secular) series leads to the renormalized fluctuation propagator

$$\epsilon_k^l \to \epsilon_k = (1 + \chi_{\mathbf{k}})(\omega - \omega_{\mathbf{k}} + i\nu_{\mathbf{k}} - i\gamma_{\mathbf{k}}^l), \tag{7.44}$$

where $\omega_{\mathbf{k}}$ and $\nu_{\mathbf{k}}$ are nonlinear functions of the $I_k(\omega)$ spectrum given by

$$\omega_{\mathbf{k}} = \omega_{\mathbf{k}}^l + \mathrm{Re}\sum_{\mathbf{k}_1}\int_{-\infty}^{+\infty}\frac{d\omega_1}{2\pi}\frac{(\mathbf{k}\times\mathbf{k}_1)^2(\chi_{\mathbf{k}-\mathbf{k}_1}-\chi_{\mathbf{k}_1})(\chi_{\mathbf{k}}-\chi_{-\mathbf{k}_1})I_{\mathbf{k}_1}(\omega_1)}{(1+\chi_{\mathbf{k}})(1+\chi_{\mathbf{k}-\mathbf{k}_1})(\omega_{\mathbf{k}}-\omega_1-\omega_{\mathbf{k}-\mathbf{k}_1}+i\nu_{\mathbf{k}-\mathbf{k}_l}+i\nu_{\mathbf{k}})}$$

(7.45)

$$\nu_{\mathbf{k}} = -\mathrm{Im}\sum_{\mathbf{k}_1}\int_{-\infty}^{+\infty}\frac{d\omega_1}{2\pi}\frac{(\mathbf{k}\times\mathbf{k}_1)^2(\chi_{\mathbf{k}-\mathbf{k}_1}-\chi_{\mathbf{k}_1})(\chi_{\mathbf{k}}-\chi_{-\mathbf{k}_1})I_{\mathbf{k}_1}(\omega_1)}{(1+\chi_{\mathbf{k}})(1+\chi_{\mathbf{k}-\mathbf{k}_1})(\omega_{\mathbf{k}}-\omega_1-\omega_{\mathbf{k}-\mathbf{k}_1}+i\nu_{\mathbf{k}-\mathbf{k}_l}+i\nu_{\mathbf{k}})}.$$

(7.46)

In writing these equations $\gamma_{\mathbf{k}}^l$ has been neglected in the denominators of Eqs. (7.45) and (7.46).

For $\tilde{\varphi}$ above the critical level determined by $\nu_k \gtrsim k_y v_d k^2 \rho_s^2$ the turbulent or eddy damping rate ν_k dominates the three wave frequency mismatch in Eqs. (7.45) and (7.46). In this finite amplitude regime the scaling and magnitude of ν_k is given by

$$\nu_k \simeq \frac{k^2|\chi_k|[kI(k)]^{1/2}}{|1+\chi_k|}$$

(7.47)

where $I(k)$ is the mean square potential per unit wavenumber defined by

$$I(k) = \sum_{\mathbf{k}_1}\int_{-\infty}^{+\infty}\frac{d\omega_1}{2\pi}\delta(|\mathbf{k}_1|-k)I_{\mathbf{k}_1}(\omega_1).$$

The spectral density $I(k)$ is called the omni-directional potential spectrum since all directions of the $\mathbf{k}$-vector are summed over.

The contribution to $\langle\varphi^2\rangle$ from scale k is $kI(k)$. The eddy damping rate ν_k given by Eq. (7.47) is smaller than the $\mathbf{E}\times\mathbf{B}$ rotation frequency $\Omega_E = k^2\varphi_k$ reduced by the mode coupling and shielding factor $|\chi_k|/|1+\chi_k|$. Formula (7.47) for ν_k shows that large-scale eddies ($|\chi_k|\ll 1$) are weakly affected by the turbulence. Thus, particles can make many rotations in the core of large-scale eddies or vortices before the eddies decay in time $1/\nu_k$ since $\Omega_E \gg \nu_k$ for $|\chi_k|\ll 1$.

The expansion parameter in the renormalized perturbation series is

$$R_k^{\mathrm{rn}} = \left|\frac{1}{\epsilon_{\mathbf{k}}}\sum_{\mathbf{k}_1}(\hat{\mathbf{z}}\cdot\mathbf{k}\times\mathbf{k}_1)(\chi_{\mathbf{k}-\mathbf{k}_1}-\chi_{\mathbf{k}_1})\varphi_{\mathbf{k}_1}\right| \leq \frac{k\nu_k|\chi_k|(\sum_{\mathbf{k}_1}k_1^2|\varphi_{\mathbf{k}_1}|^2)^{1/2}}{|1+\chi_k|[(\omega-\omega_k)^2+\nu_k^2]}$$

(7.48)

where $\epsilon_{\mathbf{k}}(\omega)$ is given by Eq. (7.44). Evaluating the denominator of (7.48) at $\omega = \omega_{\mathbf{k}}$ and using Eq. (7.47) for ν_k shows that the renormalized expansion parameter $R_k^{\mathrm{rn}} \leq 1$ for all $\tilde{\varphi}$. Thus, the renormalized perturbation expansion in R_k^{rn} appears nominally convergent. Of course strong assumptions are made when the off-diagonal and nonsecular terms at each order of the series in $\tilde{\varphi}^m$ are neglected. The conditions required to justify the neglect of these less secular, off-diagonal terms, while not well understood mathematically, are presumably the absence of the coherent, soliton-like structures of the type analyzed in Chapter 6.

7.3.1 Incoherent fluctuation source

The spectral mode coupling equation is derived from Eq. (7.12) by closing the infinite-order perturbation series with the quasinormal approximation and renormalizing:

$$|\epsilon_{\mathbf{k}}(\omega)|^2 I_{\mathbf{k}}(\omega) = S_{\mathbf{k}}(\omega)$$

$$= \tfrac{1}{2} \sum_{\mathbf{k}_1} \int_{-\infty}^{+\infty} \frac{d\omega_1}{2\pi} (\mathbf{k} \times \mathbf{k}_1)^2 |\chi_{\mathbf{k}-\mathbf{k}_1} - \chi_{\mathbf{k}_1}|^2 I_{\mathbf{k}_1}(\omega_1) I_{\mathbf{k}-\mathbf{k}_1}(\omega - \omega_1) \quad (7.49)$$

where $S_{\mathbf{k}}(\omega)$ given by Eq. (7.49) is called the incoherent source. The incoherent source $S_{\mathbf{k}}(\omega)$ arises from that part of $\langle \varphi_k^* \varphi_{k_1} \varphi_{k-k_1} \rangle$ uncorrelated with $\varphi_{\mathbf{k}}$, where $\epsilon_{\mathbf{k}}(\omega)$ is given by Eqs. (7.44)–(7.46).

In general the ω dependence of $I_{\mathbf{k}}(\omega)$ and thus of $S_{\mathbf{k}}(\omega)$ is unknown and may contain multiple poles and branch points. Assuming to the contrary, that $S_{\mathbf{k}}(\omega)$ is slowly varying in the region near $\omega_{\mathbf{k}}$, then the right-hand side of Eq. (7.49) is taken as constant which with Eq. (7.45) leads to the Lorentzian spectrum

$$I_{\mathbf{k}}(\omega) = \frac{2\nu_{\mathbf{k}} I_{\mathbf{k}}}{(\omega - \omega_{\mathbf{k}})^2 + \nu_{\mathbf{k}}^2} \quad (7.50)$$

with

$$I_{\mathbf{k}} = (2\pi)^{-1} \int_{-\infty}^{+\infty} d\omega I_{\mathbf{k}}(\omega)$$

and $\omega_{\mathbf{k}}, \nu_{\mathbf{k}}$ determined through Eqs. (7.45)–(7.46). In approximation (7.50) the exponential decay rate $\nu_{\mathbf{k}}$ of the two-time correlation function is given by the damping of the mode $\epsilon_{\mathbf{k}}(\omega) = 0$. Substituting approximation (7.50) back into Eq. (7.49) shows that in the next approximation $I_{\mathbf{k}}(\omega)$ contains branch points starting at $2\omega_{(\mathbf{k}/2)} \pm i\nu_{\mathbf{k}}$ with branch cuts extending to $\text{Im}(\omega) = \pm i\infty$.

With the Lorentzian (7.50) the ω_1 integrals in Eqs. (7.45), (7.46), and (7.49) can be performed analytically. The Lorentzian parameterization of the frequency spectrum is called a Markovianization of the spectral equations since it eliminates the time-history integrals. The approximation appears to describe adequately the observed decay of the two-time correlation functions which are essentially exponential (Horton, 1986).

7.3.2 Dynamics of the correlation function in renormalized turbulence theory

Up to this point we have assumed that the turbulent fluctuations are statistically stationary. Restoring the long-time scale $T = (t_1 + t_2)/2$ variation to the frequency integrated spectrum $I_{\mathbf{k}} \to I_{\mathbf{k}}(T)$ yields the spectral balance equation. We write the equation in terms of the modal energy

$$E(\mathbf{k}, T) = \frac{1}{2}(1 + k^2) I_{\mathbf{k}}(T)$$

and let the long-time scale $T \to t$. The Markovian (local in time) spectral equation is

$$\frac{dE_{\mathbf{k}}(T)}{dT} = (2\gamma_{\mathbf{k}}^l - 2\nu_{\mathbf{k}})E_{\mathbf{k}} + S_{\mathbf{k}}(\{E_{\mathbf{k}}\}) = 2\gamma_{\mathbf{k}}^l E_{\mathbf{k}} + T_{\mathbf{k}}(\{E_{\mathbf{k}}\}) \qquad (7.51)$$

where the turbulent modal transfer or turbulent collision operator is

$$T_{\mathbf{k}}(\{E_{\mathbf{k}}\}) = -\operatorname{Im} \sum_{\mathbf{k}} \frac{(\mathbf{k} \times \mathbf{k}_1)^2 R_{\mathbf{k}, \mathbf{k}_1, \mathbf{k}_2} (\chi_{\mathbf{k}_2} - \chi_{\mathbf{k}_1})}{(1 + k^2)(1 + k_1^2)(1 + k_2^2)} \Big[(\chi_{-\mathbf{k}_2} - \chi_{-\mathbf{k}_1}) E_{\mathbf{k}_1} E_{\mathbf{k}_2}$$

$$- (\chi_{\mathbf{k}} - \chi_{-\mathbf{k}_1}) E_{\mathbf{k}} E_{\mathbf{k}_1} + (\chi_{\mathbf{k}} - \chi_{-\mathbf{k}_2}) E_{\mathbf{k}} E_{\mathbf{k}_2} \Big] \qquad (7.52)$$

and

$$R_{\mathbf{k}, \mathbf{k}_1, \mathbf{k}_2} = [\omega_{\mathbf{k}} - \omega_{\mathbf{k}_1} - \omega_{\mathbf{k}_2} + i(\nu_{\mathbf{k}} + \nu_{\mathbf{k}_1} + \nu_{\mathbf{k}_2})]^{-1}$$

where $\omega_{\mathbf{k}}$ and $\nu_{\mathbf{k}}$ are functions of $I_{\mathbf{k}}$ through Eqs. (7.45) and (7.46).

In the limit where the linear susceptibilities $\chi_{\mathbf{k}}$ are real and that $\max(\nu_{\mathbf{k}}) \ll \Delta\omega_{\mathbf{k}}$ then the Im R gives the delta function $\pi\delta(\omega_{\mathbf{k}} - \omega_{\mathbf{k}_1} - \omega_{\mathbf{k}_2})$ and Eq. (7.51) reduces the classical weak turbulence wave-kinetic equation (Sagdeev and Galeev, 1969).

7.3.3 Properties of $T_{\mathbf{k}}(\{E\})$ — the nonlinear transfer operator

The nonlinear transfer operator $T_{\mathbf{k}}(\{E\})$ has the properties that

(i) $T_{\mathbf{k}}$ conserves energy

$$\sum_{\mathbf{k}} T_{\mathbf{k}}(\{E\}) = 0; \qquad (7.53)$$

(ii) the equipartition of energy in Eq. (7.27), $E_{\mathbf{k}} = E_{\mathrm{eq}} = E_T/N$-independent of $\mathbf{k}$, is a solution of

$$T_{\mathbf{k}}(E_{\mathrm{eq}}) = 0, \qquad (7.54)$$

(iii) the entropy production $\sigma(\{E\})$ from $T_{\mathbf{k}}$ is positive definite. The entropy of the state $\{E_{\mathbf{k}}\}$ is

$$S = \sum_{\mathbf{k}} \ln E_{\mathbf{k}}. \qquad (7.55)$$

7.4 Local Isotropic Approximation

The rate of change of entropy computed from Eq. (7.51) is

$$\sigma = \frac{dS}{dt} = \sum_{\mathbf{k}} 2\gamma_{\mathbf{k}}^l + \sigma^{nl}(\{E_{\mathbf{k}}\}) \tag{7.56}$$

where

$$\sigma^{nl}(\{E_{\mathbf{k}}\}) = \frac{-1}{3} \operatorname{Im} \sum_{\mathbf{k}_1+\mathbf{k}_2=\mathbf{k}} (\mathbf{k} \times \mathbf{k}_1)^2 R_{\mathbf{k},\mathbf{k}_1,\mathbf{k}_2} E_{\mathbf{k}_1} E_{\mathbf{k}_2} E_{\mathbf{k}}$$

$$\left| \frac{(\chi_{\mathbf{k}_2} - \chi_{\mathbf{k}_1})}{E_{\mathbf{k}}} + \frac{(\chi_{-\mathbf{k}} - \chi_{\mathbf{k}_2})}{E_{\mathbf{k}_1}} + \frac{(\chi_{\mathbf{k}_1} - \chi_{-\mathbf{k}})}{E_{\mathbf{k}_2}} \right|^2 \geq 0, \tag{7.57}$$

provided that $\nu_{\mathbf{k}} \geq 0$ for all $\mathbf{k}$. Equipartition is the state with no entropy production $\sigma(\{E_{\text{eq}}\}) = 0$. Near equipartition the equations of motion show that $d\sigma/dt < 0$, and thus equipartition is a stable attractor in the $E_{\mathbf{k}}$ state space.

In the unstable, turbulent steady state the rate of entropy production dS/dt is given by

$$\frac{dS}{dt} = \sum_{\mathbf{k}} 2\gamma_{\mathbf{k}}^l + \sigma^{nl}(\{E\}) \equiv 2\gamma_T(\mu) + \sigma^{nl}(\{E\}) = 0. \tag{7.58}$$

The conditions in Sec. 7.2.3. for the existence of a steady state require that the spectrum $E_{\mathbf{k}}$ adjust to where the nonlinear entropy production σ^{nl} balances the entropy decrease due to volume contraction $2\gamma_T(\mu) < 0$ of the flow.

For $\nu_{\mathbf{k}} \gg \gamma_{\mathbf{k}}^l$ the evolution in Eq. (7.51) proceeds first toward equipartition to minimize the entropy production, and then to the final spectral distribution determined by

$$T_{\mathbf{k}}(\{E_{\mathbf{k}}\}) = -2\gamma_{\mathbf{k}}^l E_{\mathbf{k}}. \tag{7.59}$$

The time scale for the first nonlinear adjustment is given by Eq. (7.51) and is proportional to $1/E^{1/2}$.

Solutions of the Markovianized spectral balance Eq. (7.59), even for given $\omega_{\mathbf{k}}, \nu_{\mathbf{k}}$, are not known. We now consider the scaling and magnitude of $E(\mathbf{k})$ implied by the local, isotropic approximation to the problem. Anisotropic scaling solutions reflecting more of the drift-wave structure are given by Balk and Zakharov (1990). The review of Horton and Hasegawa (1994) also analyzes the anisotropy and compares with the work of Balk *et al.* (1990).

7.4 Local, Isotropic Approximation

The numerical solutions for $E(\mathbf{k})$ show a definite k_x, k_y structure reflecting the k_x, k_y structure in $\omega_{\mathbf{k}}^l$ and $\gamma_{\mathbf{k}}^l$ shown in Fig. 7.3. At high levels of turbulence, however, the variation of $E_{\mathbf{k}}$ with $\theta = \tan^{-1}(k_y/k_x)$ is not too strong to allow the isotropic approximation of $I(k_x, k_y) \to I(k)$ and $\nu(k_x, k_y) \to \nu(k)$ as a useful first approximation. In addition the **k** spectra are broad Δk with the maximum below $k\rho_s = 1$. Finally, we assume that the integrals over k_1 of the angle-averaged coupling terms in Eq. (7.52) may be approximated by their local values at $k_1 = k$,

$$\sum_{\mathbf{k}_1} \langle F(\mathbf{k}, \mathbf{k}_1) I_{\mathbf{k}_1} \rangle_{\theta_k} \cong \sum_{\mathbf{k}_1} \langle F(\mathbf{k}, \mathbf{k}_1) \rangle_{\theta_k, \theta_{k_1}} I_{k_1} \simeq \langle F \rangle_{k_1 = k} kI(k) \qquad (7.60)$$

with $I(k)$ and $I_{\mathbf{k}}$ related through the equation following (7.47).

In the local, isotropic limit the spectral equation is

$$\frac{dE(k)}{dt} = [2\gamma^l(k) - 2\nu(k)]E(k) + S(k) \qquad (7.61)$$

with

$$\nu(k) = \frac{k^2 |\chi_k| [kI(k)]^{1/2}}{|1 + \chi_k|} \qquad (7.62)$$

and

$$S(k) = 2\frac{k^5 |\chi_k|^2 I(k) \, E_{\text{eq}}}{|1 + \chi_k|^2 \, \nu(k)} \qquad (7.63)$$

with the balance $T_k = 0$ implying $E(k) = E_{\text{eq}}$.

Coherent approximation

The simplest approximation to the steady-state solution of Eq. (7.61) occurs by balancing $\gamma_k^l = \nu_k$ neglecting the incoherent source $S(k)$. The balance occurs at

$$I(k) = I^c(k) = \frac{1}{k^3} \frac{(\chi_k'')^2}{(\chi_k')^2 + (\chi_k'')^2} \qquad (7.64)$$

where we use $\gamma_k^l = -k\chi_k''/|1 + \chi_k|$ in Eq. (7.61) and Eq. (7.47)

$$\nu_k \simeq \frac{k^2 |\chi_k| [kI(k)]^{1/2}}{|1 + \chi_k|} \qquad (7.65)$$

In this approximation the spectrum $I(k)$ vanishes where $\chi_k'' = 0$ and reduces to k^{-3} or

$$e[kI(k)]^{1/2}/T_e = (\rho_s/L_n)/(k\rho_s) = 1/kL_n \qquad (7.66)$$

in regions where $(\chi_k'')^2 \gg (\chi_k')^2$. The mixing length formula (7.66) for $I(k)$ underestimates the magnitude of $I(k)$ due to the neglect of the positive definite $S(k)$ and underestimates when compared with the $I(k)$ obtained in the simulations. The peak of the spectrum is shifted to small k. The shift to low k is relative to the k that maximizes the linear growth, and the shift is described by Eq. (7.64).

Incoherent emission

The incoherent emission $S(k) > 0$ raises the spectrum $I(k)$ above the level of the coherent spectrum $I^c(k)$ in Eq. (7.64) yielding solutions with $\nu_k \gg \gamma_k^l$.

To estimate the solution of $T_\mathbf{k}(\{E\}) = -2\gamma_\mathbf{k}^l E_\mathbf{k}$ including the incoherent emission, we introduce a local energy conserving "collision" model for the transfer operator for the wave transport problem $d_t E_\mathbf{k} = T_\mathbf{k}(\{E\})$ with the constraint $\sum_\mathbf{k} T_\mathbf{k} = 0$ and $E_\mathbf{k}(t) \geq 0$ in strict analogy with the particle transport $d_t f(\mathbf{v}, t) = C_\mathbf{v}(f, f)$ with $\int d\mathbf{v} C_\mathbf{v} = 0$ and $f(\mathbf{v}, t) \geq 0$. In the simplest relaxation-type model the $\mathbf{k}$ dependence of $\nu(\mathbf{k})$ is lost just as the $\mathbf{v}$ dependence of the particle collision frequency from $C_\mathbf{v}(f, f)$ is lost when modelling by $C \to -(f - f^M)/\tau$.

With the relaxation model for the turbulent collision operator the modal transport equation is

$$d_t E_\mathbf{k} = 2\gamma_\mathbf{k}^l E_\mathbf{k} - 2\nu_E (E_\mathbf{k} - E_\text{eq}) \qquad (7.67)$$

with $\nu_E = \nu_1 E^{1/2}$ and $E_\text{eq} = \frac{1}{N} \sum_\mathbf{k} E_\mathbf{k}$. The steady-state solution of the driven transport equation is $E_\mathbf{k} = S_E/2(\nu_E - \gamma_\mathbf{k}^l)$ where the incoherent source is $S_E = 2\nu_E E_\text{eq}$. The self-consistency condition for power balance in the steady state is

$$\sum_{i=1}^{2N} \gamma_\mathbf{k}^\ell E_\mathbf{k} = S_E \sum_\mathbf{k} \frac{\gamma_\mathbf{k}^\ell}{\nu_E - \gamma_\mathbf{k}^\ell} = 0 \qquad (7.68)$$

which determines ν_E, and thus the total energy through $\nu_E = \nu_1 E^{1/2}$. Here ν_1 is a free parameter in the relaxation transport model to be estimated through Eqs. (7.61) and (7.67).

The condition for the existence of a steady-state spectrum from the power balance equation is that $\gamma_T(\mu) = \sum_\mathbf{k} \gamma_\mathbf{k}^l < 0$ which guarantees that Eq. (7.68) has a root $\nu_E > \max(\gamma_\mathbf{k}^l)$. From this model we conclude that the turbulent damping ν_E is a function of both γ_max^l and $\gamma_T(\mu)$, which is a fact observed in the simulations.

A generalized relaxation model with $\nu_E \to \nu_E(k)$ may also be constructed. Adding a drag term $2\nu_2 \partial (E_\mathbf{k} - E_\text{eq})/\partial k$ is also possible to shift the spectrum to lower k and still maintain a positive definite entropy production. The entropy for the relaxation model is $S = -\sum_\mathbf{k} (E_\mathbf{k} - E_\text{eq})^2/2$ and the production rate is $\sigma^{nl} = \frac{dS}{dt} = 2\nu_E \sum_\mathbf{k} (E_\mathbf{k} - E_\text{eq})^2$. In the absence of $\gamma_\mathbf{k}^l$ the entropy production decays $d\sigma^{nl}/dt < 0$ monotonically to zero.

Now we consider the high-k spectral region where the growth mechanism is weak, and thus $\gamma_\mathbf{k}^\ell$ may be neglected compared with the nonlinear coupling transfer rate.

7.4.1 High-wavenumber spectral balance

At high k the linear dispersion relation Eq. (7.15) reduces to $\gamma_k^l = -\gamma_i(k) = -\nu_0 k^4/(1 + k^2) \simeq -\nu_0 k^2$ and the isotropic part of the mode coupling given by $\chi_k' = k^2$ dominates over that due to χ_k''. In this part of the spectrum the problem is similar to the 2D Navier Stokes turbulence. It is useful to introduce the energy flux $\Pi(k)$ and the enstrophy flux $Z(k)$ given by

$$\Pi(k) = \nu_k k E(k) \quad \text{and} \quad Z(k) = \nu_k k^3 E(k). \tag{7.69}$$

The analysis now parallels that of 2D Navier Stokes turbulence given, e.g., by Rose and Sulem (1978). At sufficiently high wavenumber $k > k_2$, the cascade of enstrophy is the dominant process in the mode coupling equations out to the dissipation wavenumber k_d. For a constant enstrophy flux

$$Z(k) = \eta \quad \text{for} \quad k_2 < k < k_d \tag{7.70}$$

where $\nu(k) \gg \gamma^l(k)$ for $k_2 < k < k_d$, the spectrum given by $\nu(k)$ in Eq. (7.62), and Eqs. (7.69) and (7.70) is

$$I^\eta(k) = \frac{\eta^{2/3}}{k^5}, \tag{7.71}$$

and the energy spectrum is

$$E(k) = \frac{\eta^{2/3}(1 + k^2)}{k^5} \simeq \frac{\eta^{2/3}}{k^3} \quad \text{for} \quad k^2 > k_2^2 \gtrsim 1. \tag{7.72}$$

Kraichnan (1971) shows that the k^{-3} spectrum obtained here from the local enstrophy cascade is modified by long-range triplet interactions to

$$k^{-3}/[\ln(k/k_2)]^{\frac{1}{3}}.$$

Even the $\ln k$-corrected spectrum only applies asymptotically for $k_d/k_2 \to \infty$. Low resolution simulations (Kraichnan, 1959; Rose, 1978) with $k_d/k_2 < 100$ show $E(k)$ spectra characterized better by a spectral index -4, or a still steeper exponent, rather than the k^{-3} spectrum of enstrophy cascade. However, more recent higher resolution (4096^2) simulations (Kraichnan, 1985) have essentially confirmed Kraichnan's logarithmically corrected k^{-3} enstrophy-range scaling for certain stirring and damping formulas.

For the constant enstrophy flux spectrum $I^\eta(k)$ the turbulent damping (Kraichnan, 1990) given by Eq. (7.62) and Eq. (7.72) is

$$\nu(k) = \eta^{1/3} \cong \frac{\tilde{v}_E}{l_c} \tag{7.73}$$

where the enstrophy production η is evaluated from the low-k part of the spectrum using $\eta \cong (\tilde{v}_E/l_c)(\tilde{v}_E^2/l_c^2)$ where $\tilde{v}_E^2$ is the mean square $\mathbf{E} \times \mathbf{B}$ velocity in the region $k \leq k_2$.

Finally, at the highest k the turbulent damping in Eq. (7.71) balances the bare viscosity $\gamma_k^l = -\nu_0 k^2$ to determine the upper limit k_d of the $I^\eta(k)$ spectrum

$$k_d = \left(\frac{\tilde{v}_E}{l_c \nu_0}\right)^{1/2}. \tag{7.74}$$

7.4 Local Isotropic Approximation

In the drift wave simulations typical values are $\tilde{v}_E \simeq (1-3)\,[c_s\,\rho_s/L_n]$, $l_c \simeq (1-5)\,[\rho_s]$ and $\nu_0 \gtrsim 0.1\,[\rho_s^2 c_s/L_n]$ yielding $k_d \leq 10[\rho_s^{-1}]$ and $\nu_{\max} \simeq 1[c_s/L_n]$ from Eqs. (7.73) and (7.74). In the laboratory plasma $\nu_0 \lesssim 10^{-4} c_s/L_n$ and $k_d \sim 100 \rho_s^{-1}$ which is at the limit or beyond the capability of present-day computer simulations. Typical tokamak values of $c_s = 300\,\text{km/s}$, $\rho_s/L_n = 10^{-3}\,\text{m}/10^{-1}\,\text{m} = 10^{-2}$ so that $\tilde{v}_E \simeq 3\,\text{km/s}$ and $\ell_c v_E \simeq 3\,\text{m}^2/\text{s}$.

The analysis presented in this section indicates that for determining the fluctuation level $\tilde{n} = \tilde{\varphi}$ and the transport $\langle \tilde{n} \tilde{v}_E \rangle \leq \tilde{n}\tilde{v}_E$, the most important region to resolve in the numerical simulations is the $k < k_2 \sim 1$ region of $\mathbf{k}$ space. Above k_2 the enstrophy cascade spectrum $I^\eta(k) \sim \eta^{2/3} k^{-5}$ contributes little to $\tilde{\varphi}$ and $\tilde{v}_E$.

7.4.2 Quasilinear transport from the renormalized turbulence theory spectrum

The one-field drift-wave equation implies the anomalous electron transport directly from the dissipative operator $\chi''_\mathbf{k}$, giving the out-of-phase part of the density-to-potential response according to

$$\Gamma = \langle n v_x \rangle = -D \frac{dn}{dx} \tag{7.75}$$

with

$$D = -\sum_\mathbf{k} k_y \chi''_\mathbf{k} I(\mathbf{k}) = \delta_0 \, D_{dw} \sum_\mathbf{k} k_y^2 (k^2 - c_0) I(\mathbf{k}) \tag{7.76}$$

where the units of D are

$$D_{dw} = \frac{\rho_s}{L_n} \frac{cT_e}{eB}. \tag{7.77}$$

The drift wave diffusivity in Eq. (7.77) is often called gyro-Bohm diffusion.

Using the approximate spectra in formulas (7.64) and (7.71) joined continuously at $k = k_2$ we find that

$$D = \int_{k_1}^{k_2} \frac{dk \gamma_k^e}{k^3} \frac{(\chi''_k)^2}{(\chi'_k)^2 + (\chi''_k)^2} + \int_{k_2}^{k_d} \frac{\gamma_k^e \eta^{2/3} dk}{k^5} \tag{7.78}$$

where $\gamma_k^e = -k\chi''(k)$ and $\gamma^e(k_1) = 0$. The transitional wavenumber k_2 is near unity and the second integral is negligible compared with the first. Since $\chi'_k = k^2$ and $\chi''_k = \delta_0 k_y (c_0 - k^2)$ the dominant low-k part of the first integral has $(\chi'_k)^2 \lesssim (\chi''_k)^2$ and formula (7.78) reduces to

$$D \simeq \int_{k_1}^{k_2} \frac{dk}{k^3} \gamma^e(k) \sim \frac{\langle \gamma_k^e \rangle}{\langle k^2 \rangle} \tag{7.79}$$

showing how the local isotropic model spectral formulas relates in average to the important empirical transport estimate of $\gamma_k^e/k_\perp^2$. The theory given here shifts the spectrum from the k that maximizes γ_k^e to that near marginal stability k_1. The approximation (7.79) is a lower bound to D since it uses the coherent part $I^c(k)$ which is lower than the full $I(k)$ that includes the incoherent emission. The integral also neglects the k_2 to k_d contribution, but this contribution is subdominant.

In the regime where $(\chi_k')^2 \ll (\chi_k'')^2$ the rms fluctuation level according to the spectrum in Eq. (7.64) is

$$\frac{e\widetilde{\varphi}}{T_e} \cong \frac{\rho_s}{L_n}\left(\int_{k_1}^{k_2} \frac{dk}{k^3}\frac{(\chi_k'')^2}{(\chi_k')^2+(\chi_k'')^2}\right)^{1/2} \leq \left(\frac{\rho_s}{L_n}\right)\frac{1}{\sqrt{2}\,k_1}. \tag{7.80}$$

The level in (7.80) corresponds to the mixing length level estimate with $\langle v_E^2\rangle = v_d^2$ evaluated with $\langle k_x^2\rangle = \langle k^2\rangle = k_1^2$. Again Eq. (7.80) is a lower bound for the same reasons as for Eq. (7.79) and shows that it is the long wavelength part of the spectrum that determines the rms fluctuation level.

Application of the quasilinear theory to the power balance problem in tokamaks is given in Horton and Estes (1979) and Bravenec et al. (1992).

7.4.3 Relation to the Hasegawa-Wakatani drift-wave equation

Collisional drift waves described by the two-component fluid equations are given by a cubic dispersion relation (Hinton and Horton, 1971; Scott, 1992) derived from the hydrodynamic coupling of the $e\varphi/T_e$, $\delta n/n$ and $\delta T_e/T_e$ fluctuations. For high temperatures the parallel electron thermal conductivity $\chi_{\|e}$ is large and the temperature fluctuation

$$\frac{\delta T_{ek}}{T_e} \simeq \frac{1}{i(k_\|^2\chi_{\|e})}\left[\omega_*\left(1-\frac{3}{2}\eta_e\right)-\omega\right]\left(\frac{e\varphi_k}{T_e}\right) \tag{7.81}$$

becomes small compared to the density and potential fluctuations since $k_\|^2\chi_{\|e} \gg |\omega| \sim |\omega_*|$. The drift wave system reduces to a model called the Hasegawa-Wakatani system to coupled partial differential equations for $e\varphi/T_e$ and $\delta n/n$. With $\delta T_e = 0$ the parallel electron momentum balance reduces to

$$\eta j_\| = \frac{-m_e\nu_e v_{\|e}}{e} = -\nabla_\|\varphi + \left(\frac{T_e}{e}\right)\nabla_\| \ln n \tag{7.82}$$

and the divergence of the parallel current to

$$\frac{\nabla_\| j_\|}{en} = -D_\|\partial_\|^2\left(\frac{e\varphi}{T_e} - \frac{\delta n}{n}\right) \tag{7.83}$$

where

$$D_\| = \frac{T_e}{m_e\nu_e} = \frac{v_e^2}{\nu_e}. \tag{7.84}$$

7.4 Local Isotropic Approximation

Within this isothermal model a dynamical equation for the nonadiabatic part $\delta n^{ah}(\mathbf{x}, t)$ of the electron density fluctuation

$$\frac{\delta n(\mathbf{x}, t)}{n} = \frac{e\varphi(\mathbf{x}, t)}{T_e} + \delta n^{ah}(\mathbf{x}, t) \tag{7.85}$$

follows from the electron continuity equation

$$\partial_t \delta n^{ah} + \left(\frac{cT_e}{eB}\right)[\varphi, \delta n^{ah}] - D_\| \partial_\|^2 \delta n^{ah} = -(\partial_t \varphi + v_d \partial_y \varphi). \tag{7.86}$$

Using Eq. (7.83) for $\nabla_\| j_\|$ and the nonlinear polarization drift for $\nabla_\perp \cdot \mathbf{j}_\perp$, the quasineutrality condition $\nabla \cdot \mathbf{j} = 0$ becomes

$$(1 - \rho_s^2 \nabla_\perp^2)\partial_t \varphi + \left(\frac{cT_e}{eB}\right)[\varphi, \delta n^{ah} - \rho_s^2 \nabla^2 \varphi] + v_d \partial_y \varphi = -\partial_t \delta n^{ah}. \tag{7.87}$$

The one-field drift wave model in Eq. (7.12) follows from Eq. (7.87) by taking $\delta n^{ah} = \widehat{\mathcal{L}}_{ah}\varphi$.

The two-field model (7.86)–(7.87) has the quadratic $(A\omega^2 + B\omega + C = 0)$ dispersion relation given by

$$\omega\left(1 + k_\perp^2 \rho_s^2 - \frac{i\omega k_\perp^2 \rho_s^2}{\nu_\|}\right) - k_y v_d = 0 \tag{7.88}$$

for $\gamma_i = 0$. The dynamics describes the unstable drift wave mode

$$\omega_+(\mathbf{k}) \cong \frac{k_y v_d}{(1 + k_\perp^2 \rho_s^2 - ik_\perp^2 \rho_s^2 k_y v_d/\nu_\|)} \tag{7.89}$$

and the damped diffusion mode

$$\omega_-(\mathbf{k}) \cong -\frac{i\nu_\|(1 + k_\perp^2 \rho_s^2)}{k_\perp^2 \rho_s^2}. \tag{7.90}$$

Here we are restricted to $\eta_e = 0$ due to the $\delta T_e = 0$ approximation. Energy balance and particle diffusion are given by

$$\frac{d}{dt}\frac{1}{2}\left\langle \varphi^2 + (\nabla_\perp \varphi)^2 \right\rangle = \frac{d}{dt}\sum_\mathbf{k} E_\mathbf{k} = -\left\langle \varphi \partial_t \delta n^{ah}\right\rangle = i\sum_{k=k_\omega} \omega \varphi_k^* \delta n_k^{ah} \tag{7.91}$$

and

$$\frac{\partial}{\partial t}\langle n \rangle = \frac{\partial}{\partial x}\left\langle \frac{\partial \varphi}{\partial y}\delta n^{ah}\right\rangle = -\frac{\partial}{\partial x}\sum_{k=k_\omega} k_y \varphi_k^* \delta n_k^{ah}. \tag{7.92}$$

Numerous works consider the turbulence and transport generated by the two-field collisional drift-wave model given by Eqs. (7.86)–(7.87).

Wakatani and Hasegawa (1984) report turbulence studies for the collisional regime with strong density gradients $\kappa = \rho_s/L_n \geq 1/20$. The high-collisionality condition $\nu_\| \cong 4(k_\perp \rho_s)^2 \omega_*$ is used to determine $k_\|$ as a function of $\mathbf{k}_\perp$. For this regime $\gamma_\mathbf{k}^\ell \sim \omega_\mathbf{k}^\ell$. They find broad-band turbulence with $\Delta \omega \sim \omega_*$ at fixed $\mathbf{k}$ and the frequency integrated energy spectrum $E_\mathbf{k} \simeq k_\perp^{-3}$. In both models $\langle k_\|^2 \rangle^{1/2}$ is fixed through the constant parameters. In Sec. 7.4.2 we obtain the scaling $e\widetilde{\varphi}/T_e \sim \delta \tilde{n}/n = \kappa f(\delta_0, \eta_e, \nu_0)$ where as Wakatani and Hasegawa find $e\widetilde{\varphi}/T_e \sim \delta \tilde{n}/n \sim \kappa^{1/2}$. Because of their higher fluctuation levels they obtain Bohm-like diffusion in contrast to the drift wave type diffusion given in Eq. (7.76).

Terry and Diamond (1985) apply clump theory to the electron density Eq. (7.86) to calculate the spectrum of fluctuations. To reduce the mode coupling problem they restrict the fluctuations to those given by the ballooning mode representation after introducing the toroidal drift frequency ω_{De}. They report broad turbulence with $|\varphi_{k_y}|^2 \simeq k_y^{-17/6}$ and $|\varphi_{k_y}(\omega)|^2 \simeq \omega^{-2}$ for $\omega > \Delta \omega_k \sim \omega_*$. The diffusion is $D \propto \nu_\|^{2/3}$ for low turbulence levels and $D_{dw} = \rho_s^2 c_s/L_n$ for high-turbulence levels where $e\widetilde{\varphi}/T_e \simeq \delta \tilde{n}/n \simeq 7\rho_s/L_n$.

To understand the relation between Hasegawa-Wakatani model and the nonlinear drift wave model in Eq. (7.12) it is useful to consider the renormalized turbulence theory approximation to Eq. (7.86)

$$\delta n_{\mathbf{k}\omega}^{ah} = g_\mathbf{k}^e(\omega)(\omega - k_y v_d)\varphi_{\mathbf{k}\omega} \tag{7.93}$$

where $g_\mathbf{k}^e(\omega) = (\omega + i\nu_{\perp k} + i\nu_\|)^{-1}$ with the $\nu_\| = k_\|^2 D_\|$ and $\nu_{\perp k} \simeq k_\perp^2 D_\perp$. Keeping the ω dependence of $\delta n_{\mathbf{k}\omega}^{ah}$ in Eq. (7.93) gives the second damped mode $\omega_-(\mathbf{k})$ of the two-field equations. The numerical solutions of Wakatani and Hasegawa (1984) clearly retain the ω dependence and nonlinearity of $\delta n_{\mathbf{k}\omega}^{ah}$. In a single-field model the effect of $\delta n_{\mathbf{k}\omega}^{ah}$ is reduced by noting that collisional drift modes satisfy $k_\|^2 v_e^2 > \nu_e|\omega + i\nu_{\perp k}|$ so that $g_{\mathbf{k}\omega}^e \simeq -i\nu_\|^{-1}$ and Eq. (7.93) reduces to $\delta n_{\mathbf{k}\omega}^{ah} \simeq -i(\omega - k_y v_d)\varphi_{\mathbf{k}\omega}/\nu_\|$. Retaining the ω dependence in only the numerator of $\delta n_{\mathbf{k}\omega}^{ah}$ is inconsistent, however, since it leads to a quadratic dispersion relation in which the $\omega_-(\mathbf{k})$ mode is unstable for some $\mathbf{k}$ values. Thus the consistent approximation of $\delta n_{\mathbf{k}\omega}^{ah}$ requires $\delta n_{\mathbf{k}\omega}^{ah} = (-i/\nu_\|)(\omega_\mathbf{k}^\ell - k_y v_d)\varphi_\mathbf{k} = \mathcal{L}_\mathbf{k}\varphi_{\mathbf{k}\omega}$ giving the one-field model of Sec. 7.2 for dissipative drift-wave turbulence.

7.5 Low-Order Wave Coupling

7.5.1 Three-wave coupling and strange attractors

From the mode coupling in Eq. (7.15) we see that the drift wave turbulence may be viewed as resulting from the interaction of a large number of triplet or three-wave interactions. It is instructive to consider in some detail the dynamics of a single three-wave interaction.

7.5 Low-Order Wave Coupling

The dynamics of three interacting drift waves, one of which is linearly unstable, shows that the saturated state is typically chaotic in time with a broad-band frequency spectrum. The long-time trajectory of the $d = 4$ system is on a strange attractor in the phase space. The tendency of the strange attractor to make the plasma oscillations stochastic appears to be the explanation for the broad-band frequency spectrum measured by scattering microwaves from the electron density fluctuations and by probes in the plasma confinement experiments.

The collisional drift wave, collisionless or plateau regime drift wave, and the dissipative trapped electron drift wave instabilities are given by different choices of the anti-Hermitian operator $\mathcal{L}^{ah}$ in Eq. (7.9). The procedure is described in Sec. 7.2.1.

The collisionless or plateau regime electron drift wave regime has

$$\mathcal{L}^{ah}(\mathbf{k}) = i\delta_0 k_y \left(k_\perp^2 - \tfrac{1}{2}\eta_e\right) \tag{7.94}$$

with

$$\delta_0 = \left(\frac{\pi}{2}\right)^{1/2} \left(\frac{m_e}{m_i}\right)^{1/2} \left(\frac{L_c}{L_n}\right)$$

a constant of order unity. Other regimes for the linear electron dissipation yield similar conclusions. Here L_c is the parallel correlation length for the fluctuations. In a tokamak $L_c \simeq qR$.

In the Fourier representation

$$\varphi(\mathbf{x},t) = \sum_{\mathbf{k}} \varphi_{\mathbf{k}}(t)\exp(i\mathbf{k}\cdot\mathbf{x}) \tag{7.95}$$

the nonlinearity is given by the sum over interactions of modes $\mathbf{k}_1$ and $\mathbf{k}_2$ such that

$$\mathbf{k}_1 + \mathbf{k}_2 = \mathbf{k} = -\mathbf{k}_3. \tag{7.96}$$

The lowest order set of self-consistent nonlinear interactions from mode coupling occurs for three modes $\varphi_i(t) = \varphi_{\mathbf{k}_i}(t)$ with $i = 1, 2, 3$. Introducing the amplitude $a_j(t)$ and phase $\zeta_j(t)$ according to

$$(1 + k_j^2)^{1/2}\varphi_j(t) = a_j(t)\exp[i\zeta_j(t)], \tag{7.97}$$

the three complex equations for $\varphi_j(t)$ reduce to four real equations for the amplitudes and the total phase

$$\zeta(t) = \zeta_1(t) + \zeta_2(t) + \zeta_3(t) \tag{7.98}$$

$$\frac{da_j}{dt} = \gamma_j a_j - A a_k a_\ell (F_j \cos\zeta + G_j \sin\zeta) \tag{7.99}$$

$$\frac{d\zeta}{dt} = -\Delta\omega + A \sum_{j,k,\ell} \frac{a_k a_\ell}{a_j}(F_j \sin\zeta - G_j \cos\zeta) \tag{7.100}$$

where j, k, ℓ are cyclic permutations of 1,2,3. In Eqs. (7.99) and (7.100) we have

$$\Delta\omega = \omega_1 + \omega_2 + \omega_3 \qquad (7.101)$$

and γ_j is the linear growth of the $\mathbf{k}_j$ mode. The linear frequency and growth rate are given by

$$\omega_j + i\gamma_j = k_{yj}v_d/[1 + k_j^2 + \mathcal{L}^{ah}(\mathbf{k}_j)]. \qquad (7.102)$$

For $\delta_0 < 1$ the linear frequency is

$$\omega_j \simeq \frac{k_{yj}v_d}{(1 + k_j^2)} \qquad (7.103)$$

and the linear growth rate is

$$\gamma_j \simeq \frac{\delta_0 k_{yj}^2 v_d^2 (k_j^2 - \eta_e/2)}{(1 + k_j^2)^2}. \qquad (7.104)$$

The coupling strength A is proportional to the area of the triangle formed by the $\mathbf{k}$ vectors of the interacting modes

$$A = \frac{(\mathbf{k}_1 \times \mathbf{k}_2 \cdot \hat{\mathbf{z}})}{2[(1 + k_1^2)(1 + k_2^2)(1 + k_3^2)]^{1/2}}. \qquad (7.105)$$

The coupling factors F_j and G_j are the real and imaginary parts of the symmetrized complex susceptibility arising from the $\mathbf{E} \times \mathbf{B}$ convection,

$$F_j - iG_j = \chi(\mathbf{k}_k) - \chi(\mathbf{k}_\ell) \qquad (7.106)$$

where

$$\chi(\mathbf{k}) = \mathbf{k}_\perp^2 - \mathcal{L}^{ah}(\mathbf{k}) = k_\perp^2 - i\delta_0 k_y(k_\perp^2 - \eta_e/2). \qquad (7.107)$$

The complex susceptibilities $\chi(\mathbf{k})$ in the three-mode equations are generic to the drift-wave problem and is an important generalization of earlier works (Vyshkind and Rabinovich, 1976; Wersinger et al., 1980) where the three-wave coupling coefficients are real.

These results for the formulation of the triplet interactions of drift waves are shown by the diagrams and equations in Fig. 7.8. Also shown is the generalization to the pressure gradient driven modes.

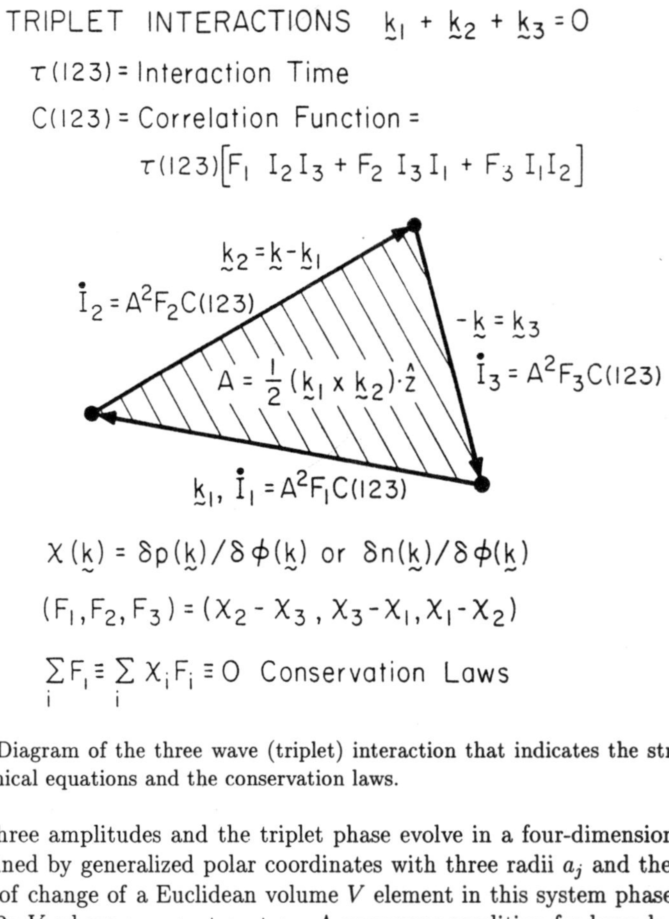

7.5 Low-Order Wave Coupling

TRIPLET INTERACTIONS $\underline{k}_1 + \underline{k}_2 + \underline{k}_3 = 0$

$\tau(123)$ = Interaction Time

$C(123)$ = Correlation Function =
$$\tau(123)\left[F_1\, I_2 I_3 + F_2\, I_3 I_1 + F_3\, I_1 I_2\right]$$

$\underline{k}_2 = \underline{k} - \underline{k}_1$

$\dot{I}_2 = A^2 F_2 C(123)$

$-\underline{k} = \underline{k}_3$

$A = \frac{1}{2}(\underline{k}_1 \times \underline{k}_2)\cdot\hat{z}$

$\dot{I}_3 = A^2 F_3 C(123)$

$\underline{k}_1,\ \dot{I}_1 = A^2 F_1 C(123)$

$X(\underline{k}) = \delta\rho(\underline{k})/\delta\phi(\underline{k})$ or $\delta n(\underline{k})/\delta\phi(\underline{k})$

$(F_1, F_2, F_3) = (X_2 - X_3,\ X_3 - X_1,\ X_1 - X_2)$

$\sum_i F_i \equiv \sum_i X_i F_i \equiv 0$ Conservation Laws

Fig. 7.8. Diagram of the three wave (triplet) interaction that indicates the structure of the dynamical equations and the conservation laws.

The three amplitudes and the triplet phase evolve in a four-dimensional phase space defined by generalized polar coordinates with three radii a_j and the angle ζ. The rate of change of a Euclidean volume V element in this system phase space is $dV/dt = 2\gamma_t V$ where $\gamma_t = \gamma_1 + \gamma_2 + \gamma_3$. A necessary condition for bounded, stable, asymptotic behavior is that $\gamma_t < 0$, giving a volume contracting flow in the phase space.

7.5.2 Integrable limit of the three-wave equations

In the limit of no dissipation ($\mathcal{L}^{ah} \to 0$) the three system vectors $\boldsymbol{\gamma}$ and $\mathbf{G}$ vanish simultaneously and the system (7.99) and (7.100) is integrable. To show integrability we introduce the action-angle variables $J_j = a_j^2/F_j$ and ζ_j, and the three-wave Hamiltonian

$$H(J,\zeta) = -\sum_{i=1,2,3} \omega_i J_i + A(F_1 F_2 F_3)^{1/2}(J_1 J_2 J_3)^{1/2} \sin(\zeta_1 + \zeta_2 + \zeta_3). \qquad (7.108)$$

Hamiltonians of the form Eq. (7.108) that depend on only one combination of the phase variables can easily be reduced to a one degree-of-freedom integrable Hamiltonian $\overline{H}(I, \theta)$. To reduce the system (7.108) we introduce the generating function $F(\zeta, I)$, for which $J = \partial F/\partial \zeta$ and $\theta = \partial F/\partial I$, where

$$F = \sum_{ij} M_{ij}\zeta_i I_j = \zeta^T M I. \qquad (7.109)$$

A simple choice is

$$M = \begin{pmatrix} 1 & 0 & 0 \\ 1 & 1 & 0 \\ 1 & 0 & 1 \end{pmatrix} \qquad (7.110)$$

which gives $\theta = \zeta^T M = (\zeta_1 + \zeta_2 + \zeta_3, \zeta_2, \zeta_3)$ and $J = MI = (I_1, I_1 + I_2, I_1 + I_3)$. Solving for M^{-1} or I gives $I_1 = J_1$, $I_2 = J_2 - J_1 = -m_{12}$ and $I_3 = J_3 - J_1 = -m_{13}$. Since I_2 and I_3 are constants of the motion, the Hamiltonian in (7.108) can be reduced to (within a constant)

$$\overline{H}(I_1, \theta_1) = -(\omega_1 + \omega_2 + \omega_3)I_1 + A(F_1 F_2 F_3)^{1/2}[I_1(I_1 - m_{12})(I_1 - m_{13})]^{1/2}\sin(\theta_1). \qquad (7.111)$$

This one degree-of-freedom Hamiltonian is integrable and the explicit solutions are obtained in terms of the Jacobi-elliptic functions summarized in Chapter 1.

The integrals may be used to express the wave energy W in terms of the actions J_i as

$$W = \frac{1}{2}\sum_{\mathbf{k}}(1 + k_\perp^2)|\varphi_{\mathbf{k}}|^2 = \frac{1}{2}\sum_j a_j^2 = \frac{1}{2}F_2(J_2 - J_1) + \frac{1}{2}F_3(J_3 - J_1) \qquad (7.112)$$

and the potential enstrophy

$$U = \frac{1}{2}\sum_{\mathbf{k}} k^2(1 + k_\perp^2)|\varphi_{\mathbf{k}}|^2 = \frac{1}{2}F_2 k_2^2(J_1 - J_2) - \frac{1}{2}F_3 k_3^2(J_1 - J_3) \qquad (7.113)$$

in terms of the actions. The plasma wave energy density W is the sum of the electron electrostatic energy density $W_e = \frac{1}{2}e\delta n_e \varphi$ in the electron distribution and the ion kinetic energy $W_i = \frac{1}{2}n_i m_i V_E^2$. The enstrophy U is related to the plasma wave-momentum density P in Eq. (7.26). From the integrals W, U and H (the interaction energy), the motion is one-dimensional and thus integrable.

7.5.3 Dissipative dynamics and the strange attractor

When dissipation is retained in the system the three vectors $\boldsymbol{\gamma}, \mathbf{F}, \mathbf{G} \neq 0$, and the dynamics is nonintegrable. For $\mathbf{G} \neq 0$ the enstrophy U is no longer conserved by the nonlinear coupling. Many numerical integrations of the dissipative equation have been performed from which we conclude that there are no isolating integrals (except possibly for special values of $\boldsymbol{\gamma}, \mathbf{F}, \mathbf{G}$). Generally, the system is nonintegrable and

7.5 Low-Order Wave Coupling

chaotic. The behavior of this system is analyzed in terms of the stability of the fixed points and limit cycles. This stability analysis delineates regions of constant, periodic, aperiodic and stochastic behavior. The fixed points of the system are the roots of $\dot{a}_j = \dot{\zeta} = 0$ obtained from Eqs. (7.99) and (7.100).

The stability $a_i(t) \sim a_0 \exp(\lambda t)$ of a fixed point $(\mathbf{a}_0, \zeta_0)$ is determined by the eigenvalues of the secular equation $\sum c_n \lambda^n = 0$ from the 4×4 matrix of the linearized equations of motion.

There are numerous ways in which the parameters may be varied to make the system pass successively from the region of a stable fixed point to a region of chaotic behavior. We describe two methods and comment on their relevance to drift waves. In one, this succession is obtained by taking $|\mathbf{F}| \ll |\mathbf{G}|$ and increasing the magnitude of the damping of the stable modes relative to the growth rate of the unstable mode (rotation of $\mathbf{G}$). In the second method, this succession occurs when the values of the growth and damping rates remain comparable in magnitude, but $|\mathbf{F}|$ is increased relative to $|\mathbf{G}|$. In this case stochastic behavior occurs for $|\mathbf{F}| \gtrsim \frac{1}{3}|\mathbf{G}|$. The first process is essentially the method applied in the studies of Wersinger et al. (1980), Wang and Masui (1981). The critical ratio Γ_c of the damping-to-growth for the onset of chaotic behavior is rather large ($\Gamma_c \sim 20$) and can occur for drift wave triplets such that $|\mathbf{k}| \ll |\mathbf{k}_1| \sim |\mathbf{k}_2|$ where the large $|k_1|, |k_2|$ modes are heavily damped due to ion-ion collisional viscosity. In contrast for triplets with comparable values of $|\mathbf{k}_i|$ the functional forms of $\gamma(\mathbf{k}_j)$ give comparable growth and damping rates $|\gamma_1| \sim |\gamma_2| \sim |\gamma_3|$. These constraints make growth rates with comparable magnitudes typical while the functional form of $\mathbf{F}$ and $\mathbf{G}$ makes $|\mathbf{F}| \sim |\mathbf{G}|$. Thus it appears that it is the complex susceptibilities which are instrumental in making the three-drift wave interaction chaotic for typical triplets in contrast to the large Γ (damping-to-growth-rate ratio) in neutral fluid models.

The time evolution of the amplitudes and phase in the stochastic regime is shown in Fig. 7.9 where the amplitudes $a_j(t)$ and cosine of the phase $\cos \zeta(t)$ are obtained by integrating Eqs. (7.99)–(7.100) for parameters typical of drift wave simulation. The amplitude and the phase appear random with the phase covering the entire range $(0, 2\pi)$ rather uniformly. After an initial transient there is saturation of the amplitudes and the total wave energy is constant in a time-average sense. The demonstration that the motion is stochastic is given by numerical evaluation of the maximal Lyapunov characteristic exponent and by the decay of the autocorrelation of $\varphi_j(t) = a_j \cos \zeta_j$. The Lyapunov exponent is positive, of order a few tenths, and the autocorrelation functions decay rapidly confirming the two important features of stochastic signals.

Recent developments in a truncated description of the mode coupling include the decimated amplitude scheme (DAS) of Kraichnan (1985). Williams et al. (1987) apply the DAS method to the Betchov model $\dot{x}_i = c_{ijk} x_j x_k$ with Gaussian random values of $\{c_{ijk}\}$ the coupling coefficients subject to the energy conservation constraint $c_{ijk} + c_{jki} + c_{kij} = 0$. They compare the 32 mode truncated solution to the full 96 mode system and to the DIA solution showing good agreement for the decay of the two-time correlation function computed from full mode coupling compared with the DIA solution.

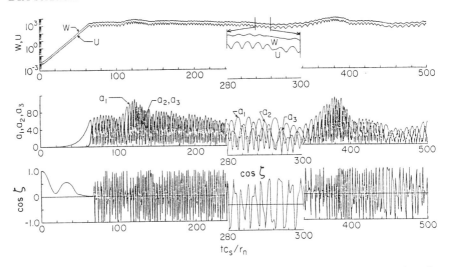

Fig. 7.9. Growth and saturation into a chaotic time series from the triplet interactions for dissipative drift waves. The time slice from $t = 280$ to $300 [L_n/c_s]$ is blown up to show the detailed oscillations of $a_i(t)$ and the relative phase $\alpha(t)$.

7.5.4 Random phase approximation

In view of the stochasticity in the three-wave interactions, we consider two important aspects of wave turbulence theory: the applicability of the random phase approximation and the frequency spectrum of the fluctuation for a fixed wavenumber **k**. Because of the random nature of the phase evolution in the three-wave interactions, the random phase approximation (RPA) appears justified. In the RPA a set of reduced equations of motion are derived for the long-time evolution of the wave spectral intensities $I_j(t)$. We compare the predictions of the RPA equations with solutions of the mode coupling equations that keep the phase $\zeta(t)$. We compare the RPA amplitudes with the exact amplitudes averaged over phases.

The RPA equations assume the separation of time scales of the random fluctuations that evolve on the time scale τ_c, the short correlation timescale seen in Fig. 7.9, from the long-time $T = |I_j/dI_j/dt|$ scale for the evolution of the average amplitudes

$$I_j(t) = \left\langle a_j^2 \right\rangle = (1 + k_j^2) \left\langle |\varphi_j|^2 \right\rangle. \tag{7.114}$$

7.5 Low-Order Wave Coupling

The random phase equations are

$$\frac{dI_j}{dt} = 2\pi\tau_{123}A^2\left\{(F_j^2 + G_j^2)I_kI_\ell + (F_jF_k + G_jG_k)I_\ell I_j + (F_jF_\ell + G_jG_\ell)I_kI_\ell\right\} + 2\gamma_j I_j \tag{7.115}$$

$$\frac{d\theta_{123}}{dt} + (\nu + i\Delta\omega)\theta_{123} = 1, \tag{7.116}$$

where $\tau_{123} = \text{Re}\,\theta_{123}$ is the maximum interaction time and $\nu = \nu_1 + \nu_2 + \nu_3$. The interaction time is, by definition, the correlation time for this 123 triplet and has the limiting value $\nu/(\Delta\omega^2 + \nu^2)$ (as $t \to \infty$). The undriven ($2\gamma_j = 0$) equilibrium or fixed points of Eq. (7.115) are $I_j = I_0$ for $\mathbf{G} \neq 0$ and $I_j = I_0/(1 + \beta k_j^2)$ for $\mathbf{G} = 0$, where I_0 and β are arbitrary constants, describing the total energy and enstrophy of the statistical equilibrium.

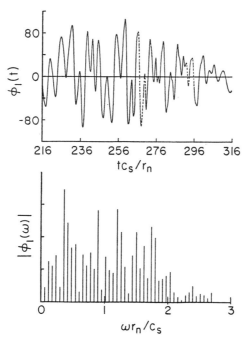

Fig. 7.10. A segment of the time series of the potential ϕ_1 of the first mode in a drift wave triplet and the associated frequency spectrum.

The kinematics and dynamics for a phase-averaged interaction triplet $\{I_j(t)\}$ are shown in Fig. 7.11.

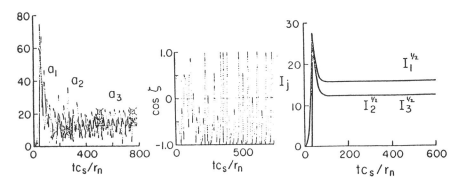

Fig. 7.11. The evolution of the intensities $I_j(t)$ obtained from the time averaged-squared amplitudes $a_j^2(t)/2$ in a triplet interaction.

Terry and Horton (1982) have compared the numerical solutions of the exact Eqs. (7.99) and (7.100) for typical drift wave parameters with numerical solution of the random phase Eq. (7.115). The comparisons show that the random phase equations faithfully reproduce the time-average behavior of the exact equations. The RPA equations give correctly the relative values of the mean amplitude at saturation and the absolute values of these amplitudes when the interaction time τ_{123} is computed accurately. Thus, the intrinsic stochasticity present in the interaction of the drift waves is a sufficient condition to justify the use of the random phase approximation.

The calculation of τ_{123} is based on the renormalized turbulence theory yielding Eq. (7.46) in the Markovian limit. For an example similar to that given here, Krommes (1982) solves the full DIA equations for τ_{123} and shows some agreement between the statistical theory and the exact numerical solutions averaged over an ensemble of Gaussian initial data [although in general closures do not perform well for low degree-of-freedom systems — e.g., see Bowman et al. (1993)]. In the full DIA τ_{123} is replaced by a non-Markovian time history integral over the propagator $g_{123}(t)$ that itself depends on the spectra $\{I_j(t)\}$. The studies indicate that replacing the time history integral with its time asymptotic value of τ_{123} gives a reliable Markovianized kinetic equation for $\{I_j(t)\}$.

The frequency spectrum for the complex amplitude components $\varphi_j(t)$ of the electric potential are shifted from the linear frequency $\omega_j = \omega(\mathbf{k}_j)$ and have Lorentzian widths of the frequency spectrum $\nu(\mathbf{k}_j)$ of the same order as the maximum frequency.

Thus we see that a broad-band stochastic oscillation of both the field amplitudes $a_j(t)$ and phases $\alpha_j(t)$ occurs from the interaction of three electron drift waves for representative values of the system parameters. In the stochastic regime the random phase approximation is found to predict the average behavior of the wave

7.5 Low-Order Wave Coupling

amplitudes. The stochastic wave interactions are sufficient to explain the broad frequency spectra observed in plasma confinement experiments.

Extensions of the three-drift wave interaction problem are made in the studies of Biskamp and He (1985) and He and Salat (1988, 1989). Biskamp and He (1985) generalize the problem to include the parallel ion acoustic oscillations as given in Sec. 6.4 and make a systematic search of the routes to chaotic dynamics in the parameter space of the problem. They find that the system goes chaotic in the manner called the Ruelle-Takens route to chaos.

As the strength of the oscillations are increased by varying a system parameter, the time behavior goes from a fixed amplitude and phase (stable fixed point) to a periodic oscillation (1D torus) to a two-frequency quasiperiodic oscillatory (2D torus) solution and then to a chaotic solution. Typically there is no regular three-frequency solution. This transition from the 2D torus to chaos is the generic route to chaos described by Ruelle and Takens (1971). This route to chaos has been observed in a number of hydrodynamic experiments as reviewed by Swinney (1983).

The classical or Landau picture for the transition to chaos is that the solutions remain quasiperiodic with a large number ($N \geq 3$) of incommensurate frequencies and their harmonics. The work of Swinney and his collaborators and others shows that the route to chaos and the transition to turbulence occur directly from the quasiperiodic 2D torus and not through the Landau picture.

In conclusion we see that the drift wave system has a wide range of dynamical behavior depending on the energy level of the wave-vortex components and the details of the dissipation (growth and damping) in the system. For the system with growth and damping characteristic of that in the tokamak plasmas the present understanding is that the drift turbulence in the interior confinement zone is in a regime of turbulence with low fluctuation levels (at or below the mixing length level) and sufficiently Gaussian-like statistics that the approach of renormalized turbulence theory with power law wavenumber spectra is the relevant theoretical model. Such anisotropic, power-law wave spectra are reviewed by Horton and Hasegawa (1994) for drift wave turbulence. Here we show one last example for this regime.

While in general large **k**-space representations, such as $(512)^2$ or $(1024)^2$, are required to faithfully represent broadband turbulent spectra, it is informative to look at a low-order **k**-space representation. In Fig. 7.12 we show a global view of a $(15)^2 - \mathbf{k}$ space representation of the drift wave turbulence as a low-order model for how the turbulence might occur in the transport zone in tokamaks. The four inserts bordering the wavenumber spectrum show fairly long time samples for the dynamics of certain **k**-modes of interest. The comparison with the Lorentzian model of the frequency power spectrum with the nonlinear frequencies and line widths are shown as well as the near Gaussianity of the pdf for the single space point plasma potential. We conclude that for this $N = 224$ degree-of-freedom model the renormalized turbulence theory yields a qualitatively correct description of the dynamics computed from the mode coupling equations. Furthermore, the qualitative picture of the turbulence appears consistent with that inferred from electromagnetic scattering experiments on the electron density fluctuations. Figure 7.13 is an example of a comparison from drift wave simulations with the electron density fluctuations reported by Mazzucato (1978) for the Princeton Large Torus (PLT) device. This type of comparison between drift wave turbulence/ion-temperature-gradient turbulence has been extensively developed over the past ten years. For example, Waltz (1983, 1985, 1986, 1988, 1990) has developed such comparisons in convincing detail. In this regime at least the basic transport mechanism seems to be established.

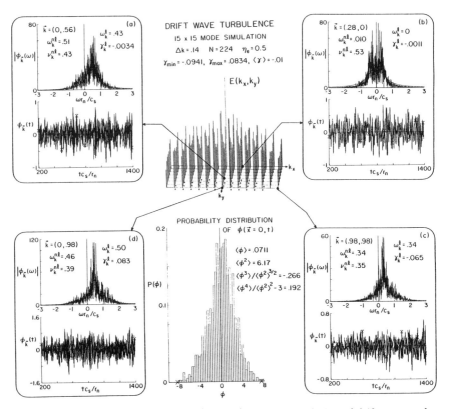

Fig. 7.12. An example of a low order (15 × 15) k space simulation of drift wave turbulence. The figure is intended to show clearly the characteristic features of the turbulent system: much larger grids with smaller Δk are required to simulate the tokamak plasma. Note the broad frequency spectrum developed for the linearly damped modes in frames (b) and (c).

7.5 Low-Order Wave Coupling

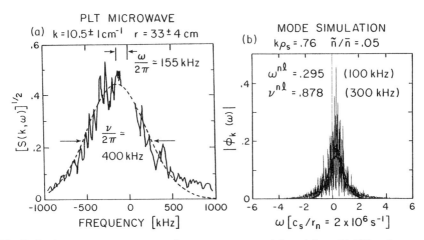

Fig. 7.13. An attempt to compare the power spectrum inferred from 10 GHz microwave scattering off the low-frequency electron density fluctuations in PLT with a drift wave simulation of Eq. (7.12).

Beyond a few general statements comparison of the turbulence models with the tokamak experiments is beyond the scope of this text. Basically there appear to be two qualitatively different regimes of turbulence in the confinement devices. In the core there is the low level, broadband fluctuations thought to be consistent with the renormalized turbulence theory. In the edge plasma and in the scrape-off-layer (SOL) plasma the fluctuations have large amplitudes, often appreciably higher kurtosis than the Gaussian kurtosis, particulary in the density fluctuations, and an intermittent transport flux. Here, a large fraction of the radial transport may well be characterized as due to "transport events" associated with coherent structures (Endler et al., 1995). While there is no general agreement on the details of this strongly nonlinear regime, the qualitative picture of coherent structures embedded in a turbulent wave background would appear most natural. Clearly, the area of research involving the role of coherent structures in turbulent transport will be an important topic in future studies of anomalous transport in plasmas. Higher resolution diagnostics coupled with more advanced theoretical models of plasma turbulence with coherent structures (Benkadda et al., 1994) provides a basis for new and exciting areas of research in plasma physics.

Bibliography

Kh. Abdullaev, I. Bogolubsky, and V. Makhankov, Phys. Lett. **56**, 427 (1974).
M.J. Ablowitz, D.J. Kaup, A.C. Newell, and H. Segur, Phys. Rev. Lett. **31**, 125 (1973).
M.J. Ablowitz, D.J. Kaup, A.C. Newell, and H. Segur, Stud. Appl. Math. **53**, 249–315 (1974).
M.J. Ablowitz and P.A. Clarkson, *Solitons, Nonlinear Evolution Equations and Inverse Scattering* (Cambridge University Press, Cambridge, 1991).
M. Abramowitz and I. Stegun, *Handbook of Mathematical Functions* (Dover, New York, 1970).
V.V. Afanasiev, R.Z. Sagdeev, and G.M. Zaslavsky, CHAOS **1**, 143 (1991).
S.V. Antipov, M.V. Nezlin, E.N. Snezhkin, and A.S. Trubnikov [Zh. Eskp. Teor. Fiz. **82**, 145 (1982)] Sov. Phys. JETP **55**, 85 (1982).
S.V. Antipov, M.V. Nezlin, E.N. Snezhkin, and A.S. Trubnikov, Sov. Phys. JETP **62**, 1097 (1985).
R.A. Antonova, B.P. Zhvaniya, D.K. Lominadze, Dzh. Nanobashvili, and V.I. Petviashvili, Pisma Zh. Eksp. Teor. Fiz. **37**, 545 (1983); [JETP Lett. **37**, 651 (1983)].
H. Aref, Ann. Rev. Fluid Mech. **15**, 345 (1983).
V.I. Arnold, Izv. Akad. Nauk. SSSR **25** 25 (1961).
V.I. Arnold, Usp. Mat. Nauk. **18**, 81 and 134 (1963).
V.I. Arnold, Dokl. Akad. Nauk. SSSR **156**, 9 (1964).
V.I. Arnold and A. Avez, *Ergodic Problems of Classical Mechanics* (Benjamin, New York, 1968).
V.I. Arnold, *Mathematical Methods of Classical Mechanics* (Springer-Verlag, New York, 1978).
N. Asano, T. Taniuti, and N. Yajima, J. Math. Phys. **10**, 2020 (1969).
R. Baierlein, *Newtonian Dynamics* (McGraw Hill, 1983) Ch. 3.
G.A. Baker *Essentials of Padé Approximations* (Academic Press, New York, 1975) pp. 92–61.

G.L. Baker and J.P. Gollub, *Chaotic Dynamics, An Introduction* (Cambridge University Press, England, 1990).
A.M. Balk, V.E. Zakharov, and S.V. Nazarenko, Sov. Phys. JETP **71**, 249 (1990).
A.M. Balk and S.V. Nazarenko, Sov. Phys. JETP **70**, 1031 (1990).
A.M. Balk, S.V. Nazarenko, and V.E. Zakharov, Phys. Lett. A **217**, (1990).
A.M. Balk, S.V. Nazarenko, and V.E. Zakharov, Phys. Lett. A **152**, 276 (1991).
A.D. Beklemishev and W. Horton, Phys. Fluids B **4**, 200 (1992).
R.P. Behringer, S.D. Meyers, and H.L. Swinney, Phys. Fluids A **3**, 1243 (1991).
G. Benettini, L. Galgani, J.-M. Strelcyn, Phys. Rev. **14**, 2338, 8 (1976).
G. Benettini, M. Casarettli, L. Galgani, and J.-M. Strelcyn, Nuovo Cimento B **44**, 183 (1978).
G. Benettini, M. Casarettli, L. Galgani, and J.-M. Strelcyn, Nuovo Cimento B **50**, 211 (1979).
T.B. Benjamin, J.L. Bona, and J.J. Mahoney, Phil. Trans. Roy. Soc. Lond. **272**, 47 (1972).
S. Benkadda, T Dudok deWit, A Verga, A. Sen, ASDEX Team, and X. Garbet, Phys. Rev. Lett. **73**, 3404 (1994).
R. Benzi, G. Paladin and A. Vulpiani, Phys. Rev. A **42**, 3654 (1988).
P. Berge, Y. Pomeau, and C. Vidal, *Order Within Chaos* (Wiley, New York, 1986).
A.R. Bishop and T. Schneider (eds.), *Solitons in Condensed Matter Physics* (Springer Verlag, New York, 1978).
D. Biskamp and K. He, Phys. Fluids **28**, 2172 (1985).
D. Biskamp and Walter (1985).
N.N. Bogolyubov and Yu.A. Mitropolsky, *Asymptotic Methods in the Theory of Nonlinear Oscillations* (Gordon and Breach, New York, 1961).
A.H. Boozer, Phys. Fluids **26**, 1288 (1983).
V. Borue, Phys. Rev. Lett. **71**, 3967 (1993).
J.C. Bowman, J.A. Krommes, and M. Ottaviani, Phys. Fluids B **5**, 3558 (1993).
R.V. Bravenec, K.W. Gentle, B. Richards, D.W. Ross, D.C. Sing, A.J. Wootton, D.L. Brower, N.C. Luhman, W.A. Peeples, C.X. Yu, T.P. Crowley, J.W. Heard, R.L. Hickok, P.M. Schock, and X.Z. Zhang, Phys. Fluids B **4**, 2127 (1992).
R. Brossier, P. Deschamps, R. Gravier, R. Pellat, and C. Renaud, Phys. Rev. Lett. **31**, 79 (1973).
R. Brossier, Nucl. Fusion **18**, 867 (1978).
D.L. Brower, W.A. Peebles, and N.C. Luhmann, Jr., Phys. Rev. Lett. **54**, 689 (1985).
J. Büchner and L.M. Zelenyĭ, Phys. Lett. A **118**, 395 (1986).
J. Büchner and L.M. Zelenyĭ, Phys. Lett. A **92**, 395 (1987).
G.R. Burkhart and J. Chen, J. Geophys. Res. **96**, A8, 14033 (1991).
B. Carreras, H.R. Hicks, and D.K. Lee, Phys. Fluids **24**, 66 (1981).
J.R. Cary and J.D. Meiss, Phys. Rev. A **24**, 2664 (1981).

G. Casati, F. Valz-Gris, I. Guarneri, Physica D **3**, 644 (1981).
D.K. Chaikovsky and G.M. Zaslavsky, Chaos **1**(4), 463 (1991).
Z. Chang, J.D. Callen, E.D. Fredrickson, R.V. Bundy, C.C. Hegna, K.M. McGuire, M.C. Zarnstorff, and jTFTR Group, Phys. Rev. Lett.**74**, 4663 (1995).
J.G. Charney, Geophys. Public. Kosjones Nors. Videnshap.-Akad. Oslo **17**, 3 (1948).
J. Chen and P.J. Palmadesso, J. Geophys. Res. **91**, 1499 (1986).
J. Chen, H.G. Mitchell, and P.J. Palmadesso, J. Geophys. Res. **95**, 15141 (1990).
J. Chen, G.R. Burkhart, and C.Y. Huang, Geophys. Res. Lett. **17**, 2237 (1990).
J. Chen, J. Geophys. Res. **97**, 15011 (1992).
M.H. Chen, Y.C. Lee, and C.S. Liu, Physica Scripta **20**, 490 (1979).
A.A. Chernikov. R.Z. Sagdeev, and G.M. Zaslavsky, Physica D **33**, 64 (1988).
A A Chernikov, T. Tel, G. Vattay, and G. M. Zaslavsky, Phys. Rev. A **40**, 4072 (1989).
A.A. Chernikov, A.I. Neishtadt, A.V. Rogalsky, and V.Z. Yakhnin, CHAOS **1**, 206 (1991).
B.V. Chirikov, Phys. Rep. **52**, 263 (1979).
Y. Couder and C. Basdevant, J. Fluid Mech. **173**, 225 (1986).
J.F. Currie, J.A. Krumhansl, A.R. Bishop, and S.E. Trullinger, Phys. Rev. A **20**, 2213 (1979).
J.F. Currie, J.A. Krumhansl, A.R. Bishop, and S.E. Trullinger, Phys. Rev. B **22**, 477 (1980).
J.H. Curry, J.R. Herring, J. Loncaric, and S.A. Orszag, J. Fluid Mech. **147**, 1 (1984).
W.D. D'haeseler, W.N.G. Hitchon, J.D. Callen, J.L. Shohet, *Flux Coordinates and Magnetic Field Structure* (Springer-Verlag, Berlin, 1991).
R.C. Davidson, *Methods in Nonlinear Plasma Theory* (Academic Press, New York, 1972).
S.P. Dawson and C.F. Fontane, Phys. Fluids **31**, 83 (1983).
D. del-Castillo-Negrete and P.J. Morrison, Phys. Fluids A **5**, 948 (1993).
D. del-Castillo-Negrete, J.M. Greene and P.J. Morrison, Physica D **90**, – (1996).
R.K. Dodd, J.C. Eilbeck, J.D. Gibbon, and H.C. Morris, *Solitons and Nonlinear Wave Equations* (Academic Press, London, 1982).
C.F. Driscoll and K.S. Fine, Phys. Fluids B **2**, 1359 (1990).
C.F. Driscoll and K.S. Fine, Phys. Rev. Lett. **74**, 4424-4427 (1995).
J.W. Dungey, Phys. Rev. Lett. **6**, 47 (1961).
T.H. Dupree, Phys. Fluids **25**, 277 (1982).
A.I. Dyachenko, S.V. Nazarenko, and V.E. Zakharov, Phys. Lett. A (1992).
J.-P. Eckman, J.P. Crutchfield, and N.H. Packard, Appl. Phys. Lett. **37**, 750–753 (1980).
M. Endler, H. Niedermeyer, L. Giannone, E. Holzhauer, A. Rudyj, G. Theimer, N. Tsois, and the ASDEX-Team, Nucl. Fusion **35**, 1307 (1995).

H. Ertel, Meteorol. Zh. **59**, 277 (1942).
D.F. Escande and F. Doveil, J. Stat. Phys. **26**, 257 (1981).
D.F. Escande, in *Long-Time Prediction in Dynamics*, eds. W. Horton, L.E. Reichl, and V.G. Szebehely (Wiley Interscience, 1983), p. 149.
D.F. Escande, Phys. Rep. **121**, 165 (1985).
A. Fasoli, F. Skiff, R. Kleiber, M.Q. Tran, and P.J. Paris, Phys. Rev. Lett. **70**, 303 (1993).
M.J. Feigenbaum, "Universal behavior in nonlinear systems," Los Alamos Science 1 (Summer 1980), 4–27 (reprinted in Physica D **7**, 16–39 (1983)).
E. Fermi, J. Pasta, and S. Ulam, in *Enrico Fermi: Collected Papers* (University of Chicago Press, 1965), Vol. II, pp. 978–988.
H. Flaschka and A.C. Newell, "Integrable systems of nonlinear evolution equations," in *Dynamics Systems: Theory and Applications*, eds. J. Moser (Springer Verlag, 1975), p. 355.
G.R. Flierl, Dyn. Atmos. & Oceans **3**, 15 (1979).
G.R. Flierl, V.D. Larichev, J.C. McWilliams, and G.M. Reznik, Dyn. of Atmos. & Oceans **5**, 1 (1980).
A.S. Fokas, V.E. Zakharov (eds.), *Important Developments in Soliton Theory* (Berlin; New York. Springer, 1993).
E. Forest and R.D. Ruth, Physica D **43**, 105–117 (1990).
V. Frisch and P.L. Sulem, Phys. Fluids **27**, 1921 (1984).
D. Fyfe and D. Montgomery, Phys. Fluids **22**, 246 (1979).
A.A. Galeev, R.Z. Sagdeev, Yu. Sigov, V.D. Shapiro, and V.I. Shevchenko, Fiz. Plasmy **1**, 10 (1975).
C.S. Gardner and J.M. Morikawa, Rept. NYU-9082, Courant Inst. of Mathematical Science, New York University (1960).
C.S. Gardner, J.M. Greene, M.D. Kruskal, and R.M. Miura, Phys. Rev. Lett. **19**, 1095 (1967) and Comm. Pure and Appl. Math. **27**, 97 (1974).
J.P. Gollub and H.L. Swinney, Phys. Rev. Lett. **41**, 448 (1978).
M.V. Goldman, Physica D **18**, 67–76 (1986).
V.A. Gordin and V.I. Petviashvili, Dokl. Akad. Nauk. SSSR **285**, 857 (1985) [Sov. Phys. Dokl **30**, 1004 (1985).
H. Grad, Phys. Fluids **10**, 137 (1967).
P. Grassberger and I. Procaccia, Phys. Rev. Lett. **50**, 346 (1983).
P. Grassberger and H. Kantz, Phys. Lett. A **114**, 167 (1985).
J.M. Greene, J. Math. Phys. **9**, 760 (1968).
J.M. Greene, J. Math. Phys. **20**, 1183 (1979).
J.M. Greene, R.S. MacKay, F. Vivaldi, and M.J. Feigenbaum, Physica 3D, 468 (1981).
J.M. Greene, Physica D **18**, 427 (1986).
J.M. Greene and J.S. Kim, Physica D **36**, 89–91 (1989).
R.W. Griffiths and E.J. Hopfinger, J. Fluid Mech. **173**, 501 (1986).

R.W. Griffiths and E.J. Hopfinger, J. Fluid Mech. **178**, 73 (1987).
A. Gruzinov, M.B. Isichenko, and J. Kalda, Sov. Phys. JETP **70**, 263 (1990).
J. Guckenheimer and P. Holmes, *Nonlinear Oscillations, Dynamical Systems, and Bifurcations of Vector Fields* (Springer-Verlag, New York, 1983).
Bai-Lin Hao, *Chaos* (World-Scientific, Singapore, 1984).
J.H. Harris, T.C. Jernigan, F.S.B. Anderson, R.D. Benson, R.J. Colchin, M.J. Cole, A.C. England, R.F. Gandy, M.A. Henderson, D.L. Hillis, R.L. Johnson, D.K. Lee, J.F. Lyon, G.H. Neilson, B.E. Nelson, J.A. Rome, M.J. Saltmarsh, C.W. Simpson, D.J. Taylor, P.B. Thompson, and J.C. Whitson, Fusion Tech. **17**, 53, (1990).
A. Hasegawa and F. Tappert, Appl. Phys. Lett. **23**, 142 (1973).
A. Hasegawa and K. Mima, Phys. Rev. Lett. **39**, 205 (1977).
A. Hasegawa and K. Mima, Phys. Fluids **21**, 87 (1978).
A. Hasegawa, T. Imamura, K. Mima, and T. Taniuti, J. Phys. Soc. Jpn. **45**, 1005 (1978).
A. Hasegawa, C.G. Maclennan, and Y. Kodama, Phys. Fluids **22**, 2122 (1979).
A. Hasegawa and M. Wakatani, Phys. Rev. Lett. **50**, 682 (1983).
A. Hasegawa, Applied Optics **23**, 3302–3309 (1984).
A. Hasegawa and M. Wakatani, Phys. Rev. Lett. **59**, 1581 (1987).
A. Hasegawa, *Optical Solitons in Fibers*, 2nd Ed. (Spring-Verlag, New York, 1990).
T. Hatori, T. Kamimura, and Y.H. Ichikawa, Physica D **14**, 193 (1985).
T. Hatori and H. Irie, Prog. Theor. Phys. **78**, 249 (1987).
T. Hatori, H. Irie, Y. Abe, and K. Urata, Prog. Theor. Phys. **98**, 83 (1989).
R.D. Hazeltine and J.D. Meiss, *Plasma Confinement* (Addison-Wesley, Redwood City, California, 1992).
K. He and A. Salat, Phys. Letts. A **132**, 175 (1988).
K. He and A. Salat, Plasma Phys. Contr. Fusion **31**, 123 (1989).
H.W. Hendel, T.K. Chu, and P.A. Politzer, Phys. Fluids **11**, 2426 (1968).
M. Henon, "Numerical Exploration of Hamiltonian Systems" in *Chaotic Behavior of Deterministic Systems*, eds. G. Iooss, R.H.G. Helleman, R. Stora (North-Holland, Amsterdam, 1983) pp. 56–129.
F.L. Hinton and W. Horton, Phys. Fluids **14**, 116 (1971).
B.G. Hong, F. Romanelli, and M. Ottaviani, Phys. Fluids B **3**, 615 (1991).
S. Horihata and M. Sato, J. Phys. Soc. Jpn. **56**, 2611 (1987).
T. Horita, H. Hata, H. Mori, and K. Tomita, Prog. Theor. Phys. **81**, 1073 (1989).
T. Horita, H. Hata, R. Ishizaki, and H. Mori, Prog. Theor. Phys. **83**, 1065 (1990).
W. Horton, Phys. Rev. Lett. **19**, 1269 (1976).
W. Horton and R.D. Estes, Nucl. Fusion **19**, 203 (1979).
W. Horton and D. Choi, Phys. Rep. **49**, 273 (1979).
W. Horton, D.-I. Choi, and W. M. Tang, Phys. Fluids **24**, 1077 (1981).
W. Horton, Plasma Phys. **23**, 1107 (1981).

W. Horton, "Drift wave turbulence and anomalous transport," *Handbook of Plasma Physics, Vol. II*, eds. M.N. Rosenbluth and R.Z. Sagdeev (Elsevier Science Publishers, 1984), pp. 383–449.
W. Horton, Plasma Phys. Contr. Fusion **27**, 937 (1985).
W. Horton, Phys. Fluids **29**, 1491 (1986).
W. Horton, J. Liu, J. Sedlak, and J. Meiss, Phys. Fluids **29**, 1004 (1986).
W. Horton, T. Tajima, and T. Kamimura, Phys. Fluids **30**, 3485 (1987).
W. Horton, Phys. Fluids **1**, 524 (1989).
W. Horton and T. Tajima, Geophys. Res. Lett. **17**, 123 (1990).
W. Horton, Phys. Rep. **192**, 1 (1990).
W. Horton, C. Liu, J. Hernandez, and T. Tajima, Geophys. Res. Lett. **18**, 1575 (1991).
W. Horton and W. Rowan, Phys. Plasmas **901** (1994).
W. Horton and T. Tajima, Geophys. Res. Lett. **18**, 15811 (1991).
W. Horton, D. Jovanović, and J. Juul Rasmussen, Phys. Fluids B **4**, 3336 (1992).
W. Horton and V. Petviashvili, "On the trapping condition for planetary vortex structures," *Research Trends in Physics: Chaotic Dynamics and Transport in Fluids and Plasmas*, eds. W. Horton, Y. Ichikawa, I Prigogine (Ed. in-Chief), G. Zaslavsky, American of Physics (New York, 1993).
W. Horton, J.Q. Dong, X.N. Su, and T. Tajima, J. Geophys. Res. **98**, 13377 (1993).
W. Horton, A. Wootton, and M. Wakatani, eds., "Transport in the self-organized relaxed state of ion temperature gradient instability," in *Ion Temperature Gradient-Driven Turbulent Transport* (AIP Conference Proceedings, No. 284, 1994).
W. Horton and A. Hasegawa, CHAOS **4**, 227–251 (1994).
J.E. Howard and S.M. Hohs, Phys. Rev. A **17**, 2237 (1990).
T. Huld, A.H. Nielsen, H.L. Pécseli, and J. Juul Rasmussen, Phys. Fluids B **3**, 1609 (1991).
Y.H. Ichikawa, T. Suzuki, and B. Fried, J. Plasma Phys. **10**, 219 (1973).
Y.H. Ichikawa, T. Kamimura, and C.F.F. Karney, Physica D **6**, 233 (1983).
Y.H. Ichikawa, T. Kamimura, and T. Hatori, Physica D **29**, 247 (1987).
Y.H. Ichikawa and Y. Abe, Prog. Theoret. Phys. Suppl. **94**, 128 (1988).
Y.H. Ichikawa, T. Kamimura, T. Hatori, and S.Y. Kim, Prog. Theor. Phys. Suppl. **98**, 1 (1989).
Y.H. Ichikawa, Y. Nomura, and T. Kamimura, Prog. Theor. Phys. Suppl. **99**, 220 (1989).
S. Ichimaru, *Plasma Physics: An Introduction to Statistical Physics of Charged Particles* (Benjamin/Cummings, Menlo Park, 1986).
M.B. Isichenko, W. Horton, D.E. Kim, E.G. Heo, and D.-I. Choi, Phys. Fluids B **4**, 3973 (1992).
M.B. Isichenko, Rev. Mod. Phys. **64**, (4), 961 (1992).

D. Jovanović and W. Horton, Phys. Fluids B **5**, 9 (1993).
D. Jovanović, P.K. Shukla, U. de Angelis, and W. Horton, Phys. Fluids B **3**, 45 (1991).
B.B. Kadomtsev and V.I. Petviashvili, Sov. Phys. Dokl. **15**, 539–541 (1970).
B.B. Kadomtsev and V.I. Petviashvili, Dokl. Akad. Nauk. SSSR **192**, 753 (1970).
B.B. Kadomtsev and V.I. Karpman, Sov. Phys. Uspekhi **14**, 40 (1971).
F. Kako and N. Yajima, J. Phys. Soc. Jpn. **49**, 2063 (1980).
F. Kako and N. Yajima, J. Phys. Soc. Jpn. **51**, 311 (1982).
T. Kakutani, H. Ono, T. Taniuti, and C.C. Wei, J. Phys. Soc. Jpn. **24**, 1159 (1968).
T. Kakutani and N. Sugimoto, Phys. Fluids **17**, 1617 (1974).
H.E. Kandrup and P.J. Morrison, Ann. Phys. **225**, 114, Sec. 5 (1993).
C.F.F. Karney, Phys. Fluids **22**, 2188 (1979).
C.F.F. Karney, Physica D **8**, 360 (1983).
C.F.F. Karney and D.J. Kaup, SIAM J. Appl. Math. **31**, 121–133 (1976).
D.J. Kaup and A.C. Newell, J. Math. Phys. **19**, 798 (1978).
L.C. Kells and S.A. Orszag, Phys. Fluids **21**, 162 (1978).
C.F. Kennel, "Cosmic Ray Acceleration: A Plasma Physicist's Perspective," in *From Particles to Plasmas*, Ed. J.W. Van Dam (Addison-Wesley, New York, 1988).
M. Khazei, J. Bulson, and K.E. Lolnngren, Phys. Fluids **25**, 759 (1982).
A.S. Kingsep, L.I. Rudakov, and R.N. Sudan, Phys. Rev. Lett. **31**, 1482 (1973).
R.G. Kleva and J.F. Drake, Phys. Fluids **27**, 1686 (1984).
A.N. Kolmogorov, Dokl. Akad. Nauk. SSR **30**, 301 (1941).
A.N. Kolmogorov, Dokl. Akad. Nauk. SSSR **98**, 527 (1954).
K. Konno, Y.H. Ichikawa, and M. Wadati, J. Phys. Soc. Jpn. **50**, 1025 (1981).
K. Konno and A. Jeffrey, J. Phys. Soc. Jpn. **52**, 1 (1983).
K. Konno and Y.H. Ichikawa, Chaos **2**, 237 (1992).
M. Kono and H. Sanuki, J. Phys. Soc. Jpn. **33**, 1731 (1972).
M. Kono and E. Miyashita, Phys. Fluids **31**, 326 (1988).
M. Kono and W. Horton, Phys. Fluids B, 3255 (1991).
D.J. Korteweg and G. DeVries, Philosophical Magazine **39**, 422–443 (1895).
N.E. Kosmatov, V.F. Shvets, and V.E. Zakharov, Physica D **52**, 16–36 (1991).
R.H. Kraichnan, Phys. Rev. **109**, 1407 (1958).
R.H. Kraichnan, J. Fluid Mech. **5**, 497 (1959).
R.H. Kraichnan, Phys. Fluids **10**, 1417 (1967).
R.H. Kraichnan, J. Fluid Mech. **59**, 745 (1967).
R.H. Kraichnan, J. Fluid Mech. **47**, 525 (1971).
R.H. Kraichnan and D. Montgomery, Rep. Prog. Phys. **43**, 547 (1980).
R.H. Kraichnan, in *Theoretical Approaches to Turbulence*, Vol. 58 of Applied Mathematical Sciences Series, edited by D.L. Dwoyer, M.Y. Hussaini, and R.G. Voigt (Springer-Verlag, New York, 1985), Chap. V, p. 91.

R.H. Kraichnan, in *Theoretical Approach to Turbulence*, eds. D.L. Dwoye, M.Y. Hussain, and R.G. Voigt (Springer-Verlag, New York, 1989) p. 91.
R.H. Kraichnan, Phys. Rev. Lett. **65**, 575 (1990).
J.A. Krommes, Phys. Fluids **25**, 1393 (1982).
J.A. Krommes, *Handbook of Plasma Physics II*, Eds. M.N. Rosenbluth and R.Z. Sagdeev (North-Holland Publishing, Amsterdam, 1984) pp. 229–239.
W.L. Kruer, *The Physics of Laser Plasma Interactions* (Addison-Wesley, Redwood City, CA, 1988).
R. Kubo, M. Toda, and N. Hashitsume, *Statistical Physics II* (Springer-Verlag, New York, 1985).
N. Kukharkin, S.A. Orszag, and V. Yakhot, Phys. Rev. Lett. **75**, 2486 (1995).
R.M. Kulsrud, Phys. Rev. **106**, 205 (1957).
G.L. Lamb, Jr. *Elements of Soliton Theory* (John Wiley & Sons, Inc., 1980).
L.D. Landau and E.M. Lifshitz, *Classical Mechanics* (Addison-Wesley, 1958).
E.W. Laedke and K.H. Spatschek, Phys. Fluids **29**, 134 (1986).
E.W. Laedke and K.H. Spatschek, Phys. Fluids **31**, 1492 (1988).
V.D. Larichev and G.K. Reznik, Fiz. Plasmy **3**, 270 (1976).
V.D. Larichev and G.M. Reznik, Oceanology **16**, 547 (1976).
V.D. Larichev and J.C. McWilliams, Phys. Fluids A **3**, 938 (1991).
Y.-T. Lau and J.M. Finn, Ap.J. **350**, 672–691 (1991).
P.D. Lax, Commun. Pure. Appl. Math. **21**, 467 (1968).
B. Legras, P. Santangelo, and R. Benzi, Europhys. Lett. **5**, 37 (1988).
A.J. Lichtenberg, M.A. Lieberman, and R.H. Cohen, Physica D **1**, 291–305 (1980).
A.J. Lichtenberg and M.A. Lieberman, Physica D **33**, 211 (1988).
A.J. Lichtenberg and M.A. Lieberman, "Regular and Chaotic Dynamics," *Applied Mathematical Sciences* (Springer-Verlag, 1991), p. 38.
A.J. Lichtenberg and M.A. Lieberman, *Regular and Chaotic Dynamics*, 2nd Edition (Springer-Verlag, New York, 1991).
R.G. Littlejohn, Phys. Fluids **24**, 1730 (1981).
R.G. Littlejohn, J. Plasma Phys. **29**, 111 (1983).
J. Liu and W. Horton, J. Plasma Phys. **29**, 1828, 36, 1 (1986).
D.W. Longcope and R.N. Sudan, Phys. Fluids B **3**, 1945 (1991).
K. Lonngren and B. Scott, *Solitons in Action* (Academic Press, New York, 1978).
E.N. Lorenz, J. Atmos. Sci. **20**, 130 (1963).
L.R. Lyons and T.W. Speiser, J. Geophys. **87**, 2276 (1982).
R. Lyons and T.W. Speiser, J. Geophys. Res. **90**, 8543 (1985).
R.S. MacKay, J.D. Meiss, and I.C. Percival, Phys. Rev. Lett. **52**, 697 (1984a).
R.S. MacKay, J.D. Meiss, and I.C. Percival, Physica D **52**, 13 (1984b).
R.S. MacKay, Phys. Lett. A **106**, 99 (1984).
R.S. MacKay, J.D. Meiss, and I.C. Percival, Physica D **13**, 55 (1984).
R.S. MacKay and J.D. Meiss, *Hamiltonian Dynamical Systems* (Adam Higler, Bristol, 1987).

V.G. Makhankov, Phys. Rep. **35**, 1–128 (1978).
M. Makino, T. Kammimura, and T. Taniuti, J. Phys. Soc. Jpn. **50**, 980 (1981).
S.V. Manakov, Sov. Phys. JETP **38**, 248 (1974).
P.C. Martin and E.D. Siggia, Phys. Rev. A **8**, 423 (1973).
T. Maxworthy and L.G. Redekopp, Icarus **29**, 261 (1976).
E. Mazzucato, Phys. Rev. Lett. **36**, 792 (1976).
E. Mazzucato, Phys. Fluids **21**, 1063 (1978).
E. Mazzucato, Phys. Rev. Lett. **48**, 1828 (1982).
J.C. McWilliams, G.R. Flierl, V.D. Larichev, and G.M. Reznik, Dyn. Atmos. & Oceans **5**, 219 (1981).
J.C. McWilliams and N.J. Zabusky, Geophys. & Astrophys. Fluid Dyn. **19**, 207 (1982).
J.C. McWilliams, J. Fluid Mech. **146**, 21 (1984).
J.C. McWilliams, G.R. Flierl, V.D. Larichev, Phys. Fluids A **2**, 547 (1990a).
J.C. McWilliams, J. Fluid Mech. **219**, 361 (1990b).
J.C. McWilliams, in *Predictability of Fluid Motions*, eds. G. Holloway and B.J. West (American Institute of Physics, Conf. Proc. No. 106, 1984), p. 205.
J.D. Meiss and W. Horton, Phys. Fluids **25**, 1838 (1982a).
J.D. Meiss and W. Horton, Phys. Rev. Lett. **48**, 1362 (1982b).
J.D. Meiss and W. Horton, Jr., Phys. Fluids **26**, 990 (1983).
J.D. Meiss, J.R. Cary, C. Grebogi, J.D. Crawford, A.N. Kaufman, and H.D. Abarbanel, Physica D **6**, 375 (1983).
J.D. Meiss, Physica D **74**, 254 (1994).
V.K. Mel'nikov, J. Math. Phys. **31**, 1106 (1990).
M.V. Melander, J.C. McWilliams, and N.J. Zabusky, J. Fluid Mech. **178**, 137 (1987a).
M.V. Melander, Phys. Fluids **30**, 2610 (1987b).
A.B. Mikhailovskii, G.D. Aburdzhaniya, O.G. Onishchenko, and S.E. Sharapov, Phys. Lett. A **104**, 94 (1984a).
A.B. Mikhailovskii, G.D. Aburdzhaniya, O.G. Onishchenko, and A.P. Churikov, Phys. Lett. A **101**, 263 (1984b).
A.B. Mikhailovskii, V.P. Lakhin, L.A. Mikhailovskaya, and O.G. Onishchenko, Sov. Phys.-JETP **59**, 1198 (1985).
A.B. Mikhailovskii, S.V. Nazarenko, S.V. Novakovskii, A.P. Churikov, and O.G. Onishchenko, Phys. Lett. A **133**, 407 (1988).
A.B. Mikhailovskii, A.M. Pukhov, and O.G. Onishchenko, Phys. Lett. A **141**, 154 (1989b).
J.W. Miles, J. Fluid Mech. **79**, 171 (1979).
N. Minorsky, *Nonlinear Oscillations* (Van Nostrand, Princeton, 1962).
E. Mjølhus, J. Plasma Physics **19**, 437 (1973).
E. Mjølhus and J. Wyller, Physica Scripta **33**, 442 (1986).

L.F. Mollenauer, J.P. Gordon, and M.N. Islam, IEEE J. Quantum Elect. **22**, 157–173 (1986).

A.S. Monin and A.M. Yaglom, *Statistical Fluid Mechanics: Mechanics of Turbulence* (MIT Press, Cambridge, Mass, 1971).

H. Mori, H. Hata, T. Horita, and T.Kobayashi, Prog. Theor. Phys. Suppl. **99**, 1–63 (1989).

G.Y. Morikawa, J. Meteorol. **17**, 148 (1960).

H.C. Morris and R.K. Dodd, Phys. Lett. A **75**, 20 (1980).

P.J. Morrison, "Poisson brackets for fluids and plasmas," in *Mathematical Methods in Hydrodynamics and Integrability in Dynamical Systems*, eds. M. Tabor and Y. Treve, (American Institute of Physics Conference Proceedings, New York, 1982), Vol. 88.

P.J. Morrison, J.D. Meiss, and J.R. Cary, Physica D **11**, 324 (1984).

J. Moser, Wachr. Acad. Wiss., Gottingen, Math. Phys. Kℓ **2**, 1 (1962).

A. Muhm, A.M. Pukhov, K.H. Spatschek, and V. Tsytovich, Phys. Fluids B **4**, 336 (1992).

S.V. Muzylev and G.M. Reznik, Phys. Fluids B **4**, 2841 (1992).

M.V. Nezlin, Sov. Phys. Usp. **29**, 807 (1986) [Usp. Fiz. Nauk. **150**, 3 (1986)].

M.V. Nezlin and E.N. Snezhkin, in *Rossby Vortices, Spiral Structures, Solitons* (Springer-Verlag, 1993), pp. 196–203.

S.V. Novakovskii, A.B. Mikhailovskii, and O.G. Onischenko, Phys. Lett. A **132**, 33 (1988).

S.P. Novikov, S.V. Manakov, L.P. Pitaevskii, and V.E. Zakharov, *Theory of Solitons. The Inverse Scattering Method* (Plenum, New York, 1984).

H. Nagashima and M. Kuwahara, J. Phys. Soc. Jpn. **50**, 3792 (1981).

Y. Nomura, Y.H. Ichikawa, and W. Horton, Phys. Rev. A **45**, 1103 (1992).

K. Nozaki, Phys. Rev. Lett. **49**, 1883 (1982).

J. Nycander and M.B. Isichenko, Phys. Fluids B **2**, 2042–2047 (1990).

J. Nycander, Phys. Fluids A **4**, 467 (1992).

Y. Ogura, J. Fluid Mech. **16**, 33 (1963).

K. Ohkuma, Y.H. Ichikawa, and Y. Abe, Opt. Lett. **12**, 516 (1987).

L. Onsager, Phys. Rev. **37**, 405 (1931).

L. Onsager, **38**, 2265 (1931).

L. Onsager, Supplto. Nuovo Cim. **6**, 279 (1949).

S.A. Orszag, J. Fluid Mech. **41**, 363 (1970).

S.A. Orszag and A.T. Patera, J. Fluid. Mech. **128**, 347 (1983).

S.A. Orszag, in *Fluid Dynamics*, edited by R. Balian and J.-L. Peube (Gordon and Breach, London, 1977), pp. 236–373, summer school lectures given at Grenoble University, 1973.

E.A. Overman and N.J. Zabusky, Phys. Fluids **25**, 1297 (1982).

S.E. Parker, W.W. Lee, and R.A. Santoro, Phys. Rev. Lett. **71**, 2042 (1993).

L.D. Pearlstein and H.L. Berk, Phys. Rev. Lett. **23**, 220 (1969).

D. Peregrine, J. Fluid Mech. **25**, 321 (1966).
V.I. Petviashvili, Fiz. Plazmy **3**, 270 (1977) [Sov. J. Plasma Phys. **3**, 150 (1977)].
J. Pedlosky, *Geophysical Fluid Dynamics* (Springer, New York, 1987).
H.L. Pécseli, J. Juul Rasmussen, and K. Thomsen, Phys. Rev. Lett. **52**, 2148 (1984).
H.L. Pécseli, J. Juul Rasmussen, and K. Thomsen, Plasma Phys. Contr. Fusion **27**, 837 (1985).
H.L. Pécseli and J. Trulsen, Phys. Fluids B **1**, 1616 (1989).
J. Pedlosky, *Geophysical Fluid Dynamics* (Springer-Verlag, New York, 1987), pp. 518–532.
V.I. Petviashvili, Sov. J. Plasma Phys. **3**, 150 (1977).
V.I. Petviashvili, JETP Lett. **32**, 619 (1980).
V.I. Petviashvili and O.A. Pokhotelov, Sov. J. Plasma Phys. **12**, 651 (1986).
V.I. Petviashvili and O.A. Pokhotelov, J. Plasma Phys. **12**, 657 (1986).
V.I. Petviashvili and O.A. Pokhotelov, *Solitary Waves in Plasmas and in the Atmosphere* (Gordon and Breach Science Publishers, Philadelphia, 1992).
S. Pierini, Dyn. Atmos. Oceans **9**, 273 (1985).
R.T. Pierrehumbert and S.E. Widnall, J. Fluid Mech. **114**, 59 (1982).
E. Pina and L.J. Lara, Physica D **26**, 369 (1987).
L.I. Piterbarg and E.I. Shulman, Phys. Lett. A **140**, 29 (1989).
A.B. Rechester and T.H. Stix, Phys. Rev. Lett. **36**, 587 (1976).
A.B. Rechester and M.N. Rosenbluth, Phys. Rev. Lett. **40**, 88 (1978).
A.B. Rechester and T.H. Stix, Phys. Rev. A **19**, 1656 (1979).
A.B. Rechester and R.B. White, Phys. Rev. Lett. **44**, 1586 (1980).
A.B. Rechester, M.N. Rosenbluth, and R.B. White, Phys. Rev. A **23**, 2664 (1981).
L.G. Redekopp, J. Fluid Mech. **82**, 725 (1977).
J.A. Robertson, W. Horton, and D-I. Choi, Phys. Fluids **30**, 1059 (1987).
H.A. Rose and P.L. Sulem, J. Physique (Paris) **39**, 441 (1978).
M.N. Rosenbluth, R.Z. Sagdeev, J.B. Taylor, and G.M. Zaslavskii, Nucl. Fusion **6**, 297 (1966).
M.N. Rosenbluth, H.L. Berk, I. Doxas, and W. Horton, Phys. Fluids **30**, 2636 (1987).
L.I. Rudakov and R.Z. Sagdeev, Sov. Phys. Dokl. **6**, 415 (1961).
O. Ruelle and F. Takens, Comm. Math. Phys. **20**, 167 (1971).
J.S. Russel, Proc. Roy. Soc. Edinb. 319 (1844).
R.Z. Sagdeev and A.A. Galeev, *Nonlinear Plasma Theory* (Benjamin, New York, 1968).
R.Z. Sagdeev, D.A. Usikov, and G.M. Zaslavsky, *Nonlinear Physics from the Pendulum to Turbulence and Chaos* (Harwood Academic Pub., 1988).
N. Sasa and J. Satsuma, J. Phys. Soc. Jpn. **60**, 409 (1991).
J. Satsuma and N. Yajima, Suppl. Prog. Theor. Phys. **55**, 284 (1974).
G. Schmidt and J. Bialek, Physica D **5**, 397 (1982).

A.C. Scott, F.Y.F. Chu, and D.W. McLaughlin, Proc. IEEE **61**, 1443–1483 (1973).
B.D. Scott, Phys. Fluids B **4**, 2468 (1992).
V.D. Shapiro, P.H. Diamond, V.B. Lebedev, G. Isoloviev, and V.I. Shevchenko, Plasma Phys. Contr. Fusion **35**, 1032 (1993).
P.K. Shukla, M.Y. Yu, H.V. Rahman, and K.H. Spatschek, Phys. Rep. C. **105**, 227 (1984).
R.E. Slusher and C.M. Surko, Phys. Rev. Lett. **40**, 400 (1978).
R.A. Smith, J.A. Krommes, and W.W. Lee, Phys. Fluids **28**, 1069 (1985).
T.H. Solomon and J.P. Gollub, Phys. Fluids **31**, 1372 (1988).
T.H. Solomon, E.R. Weeks, and H.L. Swinney, Phys. Rev. Lett. **71**, 3975 (1993).
T.H. Solomon, E.R. Weeks, and H.L. Swinney, Physica D **76**, 70 (1994).
J. Sommeria, S.D. Meyers, and Harry L. Swinney, "Experiments on vortices and Rossby waves in waves in eastward and westward jets," in *Nonlinear Topics in Ocean Physics*, Ed. A.R. Osborne (North-Holland, Amsterdam, 1991), pp. 227–269.
B.U.Ö. Sonnerup, J. Geophys. Res. **76**, 8211 (1971).
S.R. Spangler, J.P. Sheerin, and G.L. Payne, Phys. Fluids **28**, 104 (1985).
T.W. Speiser, J. Geophys. Res. **70**, 4219 (1965).
T.W. Speiser, J. Geophys. Res. **72**, 3919 (1967).
T.W. Speiser and R.F. Martin, J. Geophys. Res. **97**, 10775 (1992).
Solitons, R.K. Bullough and P.J. Caudrey (eds.), (Berlin; New York, Springer-Verlag, 1980).
K.H. Spatschek, E.W. Laedke, Chr. Marquardt, S. Musher, and H.J. Stewart, Q. Appl. Math. **1**, 262 (1943).
T.H. Stix, *Waves in Plasmas* (American Institute of Physics, 1992). H.R. Strauss, Phys. Fluids **19**, 134 (1976).
H.R. Strauss, Phys. Fluids **20**, 1354 (1977).
X.N. Su, W. Horton, and P.J. Morrison, Phys. Fluids B **3**, 921 (1991).
X.N. Su, W. Horton, and P.J. Morrison, Phys. Fluids B **4**, 1238 (1992).
R.N. Sudan and M.J. Keskinen, Phys. Rev. Lett. **38**, 1869 (1977).
R.N. Sudan and M.J. Keskinen, Phys. Fluids **22**, 2305 (1979).
R.N. Sudan and D. Pfirsch, Phys. Fluids **28**, 1702 (1985).
C.M. Surko and R.E. Slusher, Phys. Rev. Lett. **36**, 1747 (1976).
C.M. Surko and R.E. Slusher, Phys. Rev. Lett. **40**, 400 (1978).
G.E. Swaters, Phys. Fluids **29**, 1419 (1986).
H.L. Swinney, Physica D **7**, 3–15 (1983).
H.L. Swinney and W.Y. Tam, Phys. Rev. A **36**, 1374 (1987).
H.L. Swinney, J. Sommeria and S.D. Meyers, Nature **331**, 216 (1988).
M. Tabor, *Chaos and Integrability in Nonlinear Dynamics: An Introduction* (Wiley, New York, 1989).
T. Tajima and J.M. Dawson, Phys. Rev. Lett. **43**, 267 (1979).
T. Taniuti, Prog. Theor. Phys. Suppl. **55**, 1 (1974).

T. Taniuti, J. Phys. Soc. Jpn. **55**, 4253 (1986).
H. Tasso, Phys. Lett. A **24**, 618 (1967).
H. Tasso, Phys. Lett. A **96**, 33 (1983a).
H. Tasso and K. Lerbinger, Phys. Lett. A **97**, 384 (1983b).
H. Tasso, Phys. Lett. **16**, 231 (1987).
H. Tasso, Phys. Lett. A **120**, 464 (1987).
J.B. Taylor and R.J. Hastie, Plasma Phys. **10**, 479 (1968).
J.B. Taylor, "Investigation of Charged Particle Invariants," Culham Lab. Prog. Report CLM-PR 12 (1969).
P.W. Terry and W. Horton, Phys. Fluids **25**, 491 (1982).
P.W. Terry and W. Horton, Phys. Fluids **26**, 106 (1983).
P.W. Terry and P.H. Diamond, Phys. Fluids **28**, 1414 (1985).
TFTR Group, in *Plasma Physics and Controlled Nuclear Fusion Research* [Proceedings of the 6th International Conference, Berchtesgaden, 1976 (IAEA, Vienna, 1977)], Vol. I, p. 35.
TFTR Group, in *Plasma Physics and Controlled Nuclear Fusion Research* [Proceedings of the 8th International Conference, Brussels, 1980, (IAEA, Vienna, 1981)], Vol. I, p. 425.
Y.M. Treve and O.P. Manley, Physica D **4**, 319 (1982).
I. Tsukabayashi, Y. Nakamura, F. Kako, and K.E. Lonngren, Phys. Fluids **26**, 790 (1983).
V.N. Tsytovich, *Theory of Turbulent Plasma* (Consultants Bureau, 1977).
N.A. Tsyganenko, Planet Space Sci. **35**, 1347 (1987).
N. Tzoar and M. Jain, Phys. Rev. A **23**, 1266 (1981).
G.J.F. Van Heijst and J.B. Flor, "Laboratory experiments on dipole structures in a stratified fluid," in *Mesoscale/Synoptic Coherent Structures in Geophysical Turbulence*, eds. J.C.J. Nihoul and B.M. Jamart (Elsevier Science Publishers, The Netherlands, 1989), pp. 591–608.
R.K. Varma, Phys. Rev. Lett. **26**, 417 (1971).
R.K. Varma and W. Horton, Phys. Fluids **15**, 1469 (1972).
A.A. Vedenov, E.P. Velikhov, and R.Z. Sagdeev, Nucl. Fusion **I**, 82 (1961).
S.Ya. Vyshkind and M.I. Rabinovich, Zh. Eksp. Teor. Fiz. **71**, 557 (1976) [Sov. Phys. JETP **44**, 292 (1976)].
M. Wadati, J. Phys. Soc. Jpn. **32**, 1681 (1972).
M. Wadati, K. Konno, and Y.H. Ichikawa, J. Phys. Soc. Jpn. **46**, 1965 (1979a).
M. Wadati, K. Konno, and Y.H. Ichikawa, J. Phys. Soc. Jpn. **47**, 1698 (1979b).
M. Wakatani and A. Hasegawa, Phys. Fluids **27**, 611 (1984).
M. Wakatani, K. Watanabe, H. Sugama, and A. Hasegawa, Phys. Fluids B **4**, 1754 (1992).
R.E. Waltz, Phys. Fluids **26**, 169 (1983).
R.E. Waltz, Phys. Fluids **28**, 577 (1985).
R.E. Waltz, Phys. Fluids **29**, 3684 (1986).

R.E. Waltz, Phys. Fluids **31**, 1962 (1988).
R.E. Waltz, Phys. Fluids B **2**, 2118 (1990).
P.K.C. Wang and K. Masui, Phys. Lett. A **81**, 97 (1981).
H. Washimi and T. Taniuti, Phys. Rev. Lett. **17**, 996 (1966).
H. Wenk, Phys. Rev. Lett. **64**, 3027 (1990).
J.M. Wersinger, J.M. Finn, and E. Ott, Phys. Fluids **23**, 1142 (1980).
What is integrability?, V.E. Zakharov, ed. with contributions by F. Calogero, *et al.* (Berlin; New York, Springer-Verlag, 1991).
R.B. White, "Resistive Instabilities and Field Line Reconnection," *Handbook of Plasma Physics*, Vol. I, eds. M.N. Rosenbluth and R.Z. Sagdeev (Elsevier Science Publishers, 1984), pp. 611–676.
R.B. White, *Theory of Tokamak Plasmas* (North-Holland, Amsterdam, 1989).
G.B. Whitham, *Linear and Nonlinear Waves* (Wiley and Sons, New York, 1974).
G.P. Williams, J. Atmos. Sci. **35**, 1399 (1978).
G.P. Williams, J. Atmos. Sci. **42**, 1237 (1985).
G.P. Williams, E.R. Tracy, and G. Vahala, Phys. Rev. Lett. **59**, 1922 (1987).
T.S. Wolfram, *Mathematica: A System for Doing Mathematics by Computer*, 2nd Edition, (Addison-Wesley, 1991).
N. Yajima, M. Oikawa, and J. Satsuma, J. Phys. Soc. Jpn. **44**, 1711 (1978).
N. Yajima, M. Kono, and S. Ueda, J. Phys. Soc. Jpn. **52**, 3414 (1983).
V. Yakhot and S.A. Orszag, J. Sci. Comput. **1**, 3 (1986).
V. Yakhot and S.A. Orszag, Phys. Rev. Lett. **57**, 1772 (1986).
V.V. Yankov, JETP **80**, 2192 (1995).
N.J. Zabusky and M.D. Kruskal, Phys. Rev. Lett. **15**, 240–243 (1965).
N.J. Zabusky and J.C. McWilliams, Phys. Fluids **25**, 2175 (1982).
V.E. Zakharov and Faddeev, Sov. Phys. JETP **33**, 538 (1971).
V.E. Zakharov and A.B. Shabat, Sov. Phys. JETP **34**, 62 (1972).
V.E. Zakharov, Sov. Phys. JETP **35**, 908 (1972).
V.E. Zakharov, Sov. Phys. JETP **38**, 108 (1974).
V.E. Zakharov and V.S. Synakh, Zh. Eksp. Teor. Fiz. **68**, 940 (1975).
V.E. Zakharov, S.V. Manakov, S.P. Novikov, and L.P. Pitaevsky, *Theory of Solitons: the Inverse Scattering Method* (Plenum Publ. Corp., New York, 1984).
V.E. Zakharov, "Kolmogorov spectra in weak turbulence problem," *Handbook of Plasma Physics*, Vol. II, eds. M.N. Rosenbluth and R.Z. Sagdeev (Elsevier Science Publishers, 1984), pp. 3–36.
V.E. Zakharov and E.A. Kuznetsov, Physica D **18**, 455 (1986).
V.E. Zakharov and E.I. Schulman, Physica D **29**, 283 (1988).
V.E. Zakharov, V.S. L'vov, and G. Falkovich, *Kolmogorov Spectral of Turbulence* (Springer-Verlag, New York, 1992).
G.M. Zaslavsky and R.Z. Sagdeev, Sov. Phys. JETP **52**, 1081 (1967).
G.M. Zaslavsky and B.V. Chirikov, Sov. Phys.-Uspekhi **14**, 549–568 (1972).

G.M. Zaslavsky, R.Z. Sagdeev, D.A. Usikov, and A.A. Chernikov, *Weak Chaos and Quasi-Regular Patterns* (Cambridge Univ. Press, England, 1991).

G.M. Zaslavsky, CHAOS **1**, 1 (1991).

G.M. Zaslavsky, D. Stevens, and H. Weitzner, Phys. Rev. E **48**, 1683 (1993).

Index

$1\frac{1}{2}$-degree-of-freedom, 70
A-type null point, 51
B-type null point, 51
$\mathbf{E} \times \mathbf{B}$ convection, 308
$\mathbf{E} \times \mathbf{B}$ drift islands, 96
$\mathbf{E} \times \mathbf{B}$ motion in a sheared magnetic shear, 107
$\mathbf{E} \times \mathbf{B}$ motion in two low-frequency waves, 84
$\mathbf{E} \times \mathbf{B}$ rotation, 272
$\mathbf{E} \times \mathbf{B}$ rotation frequency, 296
ln k-corrected, 302
∇B drift velocity, 60, 76
$m = 2$ island, 33

A-K-N-S scheme, 197
Ablowitz-Kaup-Newell-Segur, 196
accelerator modes, 129
action integral I, 65
action-angle variables, 63
action-angle variables for $\mathbf{E} \times \mathbf{B}$ motion, 87
adiabatic invariant, 64
adiabatic invariant(s) $\oint p_\ell \, dq_\ell$, 27
adiabatic invariants, 62
Alfvén soliton, 199, 204
Alfvén waves, 185
aliasing errors, 282
amplitude and size relations, 241
amplitude dependent frequency shifts, 7
amplitude dispersion, 6
amplitude expansion, 2, 7, 59
amplitude modulation, 167
analogy between drift wave and Rossby wave, 225
anomalous transport, 254
anti-hermitian part $\mathcal{L}_{ah}$, 279
area preserving map, 117
ATF torsatron, 31

ball between an oscillation wall, 160
barotropic fluid, 238
bi-stable potentials, 60
bifurcation, 26, 61
Bohm diffusion, 306
bouncing ball, 160
broadband fluctuations, 317
Burger's equation, 189

canonical distribution function of solitary waves, 211
chaotic attractor, 283
characteristic exponents, 11
Charney equation, 223
circularly polarized Alfvén waves, 185
closure, 286
coalescence, 248
coalescence of drift wave vortices, 249
coherent approximation, 300
coherent modes, 164
coherent phase-locked harmonics, 167
coherent structures, 245, 277, 317
collective modes, 204
complex drift-wave frequency, 279
computer simulations, 244
conditions for nonlinear saturation, 284
conservation laws, 229
conservation of energy and enstrophy, 237
Coriolis force $f\mathbf{v} \times \hat{\mathbf{z}}$, 222, 224

correlation functions, 275
correlation function dynamics, 297
Couette flow, 256
coupling coefficients M_0 and M_2, 170
coupling factors, 308
crossing orbits, 66
cubic nonlinear Schrödinger equation, 187, 194
current sheet invariant, 65
curvature drift, 76
curvature vector κ, 76
cyclotron orbits, 68

damped diffusion mode, 305
decimated amplitude scheme (DAS), 311
degeneracy, 77
derivative (cubic) nonlinear Schrödinger equation, 186
derivative NLS, 186, 198
DIA equations, 314
DIA solution, 312
diffusion approximation, 99
diffusion coefficient, 129
dipole potential, 244
direct interaction approximation (DIA), 286
dissipation parameters, 279
dissipative dynamics, 310
distribution function, 210
drift wave, 85, 225, 305
drift wave diffusivity, 303
drift wave experiments, 244
drift wave mechanism, 226
drift Wave–Rossby wave analogy, 221
drift wave-ion acoustic wave, 234
driven-damped pendulum, 25

eddy damping rate ν_k, 296
effect of dissipation, 294
electromagnetic scattering experiments, 316
electromagnetic vortex, 234
electron density fluctuation, 305
electron dissipation parameters, 280
electron momentum balance, 304
energy and enstrophy of the dipole vortex, 242
energy flux, 302
energy spectrum, 302
enstrophy cascade, 302
enstrophy cascade spectrum, 303
enstrophy density, 231
enstrophy flux, 231, 302
entropy production, 298
envelope-solitary wave, 178
equipartition, 285
equipartition of energy, 285
ergodic theorem, 29
Ertel's conservation theorem, 238
Ertel's theorem, 237
exponential shielding, 295

Fermi acceleration, 160
Fermi map, 160
Fermi-Pasta-Ulam recurrence phenomena, 188
Fibonacci numbers F_n, 30
Fibonacci series, 30
field reversed configuration, 57, 60
fixed points, 26, 122
Floquet theory, 10, 11
fluctuation spectrum, 263
fusion project, 51

Gaussian statistics, 295, 315
Gelfand-Levitan equation, 192
general nonlinear equation, 186
generating function, 63
geomagnetic field, 51
geostrophic flow, 221
Gibbs canonical ensemble, 211, 285
golden mean, 28, 30
gyro-Bohm diffusion, 303

Hamilton's equations in non-canonical coordinates, 109

Hamiltonian dynamics, 23
harmonic generation in small
    amplitude oscillations, 2
Hasegawa-Mima eq., 223
Hasegawa-Wakatani drift-wave
    equation, 304
head-on collisions, 208, 253
helical/stellarator, 31
hermitian part, 279
high-wavenumber spectral balance, 301
Hill equations, 8
homoclinic orbits, 61, 62
hydrodynamic experiments, 315
hyperbolic curves, 26

impact parameter, 246
incoherent emission, 301
incoherent fluctuation source, 297
incompressible flow, 24
inelastic vortex collisions, 254
integrability conditions, 196
integrable (ring) orbits, 81
integrable degrees-of-freedom, 77
integrable limit of the three-wave
    equations, 309
integrable orbits, 54
integral invariants, 23
interaction energies, 260
interaction triplet, 313
intermittent transport, 257
interplanetary magnetic field, 51
invariant curves, 26, 28
inverse scattering problem, 191
inverse scattering transformation, 189,
    194
involution, 141
ion acoustic oscillations, 315
ion acoustic wave, 165, 181
ion inertial scale, 222
irrational rotational transform, 77
isotropic spectrum, 299

Jacobi elliptic functions, 21, 66, 310

Kadomtsev-Petviashvili equations, 163
KAM theory, 27, 47
KAM theory in laboratory plasmas, 31
KAM torii, 28, 112
KdV nonlinearities, 265
kinetic KdV eq., 166
kinetic modulation instability, 175
Kolmogorov spectrum, 302
Korteweg-de Vries equation, 167, 183,
    188, 189
Kraichnan ln k-corrected Kolmogorov
    spectrum, 302
kurtosis, 263, 291

L-H dynamics, 232
Landau picture for the transition to
    chaos, 315
Larichev-Reznik dipole vortices, 240
Lax pairs, 194
leaky barriers, 28
linear dielectric response function, 165
linear eigenmode equation, 236
local gyroradius, 56
Lorentzian frequency spectrum, 314,
    315
Lorentzian parameters, 293, 297
low frequency electrostatic fluctuations,
    65
low-frequency fluctuations, 235
low-order wave coupling, 306
Lyapunov characteristic exponent, 311

MacKay theory, 30
magnetic flux surfaces, 31
magnetic islands, 35
magnetic moment, 64
magnetospheric fields, 51
magnetotail, 51
Markovian approximation, 288
Markovian spectral equation, 295, 298
Mathieu equation, 8, 10–12, 15, 16, 26
mean-free-path λ_{mfp}, 259
measure preserving flows, 23

Melnikov-Arnold integral, 71
merger, 250
mixing length, 258, 272, 278, 295, 300
mixing length level, 304
mode coupling coefficients, 165
mode coupling equation, 165
mode statistics, 289
modified Korteweg-de Vries equation, 186, 189
modified nonlinear Schrödinger equation, 198
modulational instability, 167, 176, 200
momentum conservation, 231
momentum distribution function, 131
momentum inversion symmetry, 143
momentum transport, 232
monopoles and wake fields, 246
Moser theorems, 28
multifurcation, 155

Newton's laws of motion, 25
NLS equation, 187, 194
NLS solitons, 178
non-degenerate, 77
non-geostrophic flow $\mathbf{v}^{(2)}$, 222
non-Markovian time history integral, 314
nonlinear drift wave equation, 278
nonlinear drift wave equation in $\mathbf{k}$ space, 282
nonlinear drift wave equation in configuration space, 279
nonlinear Landau damping, 176, 179
nonlinear oscillations, 1
nonlinear phenomena, 2
nonlinear Schrödinger equation, 176
nonlinear self-binding, 247
nonlinear self-interaction, 181
nonlinear self-modulation, 169
nonlinear shifts in the oscillation frequency, 5
nonlinear transfer operator, 298
nonuniform magnetic field, 53

numerical experiments, 285

omni-directional potential spectrum, 296
onset of $\mathbf{E} \times \mathbf{B}$ diffusion, 100
onset of chaos from a normal magnetic field component, 74
onset of stochasticity in two drift waves, 91
optical pulse propagation, 164
oscillations of the pendulum, 21
outermost KAM surface, 136

packing fraction, 258, 260, 262
parametric excitation of the pendulum, 13
parametric instability, 7, 14
parametric instability threshold, 9
period-3 catastrophe, 158
period-4 orbit, 155
Petviashvili drift wave solitons, 207
phase coherent harmonics, 164
phase space, 283
phase space structures, 117
plasma beat wave accelerator of Tajima and Dawson, 14
Poincaré surface of section, 25
Poisson brackets, 236, 265
Poisson brackets $\{H, F\}$, 24
ponderomotive potential, 15, 17
potential enstrophy, 310
potential vorticity, 231
Princeton Large Torus (PLT), 316
probability distribution at a given space point, 291
probability distribution for sticking, 126
probability distribution functions for the orbits, 126
propagator $g_{123}(t)$, 314

quasi-geostrophic potential vorticity eq., 223
quasi-monochromatic wave packet, 171

Index   339

quasilinear theory, 304
quasilinear transport, 303
quasinormal approximation, 286

random phase approximation, 284, 312
random phase equations, 313
rational magnetic surfaces, 32
rationals m/n, 28
realizable, 286
recombination experiment, 251
reconnection field, 74
reconnection of the streamlines, 254
reductive perturbation analysis, 181
regularized long-wave (RLW) equation, 208
renormalization map, 18
renormalized $\mathbf{E} \times \mathbf{B}$ diffusion coefficients, 102
renormalized perturbation series, 296
renormalized turbulence theory (RNT), 286
resonance conditions, 78
resonance overlap parameter, 18
resonant perturbations, 31
Reynolds stress, 231
Rossby scale length, 222
Rossby wave, 225
rotating water tank, 221, 249, 273
rotation angle $\alpha(r)$, 27
rotation number, 28
rotational transform $t(r)$, 31
route to chaos, 315
Ruelle-Takens route to chaos, 315

safety factor, 31
scrape-off-layer (SOL), 317
second adiabatic invariant, 64
secondary waves, 20
secular series, 5, 7, 295
secular terms, 6
self-consistent field problem, 163
self-organize into a vortex turbulence, 294

sheared flow, 253, 265
short-range correlations, 294
sine-Gordon equation, 197
single particle-like states, 204
skewness, 291
skewness and kurtosis, 262
small amplitude expansion, 2, 4
small gyroradius limit, 63
solitary dipolar vortex solutions, 233
solitary monopolar structures, 232
solitary wave number density and temperature, 213
solitary wave spectrum, 209
soliton, 184, 188
soliton gas models, 204
space correlation function, 293
space inversion, 141
spectral density, 210, 296
spectrum of longitudinal waves, 120
splitting into monopoles, 269
stability of the dipole vortex, 252
stability of the periodic orbits, 145
standard map, 74, 117, 119
standard map at large values of A, 131
statistical properties, 275
statistical turbulence theory, 286
stochastic motion, 125
stochastic orbits, 40, 81, 114, 124
stochastic wave interactions, 315
stochasticity theory, 51
strange attractor, 306, 310
stream function, 24
sum of resonant intervals, 29
superposition of solitary waves, 209
surface of section, 26
symmetries, 140

tangent map, 123
temporal chaos, 284
third order nonresonant contribution, 5
three-wave coupling, 306
three-wave coupling coefficients, 308

three-wave equations, integrable limit, 309
three-wave Hamiltonian, 309
tilt instability, 245
time average, 29
tokamaks, 315
transfer of energy, 78
transient unbounded orbits, 82
transmission and reflection coefficient, 192
transport events, 317
transverse electric field, 69
trapped plasma, 257
Tsyganenko model, 51
turbulent collision operator, 298
twist condition, 27, 119
twist map, 27
two degree-of-freedom problem, 25
two degrees-of-freedom and degeneracy, 77
two longitudinal waves, 17
two-dimensional map, 27

two-point correlation function, 210
two-point field correlation function, 286
two-soliton solution, 204
two-time correlation functions, 291
two-wave problem, 21

unstable trajectories, 32
unstable-chaotic trajectories, 35

Vlasov-Poisson field equations, 164
volume contraction, 283
vortex collisions, 246
vortex rotation, 233
vortex structures, 221

Wakatani and Hasegawa, 306
wave dispersion, 246
wave energy, 310
wave-momentum density, 310
wavenumber spectrum, 293
weak turbulence theory, 295
whisker map, 73
winding number ν, 30

WITHDRAWN